GENERAL SYSTEMS THEORY: MATHEMATICAL FOUNDATIONS

This is Volume 113 in
MATHEMATICS IN SCIENCE AND ENGINEERING
A series of monographs and textbooks
Edited by RICHARD BELLMAN, *University of Southern California*

The complete listing of books in this series is available from the Publisher upon request.

General Systems Theory: Mathematical Foundations

M. D. Mesarovic and Yasuhiko Takahara

Systems Research Center
Case Western Reserve University
Cleveland, Ohio

ACADEMIC PRESS New York San Francisco London 1975
A Subsidiary of Harcourt Brace Jovanovich, Publishers

ACADEMIC PRESS, INC.
111 Fifth Avenue, New York, New York 10003

United Kingdom Edition published by
ACADEMIC PRESS, INC. (LONDON) LTD.
24/28 Oval Road, London NW1

Library of Congress Cataloging in Publication Data

Mesarović, Mihajlo D
Foundations for the mathematical theory of general systems.

(Mathematics in science and engineering,)
Bibliography: p.
1. System analysis. I. Takahara, Y., joint author.
II. Title. III. Series.
QA402.M47 003 74-1639
ISBN 0-12-491540-X

PRINTED IN THE UNITED STATES OF AMERICA

To

Gordana and Mitsue

with affection and gratitude

CONTENTS

VI. Stationarity and Time Invariance

VII. Controllability

VIII. Minimal Realization

IX. Stability

X. Interconnections of Subsystems, Decomposition, and Decoupling

XI. Computability, Consistency, and Completeness

XII. Categories of Systems and Associated Functors

PREFACE

This book reports on the development of a mathematical theory of general systems, initiated ten years ago. The theory is based on a broad and ambitious program aimed at formalizing all major systems concepts and the development of an axiomatic and general theory of systems. The present book provides the foundations for, and the initial steps toward, the fulfillment of that program. The interest in the present volume is strictly in the mathematical aspects of the theory. Applications and philosophical implications will be considered elsewhere.

The basic characteristics and the role of the proposed general systems theory are discussed in some detail in the first chapter. However, the unifying power of the proposed foundations ought to be specifically singled out; within the same framework, using essentially the same mathematical structure for the specification of a system, such diverse topics are considered and associated results proven as: the existence and minimal axioms for state-space construction; necessary and sufficient conditions for controllability of multivalued systems; minimal realization from input–output data; necessary and sufficient conditions for Lyapunov stability of dynamical systems; Goedel consistency and completeness theorem; feedback decoupling of multivariable systems; Krohn–Rhodes decomposition theorem; classification of systems using category theory.

A system can be described either as a transformation of inputs (stimuli) into outputs (responses)—the so-called input–output approach (also referred to as the causal or terminal systems approach), or in reference to the fulfillment of a purpose or the pursuit of a goal—the so-called goal-seeking or decision-making approach. In this book we deal only with the input–output approach. Originally, we intended to include a general mathematical theory of goal-seeking, but too many other tasks and duties have prevented us from carrying out that intention. In fairness to our research already completed, we ought to point out that the theory of multilevel systems which has been reported elsewhere† although aimed in a different direction, does contain the elements of a general theory of complex goal-seeking systems. For the sake of completeness we have given the basic definition of a goal-seeking system and of an open system (another topic of major concern) in Appendix II.

We have discussed the material over the years with many colleagues and students. In particular the advice and help of Donald Macko and Seiji Yoshii were most constructive. The manuscript would have remained a scribble of notes on a pile of paper if it were not for tireless and almost nondenumerable series of drafts retyped by Mrs. Mary Lou Cantini.

† M. D. Mesarovic, D. Macko, and Y. Takahara, "Theory of Hierarchical Multilevel Systems." Academic Press, New York, 1970.

GENERAL SYSTEMS THEORY: MATHEMATICAL FOUNDATIONS

Chapter I

INTRODUCTION

1. GENERAL SYSTEMS THEORY: WHAT IS IT AND WHAT IS IT FOR?

Systems theory is a scientific discipline concerned with the explanations of various phenomena, regardless of their specific nature, in terms of the formal relationships between the factors involved and the ways they are transformed under different conditions; the observations are explained in terms of the relationships between the components, i.e., in reference to the organization and functioning rather than with an explicit reference to the nature of the mechanisms involved (e.g., physical, biological, social, or even purely conceptual). The subject of study in systems theory is not a "physical object," a chemical or social phenomenon, for example, but a "system": *a formal relationship between observed features or attributes.* For conceptual reasons, the language used in describing the behavior of systems is that of information processing and goal seeking (decision making, control).

General systems theory deals with the most fundamental concepts and aspects of systems. Many theories dealing with more specific types of systems (e.g., dynamical systems, automata, control systems, game-theoretic systems, among many others) have been under development for quite some time. General systems theory is concerned with the basic issues common to all of these specialized treatments. Also, for truly complex phenomena, such as those found predominantly in the social and biological sciences, the specialized descriptions used in classical theories (which are based on special mathematical structures such as differential or difference equations, numerical or abstract algebras, etc.) do not adequately and properly represent the actual events. Either because of this inadequate match between the events and types of descriptions available or because of the pure lack of knowledge, for many truly complex problems one can give only the most general statements, which are qualitative and too often even only verbal. General systems theory is aimed at providing a description and explanation for such complex phenomena.

Our contention is that one and the same theory can serve both of these purposes. Furthermore, in order to do that, it ought to be *simple, elegant, general, and precise* (unambiguous). This is why the approach we have taken is both mathematical and perfectly general. At the risk of oversimplification, the principal characteristics of the approach whose foundation is presented in this book are:

(i) It is a *mathematical* theory of general systems; basic concepts are introduced axiomatically, and the system's properties and behavior are investigated in a precise manner.

(ii) It is concerned with the goal seeking (decision making, control) and similar representations of systems, as much as with the input–output or "transformational" (causal) representations. For example, the study of hierarchical, multilevel, decision-making systems was a major concern from the very beginning.

(iii) The mathematical structures required to formalize the basic concepts are introduced in such a way that precision is obtained without losing any generality. It is important to realize that nothing is gained by avoiding the use of a precise language, i.e., mathematics, in making statements about a system of concern. We take exception, therefore, to considering general systems theory as a scientific philosophy, but rather, consider it as a scientific enterprise, without denying, however, the impact of such a scientific development on philosophy in general and epistemology in particular. Furthermore, once a commitment to the mathematical method is made, logical inferences can be drawn about the system's behavior. Actually, the investigation of the logical consequences of systems having given properties should be of central concern for any general systems theory which cannot be limited solely to a descriptive classification of systems.

The decision-making or goal-seeking view of a system's behavior is of paramount importance. General systems theory is not a generalized circuit theory—a position we believe has introduced much confusion and has contributed to the rejection of systems theory and the systems approach in fields where goal-seeking behavior is central, such as psychology, biology, etc. Actually, the theory presented in this book can just as well be termed general cybernetics, i.e., a general theory of governing and governed systems. The term "general systems theory" was adopted at the initiation of the theory as reflecting a broader concern. However, in retrospect, it appears that the choice might not have been the happiest one, since that term has already been used in a different context.

The application of the mathematical theory of general systems can play a major role in the following important problem areas.

(a) Study of Systems with Uncertainties

On too many occasions, there is not enough information about a given system and its operation to enable a detailed mathematical modeling (even if the knowledge about the basic cause–effect relationships exists). A general systems model can be developed for such a situation, thus providing a solid mathematical basis for further study or a more detailed analysis. In this way, general systems theory, as conceived in this book, significantly extends the domain of application of mathematical methods to include the most diverse fields and problem areas not previously amenable to mathematical modeling.

(b) Study of Large-Scale and Complex Systems

Complexity in the description of a system with a large number of variables might be due to the way in which the variables and the relationships between them are described, or the number of details taken into account, even though they are not necessarily germane to the main purpose of study. In such a case, by developing a model which is less structured and which concentrates only on the key factors, i.e., a general systems model in the set-theoretic or algebraic framework, one can make the analysis more efficient, or even make it possible at all. In short, one uses a mathematically more abstract, less structured description of a large-scale and complex system. Many structural problems, such as decomposition, coordination, etc., can be considered on such a level. Furthermore, even some more traditional problems such as Lyapunov stability can be analyzed algebraically using more abstract descriptions.

The distinction between classical methods of approximation and the abstraction approach should be noted. In the former, one uses the same mathematical structure, and simplification is achieved by omitting some parts of the model that are considered less important; e.g., a fifth-order differential equation is replaced with a second-order equation by considering only the two "dominant" state variables of the system. In the latter approach, however, one uses a different mathematical structure which is more abstract, but which still considers the system as a whole, although from a less detailed viewpoint. The simplification is not achieved by the omission of variables but by the suppression of some of the details considered unessential.

(c) Structural Considerations in Model Building

In both the analysis and synthesis of systems of various kinds, structural considerations are of utmost importance. Actually, the most crucial step in the model-building process is the selection of a structure for the model of a system under consideration. It is a rather poor strategy to start investigations with a detailed mathematical model before major hypotheses are tested and a better understanding of the system is developed. Especially when the system consists of a family of interrelated subsystems, it is more efficient first to delineate the subsystems and to identify the major interfaces before proceeding with a more detailed modeling of the mechanisms of how the various subsystems function. Traditionally, engineers have used block diagrams to reveal the overall composition of a system and to facilitate subsequent structural and analytical considerations. The principal attractiveness of block diagrams is their simplicity, while their major drawback is a lack of precision. General systems theory models eliminate this drawback by introducing the precision of mathematics, while preserving the advantage, i.e., the simplicity, of block diagrams. The role of general systems theory in systems analysis can be represented by the diagram in Fig. 1.1. General

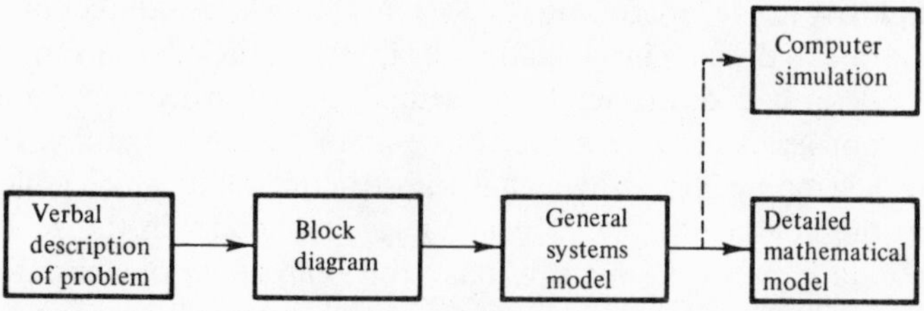

FIG. 1.1

systems models fall between block-diagram representation and a detailed mathematical (or computer) model. For complex systems, in particular, a general systems model might very well represent a necessary step, since the gulf between a block diagram and a detailed model can be too great. The fact that certain general systems techniques and results have become available to treat the systems problem on a general level makes it possible to add this step in practice.

(d) Precise Definition of Concepts and Interdisciplinary Communication

General systems theory provides a language for interdisciplinary communication, since it is sufficiently general to avoid introduction of constraints of its own, yet, due to its precision, it removes misunderstandings which can be quite misleading. (For example, the different notions of adaptation used in the fields of psychology, biology, engineering, etc., can first be formalized in general systems theory terms and then compared.) It is often stated that systems theory has to reflect the "invariant" structural aspects of different real-life systems, i.e., those that remain invariant for similar phenomena from different fields (disciplines). This similarity can be truly established only if the relevant concepts are defined with sufficient care and precision. Otherwise, the danger of confusion is too great. It is quite appropriate, therefore, to consider the mathematical theory of general systems as providing a framework for the formalization of any systems concept. In this sense, general systems theory is quite basic for the application of the "systems approach" and systems theory in almost any situation. The important point to note when using general systems theory for concept definition is that when a concept has been introduced in a precise manner, what is crucial is not whether the definition is "correct" in any given interpretation, but rather, whether the concept is defined with sufficient precision so that it can be clearly and unambiguously understood and as such can be further examined and used in other disciplines. It is in this capacity that the general systems theory offers a language for interdisciplinary communication. Such an application of general systems theory might seem trivial from the purely mathematical standpoint, but is not so from the viewpoint of managing a large team effort in which specialists from different disciplines are working together on a complex problem, as is often found in the fields of environmental, urban, regional, and other large-scale studies.

(e) Unification and Foundation for More Specialized Branches of Systems Theory

Questions regarding basic systems problems which transcend many specialized branches of systems theory (e.g., the question of state-space representation) can be properly and successfully considered on the general systems level. This will be demonstrated many times in this book. The problems of foundations are of interest for extending and making proper use of systems theory in practice, for pedagogical reasons, and for providing a coherent framework to organize the facts and findings in the broad areas of systems research.

2. FORMALIZATION APPROACH FOR THE DEVELOPMENT OF THE MATHEMATICAL THEORY OF GENERAL SYSTEMS

The approach that we have used to develop the general systems theory reported in this book is the following:†

† A comparison with some other possible approaches is given in Appendix II.

(i) The basic systems concepts are introduced via *formalization.* By this we mean that starting from a verbal description of an intuitive notion, a precise mathematical definition for the concept is given using minimal mathematical structure, i.e., as few axioms as the correct interpretation would allow.

(ii) Starting from basic concepts introduced via formalization, the mathematical theory of general systems is further developed by adding more mathematical structure as needed for the investigation of various systems properties. Such a procedure allows us to establish how fundamental some particular systems properties really are and also what is the minimal set of assumptions needed in order that a given property or relationship holds.

The starting point for the entire development is the concept of a system defined on the set-theoretic level. Quite simply and most naturally for that level, a system is defined as a relation in the set-theoretic sense, i.e., it is assumed that a family of sets is given,

$$\overline{V} = \{V_i : i \in I\}$$

where I is the index set, and a system, defined on $\overline{V}$, is a proper subset of $\times \overline{V}$,

$$S \subset \times \{V_i : i \in I\}$$

The components of S, V_i, $i \in I$, are termed the systems objects. We shall primarily be concerned with a system consisting of two objects, the input object X and the output object Y:

$$S \subset X \times Y \tag{1.1}$$

Starting a mathematical theory of general systems on the set-theoretic level is fully consistent with the stated objective of starting with the least structured and most widely applicable concepts and then proceeding with the development of a mathematical theory in an axiomatic manner.

To understand better some of the reasons for adopting the concept of a system as a set-theoretic relation, the following remarks are pertinent.

A system is defined in terms of observed features or, more precisely, in terms of the relationship between those features rather than what they actually are (physical, biological, social, or other phenomena). This is in accord with the nature of the systems field and its concern with the organization and interrelationships of components into an (overall) system rather than with the specific mechanisms within a given phenomenological framework.

The notion of a system as given in (1.1) is perfectly general. On the one hand, if a system is described by more specific mathematical constructs, e.g., a set of equations, it is obvious that these constructs define or specify a relation as given in (1.1). Different systems, of course, have different methods of specification, but they all are but relations as given in (1.1). On the other hand, in the case of the most incomplete information when the system can be described only in terms of a set of verbal statements, they still, by their linguistic function as statements, define a relation as in (1.1). Indeed, every statement contains two basic linguistic categories: nouns and functors—nouns denoting objects, functors denoting the relationship between them. For any proper set of verbal statements there exists a (mathematical) relation which represents the formal relationship between the objects denoted by nouns (technically referred to as a model for these statements). The adjective "proper" refers here, of course, to the conditions for the axioms of a set theory. In short, then, a system is always a relation, as given in (1.1), and various types of systems are more precisely defined by the appropriate methods, linguistic, mathematical, computer programs, etc.

A system is defined as a set (of a particular kind, i.e., a relation). It stands for the collection of all appearances of the object of study rather than for the object of study itself. This is necessitated by the use of mathematics as the language for the theory in which a "mechanism" (a function or a relation) is defined as a set, i.e., as a collection of all proper combinations of components. Such a characterization of a system ought not to create any difficulty since the set relation, with additional specifications, contains all the information about the actual "mechanism" we can legitimately use in the development of a formal theory.

The specification of a given system is often given in terms of some equations defined on appropriate variables. To every variable there corresponds a systems object which represents the range of the respective variable. Stating that a system is defined by a set of equations on a set of variables, one essentially states that the system is a relation on the respective systems objects specified by the variables (each one with a corresponding object as a range) such that for any combination of elements from the objects, i.e., the values for the variables, the given set of equations is satisfied.

To develop any kind of theory starting from (1.1), it is necessary to introduce more structure into the system as a relation. This can be done in two ways:

(i) by introducing the additional structure into the elements of the system objects, i.e., to consider an element $v_i \in V_i$ as a set itself with additional appropriate structure;

(ii) by introducing the structure in the object sets, V_i, $i \in I$, themselves.

The first approach leads to the (abstract) time system concept, the second to the concept of an algebraic system.

(a) Time Systems

This approach will be introduced precisely in Chapter II and will be used extensively throughout the book; therefore, a brief sketch is sufficient here.

If the elements of an object are functions, e.g., $v: T_v \to A_v$, the object is referred to as a family object or a function-generated object. Of particular interest is the case when both the domain and codomain of all the functions in the given object V are the same, i.e., any $v \in V$ is a function on T into A, $v: T \to A$. T represents the *index set* for V, while A is referred to as the *alphabet* for V. Notice that A can be of arbitrary cardinality. If the index set is linearly ordered, it is called a *time set*. This term was selected because such an index set captures the minimal property necessary for the concept of time, particularly as it relates to the time evolution and dynamic behavior of systems.

A function defined on a time set is called an *(abstract) time function*. An object whose elements are time functions is referred to as a *time object*. A system defined on time objects represents a *time system*.

Of particular interest are time systems whose input and output objects are both defined on the same sets $X \subseteq A^T$ and $Y \subseteq B^T$. The system is then

$$S \subseteq A^T \times B^T$$

(b) Algebraic Systems

An alternative way to introduce mathematical structure in a system's object V necessary for constructive specification is to define one or more operations in V so that V becomes an algebra. In the simplest case, a binary operation is given, $R: V \times V \to V$, and it is assumed that there exists a subset W of V, often of finite cardinality, such that any element in V can be obtained by the application of R on the elements of W or previously generated elements. The set W is referred to as the set of generators, or also as an *alphabet* and its elements as the *symbols*; the elements of the object V are referred to as *words*. If R is concatenation, the words are simply sequences of elements from the alphabet W.

A distinction should be noticed between the alphabet for a time object and for an algebraic object. For objects with finite alphabets, these are usually the same sets, i.e., the object, whose elements are sequences from the given set, can be viewed either as a set of time functions (on different time intervals, though) or as a set generated by an algebraic operation from the same set of symbols. When the alphabet is infinite, complications arise, and the set of generators and the codomain of the time functions are different sets, generally even of different cardinality.

In a more general situation, an algebraic object is generated by a family of operations. Namely, given a set of elements, termed primitive elements, W, and a set of operations $\bar{R} = \{R_1, \ldots, R_n\}$, the object V contains the primitive elements themselves, $W \subset V$, and any element that can be generated by a repeated application of the operations from $\bar{R}$.

We shall use primarily the time systems approach in this book because it allows a more appealing intuitive interpretation in particular for the phenomena of time evolution and state transition. Actually, it can be shown that the two approaches are, by and large, equivalent. It should be emphasized, however, that we shall be using the algebraic structure both within a general system, $S \subset X \times Y$, and a general time system, $S \subset A^T \times B^T$, although not necessarily for the problems related to time evolution.

It is interesting to note that the two methods mentioned above correspond to the two basic ways to define a set constructively: by (transfinite) induction on an ordered set and by algebraic induction. The implication and meaning of this interesting fact will not be pursued here any further.

Chapter II

BASIC CONCEPTS

In this chapter we shall introduce some basic systems notions on the set-theoretic level and establish some relationships between them. First, we shall define a general system as a relation on abstract sets and then define the general time and dynamical systems as general systems defined on the sets of abstract time functions.

In order to enable more specific definition of various types of systems, certain kinds of so-called auxiliary functions are introduced. They are abstract counterparts of relationships, often given in the form of a set of equations, in terms of which a system is defined. Auxiliary functions enable also a more detailed analysis of systems, in particular their evolution in time.

In order to define various auxiliary functions, new auxiliary objects, termed state objects, had to be introduced; the elements of such an object are termed states. The primary functions of the state, as introduced in this chapter, are:

(i) to enable a system or its restrictions, which are both in general relations, to be represented as functions;

(ii) to enable the determination of a future output solely on the basis of a given future input and the present state completely disregarding the past (the state at any given time embodies the entire past history of the system);

(iii) to relate the states at different times so that one can determine whether the state of a system has changed over time and in what way. This third requirement leads to the concept of a state space. A general dynamical system is defined in such a state space.

Some basic conditions are given for the existence of various types of auxiliary functions in general and in reference to such system properties as input completeness and linearity. A classification of systems in reference to various kinds of time invariance of certain auxiliary functions is given. Finally, some questions of time causality are considered. Two notions are introduced in this respect:

(i) A system is termed nonanticipatory if there exists a family of state objects so that the future values of any output are determined solely by the state at a previous time and the input in this time period.

(ii) A system is termed past-determined if, after a certain initial period of time, the values of any output are determined solely by the past input–output pair. Conditions are then given for time systems to be nonanticipatory or past-determined.

1. SET-THEORETIC CONCEPT OF A GENERAL SYSTEM

(a) General System, Global States, and Global-Response Function

Starting point for the entire development is provided by the following definitions.

Definition 1.1. A (general) system is a relation on nonempty (abstract) sets

$$S \subset \times\{V_i : i \in I\} \tag{2.1}$$

where $\times$ denotes Cartesian product and I is the index set. A component set V_i is referred to as a system object. When I is finite, (2.1) is written in the form

$$S \subset V_1 \times \cdots \times V_n \tag{2.2}$$

Definition 1.2. Let $I_x \subset I$ and $I_y \subset I$ be a partition of I, i.e., $I_x \cap I_y = \phi$, $I_x \cup I_y = I$. The set $X = \times\{V_i : i \in I_x\}$ is termed the input object, while $Y = \times\{V_i : i \in I_y\}$ is termed the output object. The system S is then

$$S \subset X \times Y \tag{2.3}$$

and will be referred to as an input–output system.

The form (2.3) rather than (2.2) will be used throughout this book.

Definition 1.3. If S is a function

$$S : X \to Y \tag{2.4}$$

it is referred to as a function-type (or functional) system.

Notice that the same symbol S is used both in (2.2) and (2.3) although strictly speaking the elements of the relation in (2.2) are n-tuples while those in the relation (2.3) are pairs. This convention is adopted for the sake of simplicity of notation. Which of the forms for S is used will be clear from the context in which it is used. Analogous comment applies to the use of the same symbol S in (2.3) and (2.4).

For notational convenience, we shall adopt the following conventions: The brackets in the domain of any function, e.g., $F:(A) \to B$, will indicate that the function F is only partial, i.e., it is not defined for every element in the domain A. The domain of F will be denoted by $\mathscr{D}(F) \subset A$, and the range by $\mathscr{R}(F) \subset B$. Similarly, the domain and the range of $S \subset X \times Y$ will be denoted, respectively, by

$$\mathscr{D}(S) = \{x : (\exists y)((x, y) \in S)\} \qquad \text{and} \qquad \mathscr{R}(S) = \{y : (\exists x)((x, y) \in S)\}$$

For the sake of notational simplicity, $\mathscr{D}(S) = X$ is always assumed unless stated otherwise.

Definition 1.4. Given a general system S, let C be an arbitrary set and R a function, $R:(C \times X) \to Y$, such that

$$(x, y) \in S \leftrightarrow (\exists c)[R(c, x) = y]$$

C is then a global state object or set, its elements being global states, while R is a global (systems)-response function (for S).

Theorem 1.1. Every system has a global-response function which is not partial, i.e.,

$$R: C \times X \to Y$$

PROOF. Let $F = Y^X = \{f : f : X \to Y\}$. Let $G = \{f_c : c \in C\} \subseteq F$ such that $f_c \in G \leftrightarrow f_c \subseteq S$, where C is an index set of G. Let $R: C \times X \to Y$ be such that $R(c, x) = f_c(x)$. Then we claim that $S = \{(x, y) : (\exists c)(y = R(c, x))\}$. Let $S' = \{(x, y) : (\exists c)(y = R(c, x))\}$. Let $(x, y) \in S'$ be arbitrary. Then $y = R(c, x) = f_c(x)$ for some $c \in C$. Hence, $(x, y) \in S$ because $f_c \subseteq S$. Therefore, $S' \subseteq S$. Conversely, let $(x, y) \in S$ be arbitrary. Since $\mathscr{D}(S) = X \ni x$, S is nonempty. Let $f_c \in G$. Let $\hat{f} = (f_c \setminus \{(x, f_c(x))\}) \cup \{(x, y)\}$. Then $\hat{f} \in F$ and $\hat{f} \subseteq S$. Hence, $\hat{f} = f_{c'}$ for some $c' \in C$. Consequently, $y = f_{c'}(x)$ or $(x, y) \in S'$ and hence $S \subseteq S'$. Therefore, $S = S'$. Q.E.D.

In the preceding theorem, no additional requirements are imposed either on C or R. However, if R is required to have a certain property, the global response function, although it might still exist, cannot be defined on the entire $C \times X$, i.e., R remains a partial function. Such is the case, e.g., when R

is required to be causal. Since the case when R is not a partial function is of special importance, we shall adopt the following convention:

R will be referred to as the global-response function only if it is not a partial function. Otherwise, it will be referred to as a partial global-response function.

(b) Abstract Linear System

Although many systems concepts can be defined solely by using the notion of a general system, the development of meaningful mathematical results is possible often only if additional structure is introduced. In order to avoid proliferation of definitions, we shall, as a rule, introduce specific concepts on the same level of abstraction on which mathematical results of interest can be developed; e.g., the concept of a dynamical system will be introduced only in the context of time systems. The concept of linearity, however, can be introduced usefully on the general systems level. We shall first introduce the notion of linearity, which is used as standard in this book.

Definition 1.5. Let $\mathscr{A}$ be a field, X and Y be linear algebras over $\mathscr{A}$ and let S be a relation, $S \subset X \times Y$, S is nonempty, and

(i) $s \in S \;\&\; s' \in S \rightarrow s + s' \in S$
(ii) $s \in S \;\&\; \alpha \in \mathscr{A} \rightarrow \alpha s \in S$

where $+$ is the additive operation in $X \times Y$ and $\alpha \in \mathscr{A}$.† S is then an (abstract) complete linear system.

In various applications, one encounters linear systems that are not complete, e.g., a system described by a set of linear differential equations whose set of initial conditions is not a linear space. For the sake of simplicity, we shall consider in this book primarily the complete systems, and, therefore, *every linear system will be assumed to be complete unless explicitly stated otherwise.* This is hardly a loss of generality since any incomplete linear system can be made complete by a perfectly straightforward and natural completeness procedure.

The following theorem is fundamental for linear systems theory.

Theorem 1.2. Let X and Y be linear algebras over the same field $\mathscr{A}$. $S \subset X \times Y$ is then a linear system if and only if there exists a global response function $R: C \times X \rightarrow Y$ such that:

(i) C is a linear algebra over $\mathscr{A}$;
(ii) there exists a pair of linear mappings

$$R_1 : C \rightarrow Y \qquad \text{and} \qquad R_2 : X \rightarrow Y$$

† The operation $+$ and the scalar multiplication on $X \times Y$ is defined by: $(x, y) + (\hat{x}, \hat{y}) = (x + \hat{x}, y + \hat{y})$ and $\alpha(x, y) = (\alpha x, \alpha y)$ where $(x, y), (\hat{x}, \hat{y}) \in X \times Y$ and $\alpha \in \mathscr{A}$.

such that for all $(c, x) \in C \times X$

$$R(c, x) = R_1(c) + R_2(x)$$

PROOF. The *if* part is clear. Let us prove the *only if* part. First, we shall show that there exists a linear mapping $R_2: X \to Y$ such that $\{(x, R_2(x)): x \in X\} \subset S$. Let X_s be a subspace of X and $L_s: X_s \to Y$ a linear mapping such that $\{(x, L_s(x)): x \in X_s\} \subset S$. Such X_s and L_s always exist. Indeed, let $(\hat{x}, \hat{y}) \in S \neq \phi$; then $X_s = \{\alpha\hat{x}: \alpha \in \mathscr{A}\}$ and $L_s: X_s \to Y$ such that $L_s(\alpha\hat{x}) = \alpha\hat{y}$ are the desired ones. If $X_s = X$, then L_s is the desired linear mapping. If $X_s \neq X$, then L_s can always be extended by Zorn's lemma so that $X_s = X$ is achieved. Let $\tilde{L} = \{L_p\}$ be the class of all linear mappings defined on the subspaces of X such that when the subspace X_p is the domain of L_p, $\{(x, L_p(x)): x \in X_p\} \subset S$ holds. Notice that $\tilde{L}$ is not empty. Let $\leq$ be an ordering on $\tilde{L}$ defined by: When L' and L'' are in $\tilde{L}$, then $L'' \leq L''$ iff $L'' \subseteq L''$. Since a mapping is a relation between the domain and the codomain and since a relation is a set, the above definition is proper. Let $P \subset \tilde{L}$ be an arbitrary linearly ordered subset of $\tilde{L}$. Let $L_0 = \bigcup P$, where $\bigcup P$ is the union of elements of P. We shall show that L_0 is in $\tilde{L}$. Suppose (x, y) and (x, y') are elements of L_0. Then, since $L_0 = \bigcup P$, there exist two mappings L and L' in P such that $(x, y) \in L$ and $(x, y') \in L'$. Since P is linearly ordered, e.g., $L \leq L'$, $(x, y) \in L'$ holds. Since L' is a mapping, $y = y'$; that is, L_0 is a mapping. Next, suppose (x', y') and (x'', y'') are in L_0. Then the same argument implies that $(x', y') \in L''$ and $(x'', y'') \in L''$ for some $L'' \subset P$. Since L'' is linear, $(x' + x'', y' + y'') \in L'' \subset L_0$. Furthermore, if $(x', y') \in L_0$ and $\alpha \in \mathscr{A}$, then there exists L' in P such that $(x', y') \in L'$; that is, $(\alpha x', \alpha y') \in L' \subset L_0$ holds. Hence, L_0 is a linear mapping. Finally, if $(x', y') \in L_0$, then $(x', y') \in L'$ for some L' in P. Hence, $(x', y') \in S$, or $L_0 \subset S$. Therefore, $L_0 \in \tilde{L}$. Consequently, L_0 is an upper bound of P in $\tilde{L}$. We can, then, conclude by Zorn's lemma that there is a maximal element R_2 in $\tilde{L}$. We claim that $\mathscr{D}(R_2) = X$. If it is not so, $\mathscr{D}(R_2)$ is a proper subspace of X. Then there is an element $\hat{x}$ in X such that $\hat{x}$ is not an element of $\mathscr{D}(R_2)$. Then $X' = \{\alpha\hat{x} + x: \alpha \in \mathscr{A} \;\&\; x \in \mathscr{D}(R_2)\}$ is a linear subspace which includes $\mathscr{D}(R_2)$ properly. Notice that every element x' of X' is expressed in the form $x' = \alpha\hat{x} + x$ uniquely. As a matter of fact, if $x' = \alpha\hat{x} + x = \beta\hat{x} + y$, then $(\alpha - \beta)\hat{x} = (y - x)$. If $\alpha \neq \beta$, then $\hat{x} = (\alpha - \beta)^{-1}(y - x) \in \mathscr{D}(R_2)$, which is a contradiction. By using this fact, we can define a linear mapping $L': X' \to Y$ such that $L'(\alpha\hat{x} + x) = \alpha\hat{y} + R_2(x)$, where $(\hat{x}, \hat{y}) \in S$ and $x \in \mathscr{D}(R_2)$. Then L' is linear and $\{(x', L'(x')): x' \in X'\} \subset S$, and R_2 is a proper subset of L', which contradicts the maximality of R_2. Hence, R_2 is the desired mapping. To complete the construction of R, let C be $C = \{(o, y): (o, y) \in S\}$. C is, apparently, a linear space over $\mathscr{A}$ when the addition and the scalar multiplication are defined as: $(o, y) + (o, y') = (o, y + y')$ and $\alpha(o, y) = (o, \alpha y)$, where $\alpha \in \mathscr{A}$. Let $R_1: C \to Y$ such that $R_1((o, y)) = y$. Then R_1 is a linear mapping. Let

$R(c, x) = R_1(x) + R_2(x)$. We shall show that

$$S = \{(x, y) : (\exists c)(c \in C \;\&\; y = R(c, x))\} \equiv S'$$

Suppose $(x, y) \in S$. Then $(x, R_2(x)) \in S$. Since S is linear, $(x, y) - (x, R_2(x)) = (0, y - R_2(x)) \in S$. Hence, $(\exists c)(c \in C \& y = R_1(c) + R_2(x))$; that is, $S \subseteq S'$ holds. Conversely, suppose $(x, R_1(c) + R_2(x)) \in S'$. Since $(0, R_1(c)) \in S$ and $(x, R_2(x)) \in S$ and since S is linear,

$$(x, R_2(x)) + (0, R_1(c)) = (x, R_1(c) + R_2(x)) \in S$$

Hence, $S' \subseteq S$ holds. Q.E.D.

The fundamental character of the preceding theorem is illustrated by the fact that every result on linear systems developed in this book is based on it. We can now introduce the following definition.

Definition 1.6. Let $S \subset X \times Y$ be a linear system and R a mapping $R: C \times X \to Y$. R is termed a linear global-response function if and only if

(i) R is consistent with S, i.e.,

$$(x, y) \in S \leftrightarrow (\exists c)[y = R(c, x)]$$

(ii) C is a linear algebra over the field of X and Y;
(iii) there exist two linear mappings $R_1: C \to Y$ and $R_2: X \to Y$ such that for all $(c, x) \in C \times X$

$$R(c, x) = R_1(c) + R_2(x)$$

C is referred to as the *linear global state object*. The mapping $R_1: C \to Y$ is termed the *global state response*, while $R_2: X \to Y$ is the *global input response*.

Notice the distinction between the global-response function and the linear global-response function. The first concept requires only (i), while for the second, conditions (ii) and (iii) have to be satisfied. A linear system, therefore, can have a response function which is not linear.

From Theorem 1.2 we have immediately the following proposition.

Proposition 1.1. A system is linear if and only if it has a linear global-response function.

The concept of a linear system as given by Definition 1.5 uses more than a "minimal" mathematical structure. The most abstract notion of a linear system consistent with the formalization approach is actually given by the following definition.

Definition 1.7. Let X be an (abstract) algebra with a binary operation $\cdot : X \times X \to X$ and a family of endomorphisms $\bar{\alpha} = \{\alpha : X \to X\}$; similarly, let Y has a binary operation $* : Y \times Y \to Y$ and a family $\bar{\beta} = \{\beta : Y \to Y\}$. A function system $S : X \to Y$ is a general linear system if and only if there exists a one-to-one mapping $\psi : \bar{\alpha} \to \bar{\beta}$ such that:

(i) $(\forall x, x')[S(x \cdot x') = S(x) * S(x')]$
(ii) $(\forall x)(\forall \alpha)[S(\alpha(x)) = \psi(\alpha)(S(x))]$

There could be other concepts of a linear system with the structure between that in Definitions 1.5 and 1.7, e.g., by assuming that X and Y are modules rather than abstract linear spaces. We have not considered such "intermediate" concepts in this book because for some essential results the structure of an abstract linear space is needed. Actually, one might argue that the concept of linearity based on the module structure is not satisfactory because it is neither most abstract (such as Definition 1.7) nor sufficiently rich in structure to allow proofs of basic mathematical results such as Theorem 1.2.

2. GENERAL TIME AND DYNAMICAL SYSTEMS

(a) General Time System

In order to introduce the concept of a general time system, we have to formalize the notion of time. In accordance with the strategy pronounced in Chapter I, we have to define the notion of time by using minimal mathematical structure and such that it captures the most essential feature of an intuitive notion of time. This seems a very easy task, yet the decision at this junction is quite crucial. The selection of structure for such a basic concept as the time set has important consequences for the entire subsequent developments and the richness and elegance of the mathematical results. We shall use the following notion.

Definition 2.1. A time set (for a general time system) is a linearly ordered (abstract) set. The time set will be denoted by T and the ordering in T by $\leq$.

Apparently, the minimal property of a time set is considered to be that its elements follow each other in an orderly succession. This reflects our intended usage of the concept of time for the study of the evolution of systems. No restrictions regarding cardinality are imposed on the time set. However, the time set might have some additional structure, e.g., that of an Abelian group. We shall introduce such additional assumptions when needed.

For notational convenience, T will be assumed to have the minimal element o. In other words, we assume that there exists a superset $\bar{T}$ with a linear ordering $\leq$ and a fixed element denoted by o in $\bar{T}$ such that T is defined by $T = \{t : t \geq \text{o}\}$.

We can introduce now the following definition.

Definition 2.2. Let A and B be arbitrary sets, T a time set, A^T and B^T the set of all maps on T into A and B, respectively, $X \subset A^T$ and $Y \subset B^T$. A general time system S on X and Y is a relation on X and Y, i.e., $S \subset X \times Y$. A and B are called alphabets of the input set X and output set Y, respectively. X and Y are also termed time objects, while their elements $x : T \to A$ and $y : T \to B$ are abstract time functions. The values of X and Y at t will be denoted by $x(t)$ and $y(t)$, respectively.

In order to study the dynamical behavior of a time system, we need to introduce the appropriate *time segments*. In this respect, we shall use the following notational convention.

For every t, $t' > t$,

$$T_t = \{t' : t' \geq t\}, \qquad T^t = \{t' : t' < t\}, \qquad T_{tt'} = \{t^* : t \leq t^* < t'\}$$

$$\bar{T}_{tt'} = T_{tt'} \cup \{t'\}, \qquad \bar{T}^t = T^t \cup \{t\}$$

Corresponding to various time segments, the restrictions of $x \in A^T$ will be defined as follows.

$$x_t = x \mid T_t, \qquad x^t = x \mid T^t, \qquad x_{tt'} = x \mid T_{tt'}, \qquad \bar{x}_{tt'} = x \mid \bar{T}_{tt'}$$

$$\bar{x}^t = x \mid \bar{T}^t, \qquad X_t = \{x_t : x_t = x \mid T_t \;\&\; x \in X\}$$

$$X^t = \{x^t : x^t = x \mid T^t \;\&\; x \in X\}, \qquad X_{tt'} = \{x_{tt'} : x_{tt'} = x \mid T_{tt'} \;\&\; x \in X\}$$

$$X(t) = \{x(t) : x \in X\}$$

The following conventions will also be used:

$$x_{tt} = \phi, \qquad X_{tt} = \{\phi\}$$

Based on the restriction operation, we shall introduce another operation called *concatenation*. Let $x \in A^T$ and $x^* \in A^T$. Then for any t we can define another element $\hat{x}$ in A^T:

$$\hat{x}(\tau) = \begin{cases} x(\tau), & \text{if} \quad \tau < t \\ x^*(\tau), & \text{if} \quad \tau \geq t \end{cases}$$

$\hat{x}$ is represented by $\hat{x} = x^t \cdot x_t^*$ and is called the concatenation of x^t and x_t^*.

Given a set $X \subset A^T$, the family of all restrictions of X as defined above will be denoted by $\bar{X}$, i.e.,

$$\bar{X} = \{\hat{x} : (\hat{x} = x \vee \hat{x} = x_t \vee \hat{x} = x^t \vee \hat{x} = x_{tt'} \vee \hat{x} = \bar{x}_{tt'} \vee \hat{x} = \bar{x}^t) \\ \& \, x \in X \,\&\, t, t' \in T \,\&\, t' \geq t\}$$

The restrictions in Y and the corresponding operations are defined in completely the same way as in X.

For the sake of technical convenience, we shall introduce also the following definition.

Definition 2.3. A time system $S \subset X \times Y$ is input complete if and only if

$$(\forall x)(\forall x^*)(\forall t)(x, x^* \in \mathscr{D}(S) \,\&\, t \in T \rightarrow x^t \cdot x_t^* \in \mathscr{D}(S))$$

and

$$(\forall t)(\{x(t) \mid x \in X\} = A)$$

In the succeeding discussions, every general time system is assumed input complete unless explicitly stated otherwise.

The restrictions of a time system S are defined in reference to the restrictions of inputs and outputs:

$$S_t = \{(x_t, y_t) : x_t = x \mid T_t \,\&\, y_t = y \mid T_t \,\&\, (x, y) \in S\}$$
$$S^t = \{(x^t, y^t) : x^t = x \mid T^t \,\&\, y^t = y \mid T^t \,\&\, (x, y) \in S\}$$
$$S_{tt'} = \{(x_{tt'}, y_{tt'}) : x_{tt'} = x \mid T_{tt'} \,\&\, y_{tt'} = y \mid T_{tt'} \,\&\, (x, y) \in S\}$$
$$\bar{S} = \{\hat{s} : \hat{s} = s \vee \hat{s} = s^t \vee \hat{s} = s_t \vee \hat{s} = s_{tt'}\}$$

We shall also use the following notational convention:

$$X_t = X \mid T_t, \qquad X^t = X \mid T^t, \qquad X_{tt'} = X \mid T_{tt'}$$

and completely analogous for Y, and for S, e.g.,

$$S_t = S \mid T_t, \qquad S^t = S \mid T^t, \qquad \text{and} \quad S_{tt'} = S \mid T_{tt'}$$

Definition 2.4. Let S be a time system $S \subset A^T \times B^T$. The initial state object for S and the initial systems-response function are the global state object and the global systems response of S, respectively. The initial systems response will be denoted by ρ_o, i.e., $\rho_o : C_o \times X \rightarrow Y$ such that

$$(x, y) \in S \leftrightarrow (\exists c)[\rho_o(c, x) = y]$$

We can now introduce the following definition.

Definition 2.5. Let S be a time system and $t \in T$. The state object at t, denoted by C_t, is an initial state object for the restriction S_t; i.e., it is an abstract set such that there exists a function $\rho_t : C_t \times X_t \to Y_t$ such that

$$(x_t, y_t) \in S_t \leftrightarrow (\exists c)[\rho_t(c, x_t) = y_t]$$

ρ_t is referred to as the (systems)-response function at t.

A family of all response functions for a given system, i.e.,

$$\bar{\rho} = \{\rho_t : C_t \times X_t \to Y_t \;\&\; t \in T\}$$

is referred to as a *response family* for S, while $\bar{C} = \{C_t : t \in T\}$ is a family of state objects.

Definition 2.6. Let S be a time system $S \subset X \times Y$, and ρ_t an arbitrary function such that $\rho_t : C_t \times X_t \to Y_t$. ρ_t will be termed consistent with S if and only if it is a response function at t for S, i.e.,

$$(x_t, y_t) \in S_t \leftrightarrow (\exists c)[\rho_t(c, x_t) = y_t]$$

Let

$$S_t^\rho = \{(x_t, y_t) : (\exists c)(y_t = \rho_t(c, x_t))\}$$

Then the consistency condition is expressed as

$$S_t^\rho = S_t$$

Let $\bar{\rho} = \{\rho_t : C_t \times X_t \to Y_t\}$ be a family of arbitrary functions. *Then $\bar{\rho}$ is consistent with a time system S if and only if $\bar{\rho}$ is a (systems-)response family for S, i.e., for all $t \in T$*

$$S_t^\rho = S_t = S_o^\rho \mid T_t$$

Regarding the existence of a response family, we have the following direct consequence of Theorem 1.1.

Proposition 2.1. Every time system has a response family.

A family of arbitrary maps, naturally, cannot be a response family of a time system as seen from the following theorem.

Theorem 2.1. Let $\bar{\rho} = \{\rho_t : C_t \times X_t \to Y_t \;\&\; t \in T\}$ be a family of arbitrary maps. There exists a time system $S \subset X \times Y$ consistent with $\bar{\rho}$; that is, $\bar{\rho}$ is a response family of S if and only if for all $t \in T$ the following conditions hold:

(P1) $\quad (\forall c_o)(\forall x^t)(\forall x_t)(\exists c_t)[\rho_t(c_t, x_t) = \rho_o(c_o, x^t \cdot x_t) \mid T_t]$

(P2) $\quad (\forall c_t)(\forall x_t)(\exists c_o)(\exists x^t)[\rho_t(c_t, x_t) = \rho_o(c_o, x^t \cdot x_t) \mid T_t]$

PROOF. First, we shall prove the *if* part. We have to prove that $S_t^\rho = S_o^\rho \mid T_t$ is satisfied for every $t \in T$. Let $(x_t, y_t) \in S_t^\rho$ be arbitrary. Then $y_t = \rho_t(c_t, x_t)$ for some $c_t \in C_t$. Property (P2) implies that

$$y_t = \rho_t(c_t, x_t) = \rho_o(c_o, x^t \cdot x_t) \mid T_t$$

for some $(c_o, x^t) \in C_o \times X^t$. Hence,

$$(x_t, y_t) = (x^t \cdot x_t, \rho_o(c_o, x^t \cdot x_t)) \mid T_t$$

or $(x_t, y_t) \in S_o^\rho \mid T_t$. Therefore, we have $S_t^\rho \subseteq S_o^\rho \mid T_t$. Conversely, let $(x, y) \in S_o^\rho$ be arbitrary. Then

$$y = \rho_o(c_o, x) = \rho_o(c_o, x^t \cdot x_t)$$

for some c_o. Property (P1) implies, then, that

$$y \mid T_t = \rho_o(c_o, x^t \cdot x_t) \mid T_t = \rho_t(c_t, x_t)$$

for some $c_t \in C_t$, or $(x, y) \mid T_t \in S_t^\rho$. Hence, $S_o^\rho \mid T_t \subseteq S_t^\rho$. Combining the first result with the present one, we have $S_t^\rho = S_o^\rho \mid T_t$.

Next, consider the *only if* part. Let x and c_o be arbitrary. Then $(x, \rho_o(c_o, x)) \in S_o^\rho$. Since $S_o^\rho \mid T_t \subseteq S_t^\rho$, we have that

$$(x, \rho_o(c_o, x)) \mid T_t = (x_t, \rho_o(c_o, x^t \cdot x_t) \mid T_t) \in S_t^\rho$$

or

$$\rho_o(c_o, x^t \cdot x_t) \mid T_t = \rho_t(c_t, x_t)$$

for some c_t. Hence, we have

$$(\forall t)(\forall c_o)(\forall x^t)(\forall x_t)(\exists c_t)(\rho_t(c_t, x_t) = \rho_o(c_o, x^t \cdot x_t) \mid T_t)$$

Let c_t and x_t be arbitrary. Then $(x_t, \rho_t(c_t, x_t)) \in S_t^\rho$. Since $S_t^\rho \subseteq S_o^\rho \mid T_t$, we have

$$\rho_t(c_t, x_t) = \rho_o(c_o, x^t \cdot x_t) \mid T_t$$

for some c_o and x^t. Hence

$$(\forall t)(\forall c_t)(\forall x_t)(\exists c_o)(\exists x^t)(\rho_t(c_t, x_t) = \rho_o(c_o, x^t \cdot x_t) \mid T_t) \qquad \text{Q.E.D.}$$

(b) General Dynamical System

The concept of a dynamical system is related to the way a system evolves in time. It is necessary, therefore, to establish a relationship between the values of the system objects at different times. For this purpose, the response function is not sufficient, and another family of functions has to be introduced.

Definition 2.7. A time system $S \subset X \times Y$ is a dynamical system (or has a dynamical system representation) if and only if there exist two families of mappings

$$\bar{\rho} = \{\rho_t : C_t \times X_t \to Y_t \;\&\; t \in T\}$$

and

$$\bar{\phi} = \{\phi_{tt'} : C_t \times X_{tt'} \to C_{t'} \;\&\; t, t' \in T \;\&\; t' > t\}$$

such that

(i) $\bar{\rho}$ is a response family consistent with S;
(ii) the functions $\phi_{tt'}$ in the family $\bar{\phi}$ satisfy the following conditions
(α) $\rho_t(c_t, x_t) \mid T_{t'} = \rho_{t'}(\phi_{tt'}(c_t, x_{tt'}), x_{t'})$, where $x_t = x_{tt'} \cdot x_{t'}$
(β) $\phi_{tt'}(c_t, x_{tt'}) = \phi_{t''t'}(\phi_{tt''}(c_t, x_{tt''}), x_{t''t'})$, where $x_{tt'} = x_{tt''} \cdot x_{t''t'}$

$\phi_{tt'}$ is termed the state-transition function (on $T_{tt'}$), while $\bar{\phi}$ will be referred to as the state-transition family.

$\phi_{tt'}$ is defined for $t < t'$. However, the following convention:

$$(\gamma) \quad \phi_{tt}(c_t, x_{tt}) = c_t, \qquad \text{for every} \quad t \in T$$

will be used.

Since a dynamical system is completely specified by the two families of mappings $\bar{\rho}$ and $\bar{\phi}$, the pair $(\bar{\rho}, \bar{\phi})$ itself will be referred to as a dynamical system representation or simply as a dynamical system. If a response family has a consistent state-transition family, it will be called a *dynamical systems-response family*. It will be shown that not every response family is a dynamical systems-response family.

Condition (α) represents the *consistency property of the state-transition family* (with the given response family), while (β) represents the *state-transition composition property* (also referred to as the semigroup property). Conditions (α) and (β) are rather strongly related. Actually, under fairly general conditions, property (β) is implied by (α) so that only the consistency of $\bar{\phi}$ with a response family $\bar{\rho}$ is required in order for $\bar{\phi}$ to be qualified as a state-transition family. To arrive at these conditions, we need the following definition.

Definition 2.8. Let $\bar{\rho}$ be a response family consistent with a time system S. $\bar{\rho}$ is a reduced response family if and only if for all $t \in T$

$$(\forall c_t)(\forall \hat{c}_t)[(\forall x_t)(\rho_t(c_t, x_t) = \rho_t(\hat{c}_t, x_t)) \to c_t = \hat{c}_t]$$

The reduction of $\bar{\rho}$, i.e., of associated state objects $\bar{C} = \{C_t : t \in T\}$, does not represent a significant restriction. It only requires that if the two states at any time $t \in T$ lead to the identical future behavior of the system, they ought to be recognized as being the same.

We have now the following theorem.

Theorem 2.2. Let $\bar{\rho} = \{\rho_t : C_t \times X_t \to Y_t\}$ be a response family and $\bar{\phi} = \{\phi_{tt'} : C_t \times X_{tt'} \to C_{t'}\}$ a family of functions consistent with $\bar{\rho}$, i.e., satisfying condition (α) from Definition 2.7:

$$\rho_t(c_t, x_t) \mid T_{t''} = \rho_{t''}(\phi_{tt''}(c_t, x_{tt''}), x_{t''})$$

Then if $\bar{\rho}$ is reduced, $\bar{\phi}$ has the state-transition composition property, i.e., condition (β) from Definition 2.7 is satisfied.

PROOF. From the consistency of $\bar{\phi}$, i.e., condition (α), it follows, for $t \leq t' \leq t''$,

$$\rho_t(c_t, x_t) \mid T_{t''} = \rho_{t''}(\phi_{tt''}(c_t, x_{tt''}), x_t)$$

$$\rho_t(c_t, x_t) \mid T_{t'} = \rho_{t'}(\phi_{tt'}(c_t, x_{tt'}), x_{t'})$$

$$\rho_{t'}(\phi_{tt'}(c_t, x_{tt'}), x_{t'}) \mid T_{t''} = \rho_{t''}(\phi_{t't''}(\phi_{tt'}(c_t, x_{tt'}), x_{t't''}), x_{t''})$$

Since

$$\rho_{t'}(\phi_{tt'}(c_t, x_{tt'}), x_{t'}) \mid T_{t''} = (\rho_t(c_t, x_t) \mid T_{t'}) \mid T_{t''} = \rho_t(c_t, x_t) \mid T_{t''}$$

we have

$$\rho_{t''}(\phi_{tt''}(c_t, x_{tt''}), x_{t''}) = \rho_{t''}(\phi_{t't''}(\phi_{tt'}(c_t, x_{tt'}), x_{t't''}), x_{t''})$$

for every $x_{t''} \in X_{t''}$. Since $\{\rho_t\}$ is reduced, we have

$$\phi_{tt''}(c_t, x_{tt''}) = \phi_{t't''}(\phi_{tt'}(c_t, x_{tt'}), x_{t't''}) \qquad \text{Q.E.D.}$$

It should be pointed out that in the definition of a time system, both input and output objects are defined on the same time set. This is obviously not the most general case (e.g., an output can be defined as a point rather than as a function). We have selected this approach because it provides a convenient framework for the results of traditional interest in systems theory (e.g., the realization theory presented in Chapter III). The properties and behavior of systems that have input and output objects defined on different time sets can be derived from the more complete case we are considering in this book.

(c) General Dynamical Systems in State Space

The concept of a state object as introduced so far has a major deficiency because there is no explicit requirement that the states at any two different times are related; i.e., it is possible, in general, that for any $t \neq t'$, $C_t \cap C_{t'} = \phi$.

To use the potential of the state concept fully, the states at different times ought to be represented as related in an appropriate manner. It should be possible, e.g., to recognize when the system has returned into the "same" state it was before, or has remained in the same state, i.e., did not change at all. In short, the equivalence between states at different times ought to be recognized. What is needed then is a set C such that $C_t = C$ for every $t \in T$. Such a set would represent a state space for the system. At any time the state of the system is then an element of the state space, and dynamics (i.e., change in time) of the system for any given input can be represented as mapping of the state space into itself.

This consideration leads to the following definition.

Definition 2.9. Let S be a time system $S \subset X \times Y$ and C an arbitrary set. C is a state space for S if and only if there exist two families of functions $\bar{\rho} = \{\rho_t : C \times X_t \to Y_t\}$ and $\bar{\phi} = \{\phi_{tt'} : C \times X_{tt'} \to C\}$ such that

(i) for all $t \in T$, $S_t \subset S_t^{\rho}$ and $S_0^{\rho} = \{(x, y) : (\exists c)(y = \rho_0(c, x))\} = S$
(ii) for all $t, t', t'' \in T$
 (α) $\rho_t(c, x_t) \mid T_{t'} = \rho_{t'}(\phi_{tt'}(c, x_{tt'}), x_{t'})$
 (β) $\phi_{tt'}(c, x_{tt'}) = \phi_{t''t'}(\phi_{tt''}(c, x_{tt''}), x_{t''t'})$
 (γ) $\phi_{tt}(c, x_{tt}) = c$

where $x_t = x_{tt'} \cdot x_{t'}$ and $x_{tt'} = x_{tt''} \cdot x_{t''} \cdot x_{t'}$. S is then a dynamical system in the state space C.

Notice that, in general, S_t is a proper subset of S_t^{ρ}. This is so because $\bar{\rho}$ is defined on the entire state space C, while the system might not accept all states at any particular time; i.e., the set of possible states might be restricted at a specified instant of time. This leads to the following definition.

Definition 2.10. A dynamic system in a state space C is full if and only if it has $S_t = S_t^{\rho}$ for all $t \in T$.

Many full dynamical systems, e.g., those that are also linear and time invariant, will be considered in this book. However, in general, a system need not be full; e.g., even a finite automaton is, in general, not a full dynamical system. A simple example of such a case is given by a Mealy type automaton specified by:

Input alphabet: $\{1\}$, Output alphabet: $\{1, 2\}$, State set: $\{1, 2\}$

The state transition and the output function of the system are given by the state-transition diagram in Fig. 2.1.

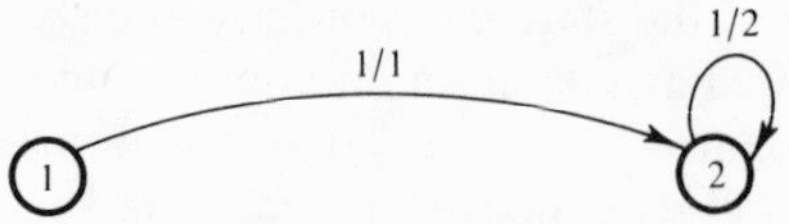

FIG. 2.1

Throughout this book when a state space is used to specify a system, the response-function consistency is assumed to mean condition (i) of Definition 2.9 and not condition (i) of Definition 2.7. The fact that a linear time-invariant system is a full dynamical system is an important property for the class of systems that is responsible for many of their convenient properties (see Chapter VIII).

3. AUXILIARY FUNCTIONS AND SOME BASIC CLASSIFICATION OF SYSTEMS

(a) Auxiliary Functions

Definition of a system as a relation, $S \subset X \times Y$, or of a time system as a relation on time objects $S \subset A^T \times B^T$, provides a starting point for the development of a general theory of systems. In order to be able to study the behavior and properties of a given system or to investigate more specific characterization of various systems, it is necessary to introduce additional concepts.

A system is, in general, a relation, and one cannot determine the output of the system when a given input is applied. In order to be able to do that, the concept of an initial state object is introduced so that the output can be determined whenever an input-initial state pair is given; i.e., the system is represented as a function, termed initial-response function, ρ_0. This, however, is not enough for an efficient specification of the outputs. For instance, when S is a time system and ρ_0 is a given initial-response function, for any given $x \in X$ the output is given by $\rho_0(c_0, x) = y$ where $x: T \to A$ and $y: T \to B$. ρ_0 is often a function of a very high cardinality as, e.g., when T is an infinite set it specifies the output only in abstract. What is needed for an efficient specification of the output is to find some simpler functions, preferably of lower cardinality, which can be used to characterize the systems behavior, e.g., by defining ρ_0 recursively. For the class of time systems, these functions can be obtained in a convenient way by considering the restrictions on subsets of T. There is a whole collection of such functions which we shall term the *auxiliary functions*. We shall consider functions that have been traditionally used in the definition of specific types of systems. We have already seen two auxiliary functions: the response function $\rho_t: C_t \times X_t \to Y_t$ and the state-transition function $\phi_{tt'}: C_t \times X_{tt'} \to C_{t'}$. We shall introduce in this section some additional auxiliary functions, which will be used subsequently to characterize various systems properties and to provide more efficient specification of the systems-response function.

(i) *Output-Generating Function*

One way to provide a more effective method to specify the systems response is to give a "procedure" in terms of a function which will generate the value of the output at any given time; i.e., for any given x^t and c_0, the required function will give $y(t)$. If such a function is given, the output $y: T \to B$ is defined for any given pair (x, c_0) in the sense that $y(t)$, the value of that output at any time $t \in T$, can be determined. This leads to the following definition.

Definition 3.1. Given a response family $\bar{\rho}$ for a time system, let $\mu_{tt'}$ be a relation

$$\mu_{tt'} \subset C_t \times \bar{X}_{tt'} \times Y(t')$$

such that

$$(c_t, \bar{x}_{tt'}, y(t')) \in \mu_{tt'} \leftrightarrow (\exists x_t)(\exists y_t)[y_t(t') = y(t') \;\&\; y_t = \rho_t(c_t, x_t) \;\&\; \bar{x}_{tt'} = x_t \mid \bar{T}_{tt'}]$$

When $\mu_{tt'}$ is a function such that

$$\mu_{tt'}: C_t \times \bar{X}_{tt'} \to Y(t')$$

it is termed an output-generating function on $\bar{T}_{tt'}$. $\bar{\mu} = \{\mu_{tt'}: t, t' \in T \;\&\; t' \geq t\}$ is termed an output-generating family (Fig. 3.1).

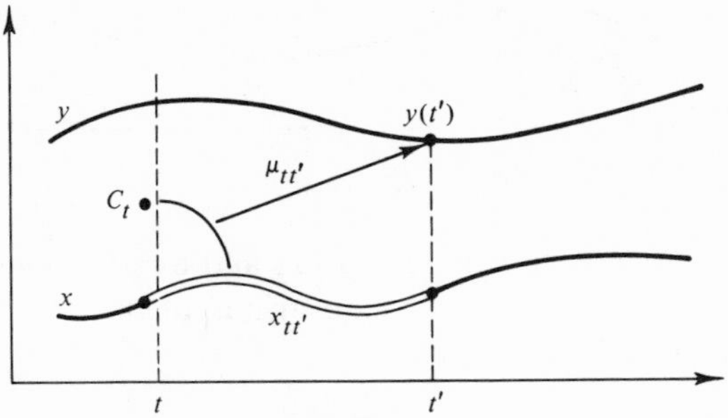

FIG. 3.1

Notice that the input restriction in the domain of $\mu_{tt'}$ is defined on $\bar{T}_{tt'} = T_{tt'} \cup \{t'\}$ rather than on $T_{tt'}$.

Apparently, the relation $\mu_{tt'}$ exists and is well defined for any general time system. However, the existence of an output-generating function, i.e., $\mu_{tt'}$ as a function, requires certain conditions to be satisfied.

(ii) *Output Function*

Time evolution of a dynamical system is customarily described in terms of the state transition, and it is of interest to relate the changes in states to the changes in outputs; specifically, the state at any time $t \in T$ ought to be related with the value of the output at that time. This leads to the following definition.

Definition 3.2. Let S be a time system with the response family $\bar{\rho}$ and λ_t a relation

$$\lambda_t \subset C_t \times X(t) \times Y(t)$$

such that

$$(c_t, x(t), y(t)) \in \lambda_t \leftrightarrow (\exists x_t)(\exists y_t)[y_t = \rho_t(c_t, x_t) \;\&\; x(t) = x_t(t) \;\&\; y(t) = y_t(t)]$$

When λ_t is a function such that

$$\lambda_t : C_t \times X(t) \to Y(t)$$

it is termed an output function at t, while $\bar{\lambda} = \{\lambda_t : t \in T\}$ is an output-function family (Fig. 3.2).

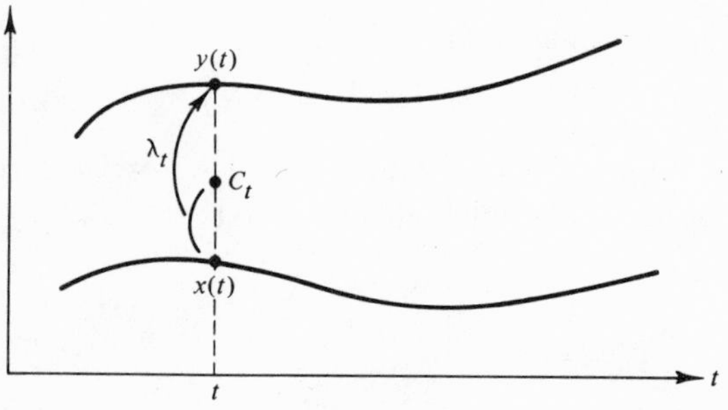

FIG. 3.2

Apparently, λ_t is a well-defined relation and exists for any general time system. The conditions for the existence of an output function will be presented in the next section.

(iii) *State-Generating Function*

For a dynamical system, the state at any time t is determined by the initial state c_0 and the initial input restriction x^t. However, for certain classes of systems, there exists a time $\hat{t} \in T$ such that the state at any subsequent time is determined solely by the past input and output restrictions; i.e., no reference to the state is needed. This leads to the following definition.

Definition 3.3. Let $\bar{\rho}$ be a response family for a time system S and η^t a relation

$$\eta^t \subset X^t \times Y^t \times C_t$$

such that

$$(x^t, y^t, c_t) \in \eta^t \leftrightarrow (\forall x_t)(\forall y_t)[(x^t \cdot x_t, y^t \cdot y_t) \in S \rightarrow y_t = \rho_t(c_t, x_t)]$$

When η^t is a function such that

$$\eta^t : X^t \times Y^t \rightarrow C_t$$

it is termed a state-generating function at t, while $\bar{\eta} = \{\eta^t : X^t \times Y^t \rightarrow C_t \,\&\, t \in T\}$ is termed a state-generating family (Fig. 3.3).

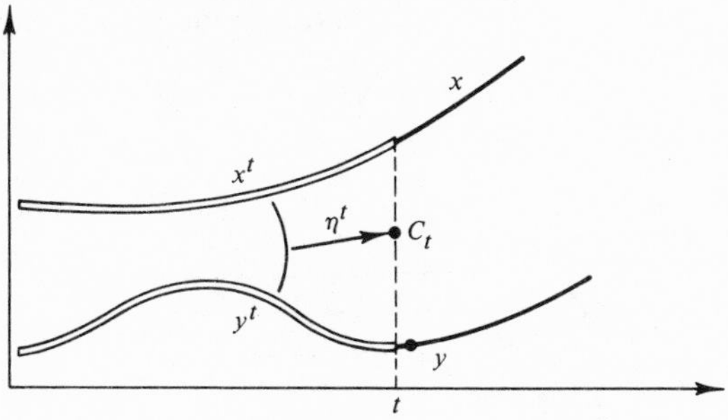

FIG. 3.3

Again, although η^t is always defined, the existence of a state-generating family requires certain conditions which, in this case, are of a more special kind.

(b) Some Classification of Time Systems

The auxiliary functions for any $t \in T$ are, in general, different. However, when some of them are the same for all $t \in T$ or are obtained from the same function by appropriate restrictions, various forms of time invariance can be introduced.

(i) *Static and Memoryless Systems*

The first type of time invariance refers to the relationship between the system objects at any given time and is intimately related with the response function.

Definition 3.4. A system S is static if and only if there exists an initial response function $\rho_0 : C_0 \times X \rightarrow Y$ for S such that for all $t \in T$

$$(\forall c_0)(\forall x)(\forall \hat{x})[x(t) = \hat{x}(t) \rightarrow \rho_0(c_0, x)(t) = \rho_0(c_0, \hat{x})(t)]$$

In other words, the system is static if and only if for any $t \in T$ there exists a map $K_t : C_o \times X(t) \to Y(t)$ such that

$$(x, y) \in S \leftrightarrow (\exists c_o \in C_o)(\forall t)(y(t) = K_t(c_o, x(t)))$$

Any time system that is not static is termed a dynamic system.

Intuitively, a system is static if the value of its output at any time t depends solely on the current value of the input and the state from which the evolution has initially started; i.e., if $x(t)$ becomes constant over a period of type, $y(t)$ becomes constant too. On the other hand, the output of a dynamic system depends not only on the current value of the input but also on the past "history" of that input as well. Notice that, in general, reference to the initial state had to be made (Fig. 3.4).

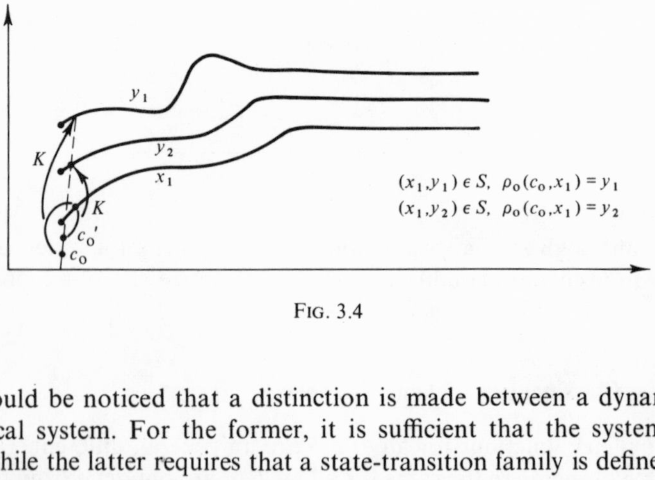

FIG. 3.4

It should be noticed that a distinction is made between a dynamic and dynamical system. For the former, it is sufficient that the system is not static, while the latter requires that a state-transition family is defined. This, perhaps, is not the most fortunate choice of terminology; however, it has been selected because it corresponds to the common usage in the already established specialized theories.

A related notion to a static system is the following definition.

Definition 3.5. A time system S is memoryless if and only if it is a static system such that

$$(\forall x)(\forall \hat{x})(\forall c_o)(\forall \hat{c}_o)[x(t) = \hat{x}(t) \to \rho_o(c_o, x)(t) = \rho_o(\hat{c}_o, \hat{x})(t)]$$

or, in terms of the mappings $K_t : C_o \times X(t) \to Y(t)$,

$$(\forall c_o)(\forall \hat{c}_o)(\forall x)(\forall \hat{x})[x(t) = \hat{x}(t) \to K_t(c_o, x(t)) = K_t(\hat{c}_o, \hat{x}(t))]$$

i.e., there exists a mapping $K_t^*: X(t) \to Y(t)$ such that $K_t^*(x(t)) = K_t(c_0, x(t))$ (Fig. 3.5).

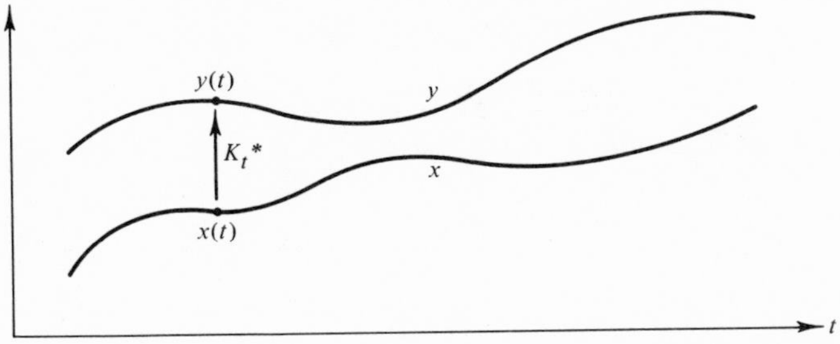

FIG. 3.5

Apparently, a memoryless system is completely characterized by the map $K_t^*: A \to B$. A system that does not satisfy Definition 3.5 is termed a *system with memory*.

(ii) *Time-Invariant Dynamical Systems*

The second kind of time invariancy refers to how system responses at two different times compare. To introduce appropriate concepts, we shall assume for this kind of time invariancy that the time set T is a right segment of a linearly ordered Abelian group $\bar{T}$ whose group operation (addition) will be denoted by $+$. More precisely, $T = \{t : t \geq o\}$, where o is the identity element of $\bar{T}$ and the addition is related with the linear ordering as

$$t \leq t' \leftrightarrow t' - t \geq o$$

The time set T defined above will be referred to as the *time set for stationary systems*.

For each $t \in \bar{T}$, let $F^t: \bar{X} \to \bar{X}$ denote an operator such that

$$(\forall t')[F^t(x)(t') = x(t' - t)]$$

Notice that F^t is defined for $t < o$ as well as $t \geq o$ and that whether or not F^t is meaningful depends on its argument. In general, $F^t(x_{t't''}) \in X_{(t'+t)(t''+t)}$ holds whenever $F^t(x_{t't''})$ is defined.

F^t is termed the *shift operator*; it simply shifts a given time function for the time interval indicated by the superscript, leaving it otherwise completely unchanged (Fig. 3.6). We shall use the same symbol F^t for the shift operator in $\bar{Y}$ and define F^t also in $\bar{S}$, $F^t: \bar{S} \to \bar{S}$, such that

$$F^t(x, y) = (F^t(x), F^t(y))$$

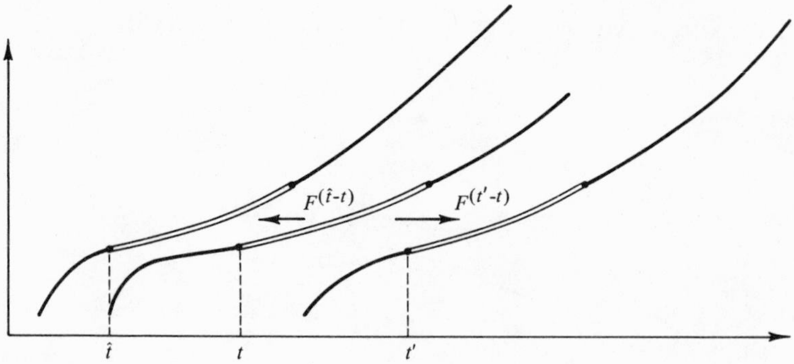

FIG. 3.6

We can introduce now the following definition.

Definition 3.6. A time system defined on the time set for stationary systems is fully stationary if and only if (Fig. 3.7)

$$(\forall t)[t \in T \rightarrow F^t(S) = S_t]$$

and stationary if and only if

$$(\forall t)(\forall t' \geq t)(t, t' \in T \rightarrow S_{t'} \subset F^{t'-t}(S_t))$$

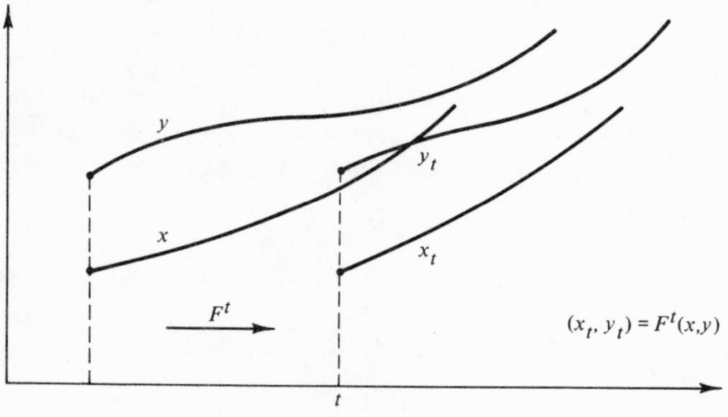

FIG. 3.7

Apparently, if a system is fully stationary, $F^{-t}(S_{t'} \mid T_t) = F^{-t'}(S_{t'})$ for any $t \geq t' \in T$; i.e., starting from any given time, its future evolution is the same except for the shift for the appropriate time interval.

When some given input and output objects X and Y satisfy the condition

$$(\forall t)(X_t = F^t(X)) \quad \text{and} \quad (\forall t)(Y_t = F^t(Y))$$

they will be referred to as the objects for a stationary system.

When the systems response for a time system is given, we have the following definition.

Definition 3.7. A (systems-) response family is time invariant if and only if for any $t \in T$ there exists a one-to-one correspondence $G_{ot}: C_o \to C_t$ such that

$$(\forall t)(\forall c_t)(\forall x_t)[\rho_t(c_t, x_t) = F^t(\rho_o(G_{ot}^{-1}(c_t), F^{-t}(x_t)))]$$

Apparently, a systems response is time invariant if and only if for any state at any time the output can be obtained from the initial response function by a shift in time (Fig. 3.8).

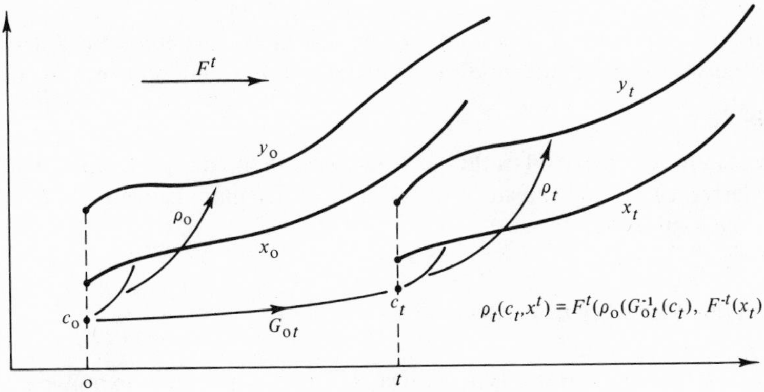

FIG. 3.8

Completely analogously, we have the following definition.

Definition 3.8. A dynamical system is time invariant if and only if

(i) its systems response family is time invariant;
(ii) for all $t, t' \in T$, $t' > t$,

$$(\forall t)(\forall t')(\forall c_t)(\forall x_{tt'})[\phi_{tt'}(c_t, x_{tt'}) = G_{\hat{t}t'}(\phi_{o\hat{t}}(G_{ot}^{-1}(c_t), F^{-t}(x_{tt'})))$$

where $t' = \hat{t} + t$ and $G_{tt'} = G_{ot'} \circ G_{ot}^{-1}$.

Apparently, for a time-invariant system, the state-transition function at any time can be obtained from the initial-response function by applying the shift operator.

Regarding the relationship between the stationarity and time invariance, the following is an immediate consequence of the definitions.

Proposition 3.1. A system with a time-invariant response (and therefore a time-invariant dynamical system) is stationary.

4. CAUSALITY

Causality refers essentially to the ability to predict the outcome or consequences of some events in the future; in other words, one has a causal representation of a phenomenon if one can recognize the associated causes and effects and has a way to describe how the causes result in effects.

Given a general system $S \subset V_1 \times \cdots \times V_n$, causality is introduced in the description of S in the three steps:

(i) It is established which objects are inputs and which are outputs.

(ii) It is recognized which output objects depend explicitly on other outputs while only implicitly on the system inputs.

(iii) A description of the time evolution of the system is provided such that the value of the output at any time depends solely on the past, i.e., the preceding state-input pair.

We shall be concerned in this section with the third aspect, which can also be referred to as time causality; it is the most pertinent causality concept for the time systems study.

(a) Time Causality Concepts

There are two causality-type concepts. The first is based on a given initial state object or rather a given initial-response function and is also referred to as *nonanticipation*. The second is given solely in terms of the inputs and outputs and is referred to as *past-determinacy*. The adjective "causal" will be used as generic covering both cases.

(i) *Nonanticipation*

Let ρ_0 be an initial-response function for a time system S. We have then the following definition.

Definition 4.1. An initial systems-response function $\rho_0 : C_0 \times X \to Y$ for a time system $S \subset X \times Y$ is nonanticipatory if and only if

$$(\forall t)(\forall c_0)(\forall x)(\forall \hat{x})[x \mid \bar{T}^t = \hat{x} \mid \bar{T}^t \to \rho_0(c_0, x) \mid \bar{T}^t = \rho_0(c_0, \hat{x}) \mid \bar{T}^t]$$

In a nonanticipatory system, the changes in the output cannot precede—"anticipate"—the changes in the input (Fig. 4.1). For certain classes of systems, the concept of nonanticipation can be strengthened as follows.

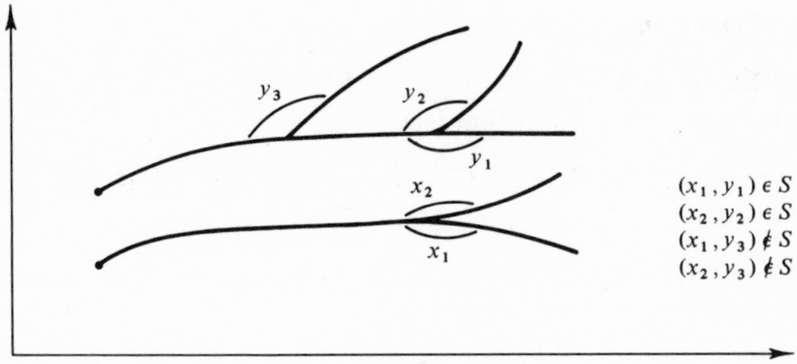

FIG. 4.1

Definition 4.2. An initial systems-response function $\rho_o : C_o \times X \to Y$ is strongly nonanticipatory if and only if

$$(\forall t)(\forall c_o)(\forall x)(\forall \hat{x})[x \mid T^t = \hat{x} \mid T^t \to \rho_o(c_o, x) \mid \bar{T}^t = \rho_o(c_o, \hat{x}) \mid \bar{T}^t]$$

Notice that Definitions 4.1 and 4.2 refer to time systems rather than to dynamical systems.

The difference between the nonanticipatory and strongly nonanticipatory response functions is that in the latter the present value of the output, $y(t)$, does not depend upon the present value of the input, $x(t)$, since the restriction in the antecedent in Definition 4.2 is on T^t while in Definition 4.1 is on $\bar{T}^t = T^t \cup \{t\}$; however, in both cases the output is restricted to $\bar{T}^t$.

It should also be pointed out that ρ_o is defined as a full function. This is an important restriction in the case of nonanticipation because it prevents some systems of having a nonanticipatory systems-response function. We shall introduce therefore the following definition.

Definition 4.3. Let $R \subset C_o \times X$ and $\rho_o : (R) \to Y$. ρ_o is an incomplete nonanticipatory initial systems response of S if and only if

(i) ρ_o is consistent with S, i.e.,

$$(x, y) \in S \leftrightarrow (\exists c_o)[\rho_o(c_o, x) = y \;\&\; (c_o, x) \in R]$$

(ii) $$(\forall t)(\forall c_o)(\forall x)(\forall \hat{x})[(c_o, x) \in R \;\&\; (c_o, \hat{x}) \in R \;\&\; x \mid \bar{T}^t = \hat{x} \mid \bar{T}^t \to \rho_o(c_o, x) \mid \bar{T}^t = \rho_o(c_o, \hat{x}) \mid \bar{T}^t]$$

Definition 4.4. We shall say that the system S is (strongly) nonanticipatory if and only if it has a complete (strongly) nonanticipatory initial-response function.

(ii) *Past-Determinacy*

Definition 4.5. A time system $S \subset A^T \times B^T$ is past-determined from $\hat{t}$ if and only if there exists $\hat{t} \in T$ such that (see Fig. 4.2)

(i) $(\forall(x, y) \in S)(\forall(x', y') \in S)(\forall t \geq \hat{t})([(x^{\hat{t}}, y^{\hat{t}}) = (x'^{\hat{t}}, y'^{\hat{t}}) \,\&\, x_{\hat{t}t} = x'_{\hat{t}t}] \rightarrow \bar{y}_{\hat{t}t} = \bar{y}'_{\hat{t}t})$

(ii) $(\forall(x^{\hat{t}}, y^{\hat{t}}))(\forall x_{\hat{t}})(\exists y_{\hat{t}})((x^{\hat{t}}, y^{\hat{t}}) \in S^{\hat{t}} \rightarrow (x^{\hat{t}} \cdot x_{\hat{t}}, y^{\hat{t}} \cdot y_{\hat{t}}) \in S)$

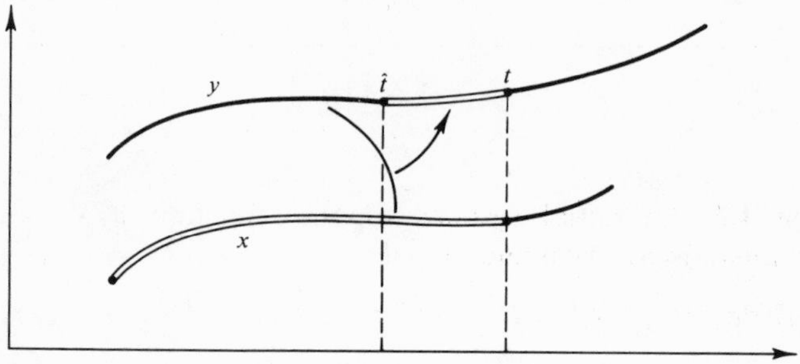

FIG. 4.2

Past-determinacy means that there exists $\hat{t} \in T$ such that for any $t \geq \hat{t}$, the future evolution of the system is determined solely by the past observations, and there is no need to refer to an auxiliary set as, e.g., the initial state object.

Condition (ii), which will be referred to as the *completeness property*, is introduced as a mathematical convenience.

(b) Existence of Causal-Response Family

Theorem 4.1. Every time system has an incomplete nonanticipatory initial systems response.

PROOF. Let $\equiv \subseteq S \times S$ be a relation such that $(x, y) \equiv (x', y')$ if and only if $y = y'$. Then, apparently, $\equiv$ is an equivalence relation. Let $S/\equiv\; = \{[s]\} \equiv C_o$, where $[s] = \{s^* \mid s^* \equiv s \,\&\, s^* \in S\}$.† Let $\rho_o: C_o \times X \rightarrow Y$ such that

$$\rho_o([s], x) = \begin{cases} y & \text{if } (x, y) \in [s] \\ \text{undefined otherwise} \end{cases}$$

† In this book the following convention will be used for a quotient set. Let E be an equivalence relation on a set X. Then the quotient set X/E will be represented by $X/E = \{[x]\}$, where $[x]$ is the equivalence class of x, i.e.,

$$[x] = \{x^* : (x, x^*) \in E \,\&\, x^* \in X\}$$

and [ ] will be considered as the natural mapping, i.e., $[\;]: X \rightarrow X/E$.

Notice that ρ_o is properly defined, because if $(x, y) \in [s]$ and $(x, y') \in [s]$, then $y = y'$. In general, ρ_o is a partial function. First, we shall show that

$$S = \{(x, y):(\exists c)(c \in C_o \;\&\; y = \rho_o(c, x))\} \equiv S'$$

If $(x, y) \in S$, then $\rho_o([(x, y)], x) = y$ by definition. Hence, $(x, y) \in S'$, or $S \subseteq S'$. Conversely, if $(x, y) \in S'$, then $y = \rho_o([s], x)$ for some $[s] \in C_o$. Then $(x, y) \in [s]$ follows from the definition of ρ_o. Hence, $S' \subseteq S$. Furthermore, if the values of $\rho_o([s], x)$ and $\rho_o([s], x')$ are defined, then $\rho_o([s], x) = \rho_o([s], x')$ for any x and x'. Hence, condition (ii) in Definition 4.3 is trivially satisfied. Q.E.D.

Theorem 4.1 cannot be extended for the full initial systems-response function, i.e., when ρ_o is a full function. There are time systems that do not have a (complete) nonanticipatory initial systems response as defined in Definition 4.1; in other words, requirement for the initial response to be a full function prevents a causal representation of the systems in the sense of nonanticipation. Such systems can either be considered to be essentially noncausal or it can be assumed that only an incomplete description of the system is available and that noncausality is due to having only partial information. This can be best shown by an example as given in Fig. 4.3.

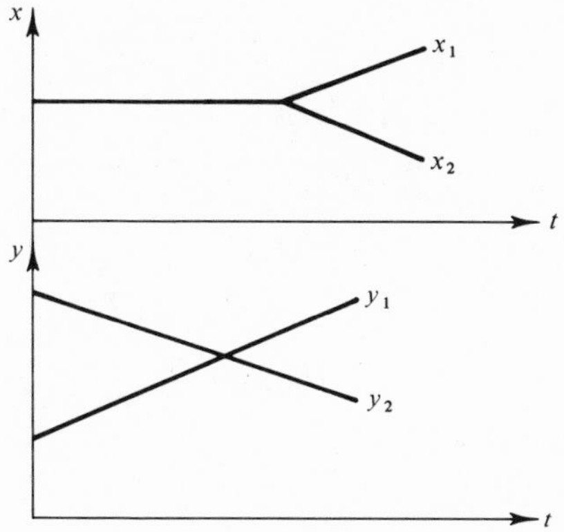

FIG. 4.3

Consider a system S which has only two elements, $S = \{(x_1, y_1), (x_2, y_2)\}$, which are as shown in Fig. 4.3. Since the initial segments of both x_1 and x_2 are the same, while those of y_1 and y_2 are different, the initial state object ought to have at least two elements if we want to have a nonanticipatory initial-response function. Let $C_o = \{c, c'\}$ and $\rho_o(c, x_1) = y_1$, $\rho_o(c', x_2) = y_2$. If ρ_o is a full function, (c, x_2) is also in the domain of ρ_o. Therefore, either $\rho_o(c, x_2) = y_1$ or $\rho_o(c, x_2) = y_2$. But $\rho_o(c, x_2) = y_1$ implies $(x_2, y_1) \in S$, i.e., ρ_o is not consistent with S, while $\rho_o(c, x_2) = y_2$ violates the nonanticipation condition in Definition 4.1 since the initial segments of x_1 and x_2 are the same, while y_1 and y_2 are not. The system S does not have, therefore, a complete nonanticipatory response function.

Definition 4.6. A response family $\bar{\rho} = \{\rho_t : t \in T\}$ is called nonanticipatory if and only if every ρ_t is an initial nonanticipatory response function of S_t.

Theorem 4.2. A time system has a nonanticipatory response family if and only if it has an initial nonanticipatory response function.

PROOF. The *only if* part is obvious. Let us consider the *if* part. Let $\rho_o : C_o \times X \to Y$ be an initial nonanticipatory systems response. Let $C_t = C_o \times X^t$ and $\rho_t : C_t \times X_t \to Y_t$ such that if $c_t = (c_o, \hat{x}^t)$, then

$$\rho_t(c_t, x_t) = \rho_o(c_o, \hat{x}^t \cdot x_t) \mid T_t, \qquad \text{for} \quad t \in T$$

Then the consistency conditions in Theorem 2.1 are trivially satisfied, i.e., $\bar{\rho} = \{\rho_t : t \in T\}$ is a response family. Furthermore, suppose $x_t \mid \bar{T}_{tt'} = x_t' \mid \bar{T}_{tt'}$ for $t' \geq t$. Then if $c_t = (c_o, \hat{x}^t)$,

$$\rho_t(c_t, x_t) \mid \bar{T}_{tt'} = \rho_o(c_o, \hat{x}^t \cdot x_t) \mid \bar{T}_{tt'}$$

and

$$\rho_t(c_t, x_t') \mid \bar{T}_{tt'} = \rho_o(c_o, \hat{x}^t \cdot x_t') \mid \bar{T}_{tt'}$$

Since ρ_o is nonanticipatory and $\hat{x}^{t'} \cdot x_t \mid \bar{T}^{t'} = \hat{x}^t \cdot x_t' \mid \bar{T}^{t'}$, we have

$$\begin{aligned}
\rho_t(c_t, x_t) \mid \bar{T}_{tt'} &= (\rho_o(c_o, \hat{x}^t \cdot x_t) \mid \bar{T}^{t'}) \mid \bar{T}_{tt'} \\
&= (\rho_o(c_o, \hat{x}^t \cdot x_t') \mid \bar{T}^{t'}) \mid \bar{T}_{tt'} \\
&= \rho_o(c_o, \hat{x}^t \cdot x_t') \mid \bar{T}_{tt'} \\
&= \rho_t(c_t, x_t') \mid \bar{T}_{tt'}
\end{aligned}$$

Hence, ρ_t is nonanticipatory. Q.E.D.

(c) Causality and Output Functions

Conditions for the existence of an output function and an output-generating function can be given directly in terms of the nonanticipation of systems response.

Dependence of the output function upon nonanticipation is clearly indicated by the following proposition.

Proposition 4.1. A time system has an output-function family $\bar{\lambda} = \{\lambda_t : C_t \times X(t) \to Y(t)\}$ if the system is nonanticipatory.

PROOF. Since the system is nonanticipatory, there exists, by Theorem 4.2, a nonanticipatory response family $\bar{\rho} = \{\rho_t : t \in T\}$. Let λ_t be defined for the nonanticipatory response family. Suppose $(c_t, x(t), y(t)) \in \lambda_t$ and $(c_t, x'(t), y'(t)) \in \lambda_t$ where $x(t) = x'(t)$. Since

$$x_t \mid \bar{T}_{tt} = x(t) = x'(t) = x_t' \mid \bar{T}_{tt}$$

and since ρ_t is nonanticipatory, we have that

$$\rho_t(c_t, x_t) \mid \bar{T}_{tt} = \rho_t(c_t, x_t') \mid \bar{T}_{tt}$$

or $y(t) = y'(t)$. Hence λ_t is a mapping such that $\lambda_t : C_t \times X(t) \to Y(t)$. Q.E.D.

The concept of an output function illustrates one of the important roles of the concept of state: If a state is given and the system is nonanticipatory, all information about the past of the system, necessary to specify the present value of output, is contained in the state itself.

Dependence of the output-generating function upon the nonanticipation of the system is quite similar to the dependence of the output function and is given by the following proposition.

Proposition 4.2. A time system has an output-generating family $\bar{\mu} = \{\mu_{tt'} : C_t \times X_{tt'} \to Y_t(t')\}$ if the system is nonanticipatory.

PROOF. The proof is similar to that of Proposition 4.1. Q.E.D.

From Proposition 4.1 it is obvious that the output of a nonanticipatory time system can be determined solely by the present state and the present value of the input. For some systems, however, the present output depends solely upon the present state and does not depend upon the present value of the input. For the analysis of these systems, a somewhat stronger notion, namely, that of strong nonanticipation, is needed.

Proposition 4.3. A time system has an output function such that for all $t \in T$

$$(\forall x(t))(\forall \hat{x}(t))[c_t = \hat{c}_t \to \lambda_t(c_t, x(t)) = \lambda_t(\hat{c}_t, \hat{x}(t))]$$

if it is a strongly nonanticipatory system.

PROOF. Since the system is strongly nonanticipatory, there exists a strongly nonanticipatory initial system response ρ_0. Then the same procedure as used in Proposition 4.1 proves the statement. Q.E.D.

When the system is strongly nonanticipatory, for every $t \in T$ there apparently exists a map

$$K_t : C_t \to B$$

such that

$$(\forall x(t))[\lambda_t(c_t, x(t)) = K_t(c_t)]$$

(d) Existence of Past-Determinacy

First of all, a distinction between past-determinacy and functionality ought to be made. A system can be functional (i.e., $S: X \rightarrow Y$), yet it does not have to be past-determined and vice versa. Notice also that for a functional system (i.e., when there is only one initial state), the concepts of strong nonanticipation and the past-determinacy from o coincide.

For any $(x, y) \in S$, let $S(x^t, y^t)$ be the set

$$S(x^t, y^t) = \{(x_t, y_t) : (x^t \cdot x_t, y^t \cdot y_t) \in S\}$$

where $S(x^t, y^t)$ is the family of "successors" to (x^t, y^t). We now have the following proposition.

Proposition 4.4. A system is past-determined from $\hat{t}$ if and only if for all $t \geq \hat{t}$, $S(x^t, y^t)$ is a strongly nonanticipatory function

$$S(x^t, y^t): X_t \rightarrow Y_t$$

PROOF. First, let us consider the *only if* part. Suppose $(x_t, y_t) \in S(x^t, y^t)$ and $(\hat{x}_t, \hat{y}_t) \in S(x^t, y^t)$. Suppose $x_t \mid T_{tt'} = \hat{x}_t \mid T_{tt'}$ for $t' > t$. Then

$$x^t \cdot x_t \mid T^{\hat{t}} = x^t \cdot \hat{x}_t \mid T^{\hat{t}}$$

and

$$y^t \cdot y_t \mid T^{\hat{t}} = y^t \cdot \hat{y}_t \mid T^{\hat{t}} \;\&\; x^t \cdot x_t \mid T_{\hat{t}t'} = x^t \cdot \hat{x}_t \mid T_{\hat{t}t'}$$

hold. Since the system is past-determined, $y^t \cdot y_t \mid \overline{T}_{\hat{t}t'} = y^t \cdot \hat{y}_t \mid \overline{T}_{\hat{t}t'}$ holds, which implies $y_t \mid \overline{T}_{tt'} = \hat{y}_t \mid \overline{T}_{tt'}$. Since $t' > t$ is arbitrary, we have that if $x_t = \hat{x}_t$, then $y_t = \hat{y}_t$, which implies $S(x^t, y^t)$ is functional. Hence, $S(x^t, y^t)$ is a strongly nonanticipatory function. Conversely, suppose $(x, y) \in S$ and $(\hat{x}, \hat{y}) \in S$. Furthermore, suppose $(x^{\hat{t}}, y^{\hat{t}}) = (\hat{x}^{\hat{t}}, \hat{y}^{\hat{t}})$ and $x_{\hat{t}t} = \hat{x}_{\hat{t}t}$. Since $S(x^{\hat{t}}, y^{\hat{t}})$ is a strongly nonanticipatory function, $\bar{y}_{\hat{t}t} = \bar{\hat{y}}_{\hat{t}t}$. Q.E.D.

Past-determinacy is related to the state-generating family $\bar{\eta}$. In this respect, we have the following proposition.

Proposition 4.5. A system is past-determined from $\hat{t}$ if and only if for any $t \geq \hat{t}$, it has a state-generating function η^t such that ρ_t is strongly nonanticipatory.

PROOF. First, we shall consider the *only if* part. Suppose S is past-determined. Let $C_t = S^t$ for $t \geq \hat{t}$. Then the desired result immediately follows from Proposition 4.4, where the state-generating function $\eta^t : S^t \to C_t$ is the identity function and $\rho_t : C_t \times X_t \to Y_t$ is $\rho_t(c_t, x_t) = S(c_t)(x_t)$. Conversely, suppose S has a state-generating function $\eta^t : S^t \to C_t$ for $t \geq \hat{t}$. Suppose $(x^t, y^t) = (\hat{x}^t, \hat{y}^t) \in S^t$ and $x_{tt'} = \hat{x}_{tt'}$ for $t' \geq t \geq \hat{t}$. Let $c_{\hat{t}} = \eta^{\hat{t}}(x^{\hat{t}}, y^{\hat{t}}) = \eta^{\hat{t}}(\hat{x}^{\hat{t}}, \hat{y}^{\hat{t}})$. Since $x^t \cdot x_{tt'} = \hat{x}^t \cdot \hat{x}_{tt'}$ and since $\rho_{\hat{t}}$ is strongly nonanticipatory, by assumption, the following holds:

$$\rho_{\hat{t}}(c_{\hat{t}}, x_{\hat{t}}) \mid \overline{T}_{\hat{t}t'} = \rho_{\hat{t}}(c_{\hat{t}}, \hat{x}_{\hat{t}}) \mid \overline{T}_{\hat{t}t'}$$

where

$$x^t \cdot x_{tt'} \mid T_{\hat{t}t'} = x_{\hat{t}} \mid T_{\hat{t}t'}, \quad \text{and} \quad \hat{x}^t \cdot \hat{x}_{tt'} \mid T_{\hat{t}t'} = \hat{x}_{\hat{t}} \mid T_{\hat{t}t'}$$

Hence the system is past-determined. Q.E.D.

Chapter III

GENERAL REALIZATION THEORY

Realization theory for the class of dynamical systems is concerned with existence of a dynamical representation for an appropriately defined time system. Usually, there is given a family of response-type functions $\bar{\rho}$, and the realization problem consists in determining whether there exists a family of state-transition functions $\bar{\phi}$ and a time system S so that $(\bar{\rho}, \bar{\phi})$ is a dynamical realization of S. Problems of this type are considered in this chapter.

Since some basic factors of the realization theory can be proven for a system whose dynamics is described solely in reference to a family of state objects rather than a state space, a portion of the realization theory is considered in that framework first. This arrangement is in accord with the formalization approach proclaimed in Chapter I, namely, to consider any given problem with a system having minimal structure.

In Section 1, conditions for the consistency and realizability of a family of response functions are given which are nontrivial only if the response family consists of full functions, i.e., partial functions are not accepted as the elements of $\bar{\rho}$. In Section 2, it will be shown that nonanticipation is necessary and sufficient for a decomposition of a time system in two families of functions: the state-transition functions and the output functions. Furthermore, when the system is strongly nonanticipatory, the output functions can be defined solely on the state objects. Since the output function is static, the entire dynamics of the system is described then by the state-transition family. Because of this important and convenient property, the representation of a system by the pair $(\bar{\phi}, \bar{\lambda})$ is referred to as canonical. Furthermore, it is shown how the states can be characterized by an equivalence relation defined on the input and the initial-states pairs.

In Section 3, the relationship between the auxiliary functions (and various systems representations) is shown to be a commutative diagram (Fig. 3.4). This procedure provides an important insight for the nature of the state space, the origin of that concept, and the reasons for its importance. Often, the state space is given as a primary concept in the very definition of a dynamical system. By showing how the state space can be constructed from a given set of input–output pairs, we have justified our viewpoint that the state space is a secondary, i.e., derived, concept. It is shown how the state space can be generated by appropriate equivalence relations (Fig. 3.9). A procedure (based on the results developed in the preceding chapters) to construct the state space representation when a time system (i.e., the input–output pairs) is given is then outlined (Fig. 3.10).

1. REALIZABILITY AND DYNAMICAL REPRESENTATION

A pair of functions $(\bar{\rho}, \bar{\phi})$ is a dynamical system representation of general time system S

$$S \subset A^T \times B^T \tag{3.1}$$

if the appropriate conditions (as given in Chapter II) are satisfied so that $\bar{\rho}$ is consistent with S and $\bar{\phi}$ is a family of state-transition functions.

The time system is defined in (3.1) as a set, and as such it has to be specified by a defining function. This is often done by giving the systems-response family $\bar{\rho} = \{\rho_t : C_t \times X_t \to Y_t \;\&\; t \in T\}$ either directly or in terms of an output-generating function. One encounters then the situation in which there is given a family of response-type functions $\bar{\rho}$, and the question arises whether there indeed exists a dynamical system for which $\bar{\rho}$ is a response family. This question can be answered in two steps:

(i) Given $\bar{\rho}$, is there a system S such that $\bar{\rho}$ is consistent with S?
(ii) Given $\bar{\rho}$, which is a response family for a system S, is there a family of transition functions $\bar{\phi}$ so that $(\bar{\rho}, \bar{\phi})$ is a dynamical representation of S?

These two problems will be considered in this section.

Definition 1.1. Given a family of functions $\bar{\rho} = \{\rho_t : C_t \times X_t \to Y_t \;\&\; t \in T\}$, we shall say that $\bar{\rho}$ has a dynamical realization or, simply, is realizable if and only if there exist a time system S and a family of functions $\bar{\phi} = \{\phi_{tt'} : C_t \times X_{tt'} \to C_{t'}\}$ so that $\bar{\rho}$ is consistent with S and $(\bar{\rho}, \bar{\phi})$ is a dynamical representation of S.

Realizability of a set of maps $\bar{\rho} = \{\rho_t : C_t \times X_t \to Y_t\}$ depends upon the properties of those maps and the sets involved. Recall that the assumptions made for $\bar{\rho}$ were the following:

(i) C_t is an arbitrary set.
(ii) $X_t \subset A^{T_t}$ and $Y_t \subset B^{T_t}$, and furthermore

$$(\forall x^t)(\forall x_t)[x^t \cdot x_t \in X]$$

(iii) ρ_t is a full function, i.e., it is defined on the entire set $C_t \times X_t$.

(a) Consistency of $\bar{\rho}$

The first realizability question concerned with the consistency of the given maps $\bar{\rho}$ with a time system S has already been answered by Theorem 2.1, Chapter II. It has been shown there that when a family of arbitrary maps $\bar{\rho}$ satisfies conditions (i), (ii), and (iii), there exists a time system S consistent with $\bar{\rho}$ if and only if for all $t \in T$

(P1) $$(\forall c_0)(\forall x^t)(\forall x_t)(\exists c_t)[\rho_t(c_t, x_t) = \rho_0(c_0, x^t \cdot x_t) \mid T_t]$$

(P2) $$(\forall c_t)(\forall x_t)(\exists c_0)(\exists x^t)[\rho_t(c_t, x_t) = \rho_0(c_0, x^t \cdot x_t) \mid T_t]$$

Let us briefly discuss the conceptual meaning of properties (P1) and (P2). First of all, observe that the condition

$$\rho_t(c_t, x_t) = \rho_0(c_0, x^t \cdot x_t) \mid T_t$$

could be interpreted to mean that the state c_t "properly" connects the previous history represented by (c_0, x^t), and the future evolution of the system represented by x_t, in the sense that the output starting from c_t, i.e., $y_t = \rho_t(c_t, x_t)$, is precisely the remainder of the output $y_0 = \rho_0(c_0, x^t \cdot x_t)$, i.e., $y_0 \mid T_t = \rho_0(c_0, x^t \cdot x_t) \mid T_t = y_t$. Condition (P1), which can also be written in the form

$$(\forall (c_0, x^t))(\forall x_t)(\exists c_t)[\rho_t(c_t, x_t) = \rho_0(c_0, x^t \cdot x_t) \mid T_t]$$

then means that for any past history (c_0, x^t) and for any future input x_t, there exists a state c_t at t which connects them properly. Condition (P2), which can also be written in the form

$$(\forall (c_t, x_t))(\exists (c_0, x^t))[\rho_t(c_t, x_t) = \rho_0(c_0, x^t \cdot x_t) \mid T_t]$$

then means that for any (c_t, x_t), i.e., the future behavior of the system, there exists a past history, i.e., an initial state and an initial input segment, so that this history leads up to c_t, which properly connects (c_0, x^t) and x_t.

(b) Realizability of $\bar{\rho}$

Theorem 1.1. Let $\bar{\rho}$ be a family of maps satisfying conditions (i) to (iii). Then $\bar{\rho}$ is realizable (i.e., there exists a state-transition family $\bar{\phi}$ so that $(\bar{\rho}, \bar{\phi})$ is a dynamical representation of a time system S) if and only if for all $t, t' \in T, t' \geq t$, $\bar{\rho}$ has the properties:

(P3) $$(\forall c_t)(\forall x_{tt'})(\exists c_{t'})(\forall x_{t'})[\rho_{t'}(c_{t'}, x_{t'}) = \rho_t(c_t, x_{tt'} \cdot x_{t'}) \mid T_{t'}]$$

(P4) $$(\forall c_t)(\forall x_t)(\exists c_0)(\exists x^t)[\rho_t(c_t, x_t) = \rho_0(c_0, x^t \cdot x_t) \mid T_t]$$

PROOF. We shall consider the *only if* part first. Let $\bar{\rho}$ be realizable and let c_t and $x_{tt'}$ be arbitrary. By consistency of $\bar{\phi}$,

$$\rho_{t'}(\phi_{tt'}(c_t, x_{tt'}), x_{t'}) = \rho_t(c_t, x_{tt'} \cdot x_{t'}) \mid T_{t'}, \qquad \text{for every} \quad x_{t'}$$

and since there exists $c_{t'}$ such that $c_{t'} = \phi_{tt'}(c_t, x_{tt'})$, (P3) holds. Since (P4) is equal to (P2) in Theorem 2.1, Chapter II, and since $\bar{\rho}$ is consistent with the time system S, i.e., $S_t{}^\rho = S_0{}^\rho \mid T_t$, (P4) also holds.

Consider now the *if* part. Let (P3) and (P4) be satisfied. Since (P3) and (P4) imply (P1) and (P2) of Theorem 2.1, Chapter II, the system consistency is naturally satisfied by $\bar{\rho}$. For each $t \in T$, let $E_t \subset C_t \times C_t$ be such that

$$(c_t, c_t') \in E_t \leftrightarrow (\forall x_t)(\rho_t(c_t, x_t) = \rho_t(c_t', x_t))$$

E_t is obviously an equivalence relation for each t. Let $\hat{\rho}_t : (C_t/E_t) \times X_t \to Y_t$ be such that

$$\hat{\rho}_t([c_t], x_t) = \rho_t(c_t, x_t)$$

where $\hat{\rho}_t$ is apparently well defined. Now, we claim that there exists a family of mappings $\{f_{tt'}\}$:

$$f_{tt'} : (C_t/E_t) \times X_{tt'} \to C_{t'}/E_{t'}$$

such that

$$\hat{\rho}_{t'}(f_{tt'}([c_t], x_{tt'}), x_{t'}) = \hat{\rho}_t([c_t], x_{tt'} \cdot x_{t'}) \mid T_{t'} \tag{3.2}$$

for every $[c_t]$, $x_{tt'}$, and $x_{t'}$. Let

$$f_{tt'} \subset (C_t/E_t \times X_{tt'}) \times (C_{t'}/E_{t'})$$

be such that

$$(([c_t], x_{tt'}), [c_{t'}]) \in f_{tt'} \leftrightarrow (\forall x_{t'})(\hat{\rho}_{t'}([c_{t'}], x_{t'}) = \hat{\rho}_t([c_t], x_{tt'} \cdot x_{t'}) \mid T_{t'})$$

(P3) implies that $f_{tt'} \neq \phi$. Suppose $(([c_t], x_{tt'}), [c_{t'}]) \in f_{tt'}$, and $(([c_t], x_{tt'}), [\hat{c}_{t'}]) \in f_{tt'}$. Then we have that

$$\hat{\rho}_{t'}([c_{t'}], x_{t'}) = \hat{\rho}_{t'}([\hat{c}_{t'}], x_{t'}), \qquad \text{for every} \quad x_{t'}$$

which implies that $[c_{t'}] = [\hat{c}_{t'}]$. Furthermore, (P3) implies that $\mathscr{D}(f_{tt'}) = (C_t/E_t \times X_{tt'})$. Hence, $f_{tt'}$ is a mapping such that

$$f_{tt'}:(C_t/E_t \times X_{tt'}) \to C_{t'}/E_{t'}$$

and

$$\hat{\rho}_{t'}(f_{tt'}([c_t], x_{tt'}), x_{t'}) = \hat{\rho}_t([c_t], x_{tt'} \cdot x_{t'}) \mid T_{t'} \qquad \text{for every} \quad x_{t'}$$

Then

$$f_{t't''}(f_{tt'}([c_t], x_{tt'}), x_{t't''}) = f_{tt''}([c_t], x_{tt'} \cdot x_{t't''}) \tag{3.3}$$

holds, because since $\{f_{tt'}\}$ is a state-transition family consistent with the reduced systems-response family $\{\hat{\rho}_t\}$, Theorem 2.2, Chapter II, is applicable to $\{f_{tt'}\}$. Finally, let $\mu_t : C_t/E_t \to C_t$ be such that $\mu_t([c_t]) \in [c_t]$. μ_t can be any function that satisfies $\mu_t([c_t]) \in [c_t]$. Let

$$\phi_{tt'}:C_t \times X_{tt'} \to C_{t'}$$

such that

$$\phi_{tt'}(c_t, x_{tt'}) = \mu_{t'}(f_{tt'}([c_t], x_{tt'})) \tag{3.4}$$

Then we can easily show

$$\rho_{t'}(\phi_{tt'}(c_t, x_{tt'}), x_{t'}) = \rho_t(c_t, x_{tt'} \cdot x_{t'}) \mid T_{t'}$$

and

$$\phi_{t't''}(\phi_{tt'}(c_t, x_{tt'}), x_{t't''}) = \phi_{tt''}(c_t, x_{tt'} \cdot x_{t't''})$$

because

$$\begin{aligned}
\rho_{t'}(\phi_{tt'}(c_t, x_{tt'}), x_{t'}) &= \rho_{t'}(\mu_{t'}(f_{tt'}([c_t], x_{tt'})), x_{t'}) && \text{[from Eq. (3.4)]} \\
&= \hat{\rho}_{t'}(f_{tt'}([c_t], x_{tt'}), x_{t'}) && \text{[since } \mu_t([c_t]) \in [c_t]\text{]} \\
&= \hat{\rho}_t([c_t], x_{tt'} \cdot x_{t'}) \mid T_{t'} && \text{[from Eq. (3.2)]} \\
&= \rho_t(c_t, x_{tt'} \cdot x_{t'}) \mid T_{t'} \\
\phi_{t't''}(\phi_{tt'}(c_t, x_{tt'}), x_{t't''}) &= \phi_{t't''}(\mu_{t'}(f_{tt'}([c_t], x_{tt'})), x_{t't''}) \\
&= \mu_{t''}(f_{t't''}(f_{tt'}([c_t], x_{tt'}), x_{t't''})) \\
&= \mu_{t''}(f_{tt''}([c_t], x_{tt'} \cdot x_{t't''})) && \text{[from Eq. (3.3)]} \\
&= \phi_{tt''}(c_t, x_{tt'} \cdot x_{t't''})
\end{aligned}$$

Q.E.D.

Consider briefly the conceptual meaning of condition (P3). The condition can be expressed in the following equivalent form which holds for all $t \in T$:

(P′3) $\qquad (\forall(c_t, x_{tt'}))(\exists c_{t'})(\forall x_{t'})[\rho_{t'}(c_{t'}, x_{t'}) = \rho_t(c_t, x_{tt'} \cdot x_{t'}) \mid T_{t'}]$

Interpretation of (P′3) is then as follows: For any state–input pair $(c_t, x_{tt'})$, there exists a state $c_{t'}$ at t' such that further time evolution for any subsequent input segment is consistent with the given state–input pair.

Condition (P′3) requires, therefore, that the state object at any time t', $c_{t'}$, "properly" connects preceding and following segments, $x_{tt'}$ and $x_{t'}$, as to be consistent with ρ_t. In particular, for any initial system's evolution, as specified by $(c_t, x_{tt'})$, there exists a state $c_{t'}$, which properly accounts for the continuation of that evolution.

(c) Dynamical Representation of a Time System

It has been shown in Section 2, Chapter II, that every time system has a response family $\bar{\rho} = \{\rho_t : C_t \times X_t \to Y_t\}$ if no constraints on $\bar{\rho}$ are assumed (Proposition 2.1, Chapter II). A similar fact is true for dynamical representation of a time system, i.e., when a family of state-transition functions $\bar{\phi}$ consistent with $\bar{\rho}$ and S is required to exist.

Theorem 1.2. Every time system S has a dynamical representation, i.e., there exist two families of mappings $(\bar{\rho}, \bar{\phi})$ consistent with S.

PROOF. Let $\rho_0 : C_0 \times X \to Y$ be an initial systems response of S. Let $C_t = C_0 \times X^t$ and $\rho_t : C_t \times X_t \to Y_t$ such that if $c_t = (c_0, x^t)$, then

$$\rho_t(c_t, x_t) = \rho_0(c_0, x^t \cdot x_t) \mid T_t$$

Then $\bar{\rho} = \{\rho_t\}$ satisfies the conditions of Theorem 1.1 and hence $\bar{\rho}$ is realizable. Q.E.D.

Consider now the effect of causality requirements. Theorem 4.2, Chapter II, has indicated that a time system has a nonanticipatory response family if and only if it has a nonanticipatory initial systems response. This result can be easily extended to dynamical systems.

Theorem 1.3. Let $(\bar{\rho}, \bar{\phi})$ be a dynamical system such that for every $t \in T$, ϕ_{ot} is an onto map. Then $\bar{\rho}$ is a nonanticipatory systems-response family if and only if ρ_0 is an initial nonanticipatory systems response.

PROOF. Let us consider the *if* part first. Suppose ρ_0 is nonanticipatory. Let x_t, x_t', and c_t be arbitrary elements. Since ϕ_{ot} is an onto map, for some $(\hat{c}_0, \hat{x}^t)$

$$\rho_t(c_t, x_t) = \rho_t(\phi_{ot}(\hat{c}_0, \hat{x}^t), x_t) = \rho_0(\hat{c}_0, \hat{x}^t \cdot x_t) \mid T_t$$

and similarly

$$\rho_t(c_t, x_t') = \rho_0(\hat{c}_0, \hat{x}^t \cdot x_t') \mid T_t$$

Since

$$(\hat{x}^t \cdot x_t) \mid \bar{T}^{t'} = (\hat{x}^t \cdot x_t') \mid \bar{T}^{t'}$$

when $x_t \mid \bar{T}_{tt'} = x_t' \mid \bar{T}_{tt'}$ we have that

$$\rho_o(\hat{c}_o, \hat{x}^t \cdot x_t) \mid \bar{T}^{t'} = \rho_o(\hat{c}_o, \hat{x}^t \cdot x_t') \mid \bar{T}^{t'}$$

Therefore, if $x_t \mid \bar{T}_{tt'} = x_t' \mid \bar{T}_{tt'}$

$$\rho_t(c_t, x_t) \mid \bar{T}_{tt'} = \rho_o(\hat{c}_o, \hat{x}^t \cdot x_t) \mid \bar{T}_{tt'} = \rho_o(\hat{c}_o, \hat{x}^t \cdot x_t') \mid \bar{T}_{tt'} = \rho_t(c_t, x_t') \mid \bar{T}_{tt'}$$

The *only if* part is apparent from the definition. Q.E.D.

We have now the following consequence for a dynamical system representation.

Theorem 1.4. A time system has nonanticipatory dynamical system representation if and only if it is nonanticipatory.

PROOF. If a time system has a nonanticipatory dynamical system representation, then it is nonanticipatory by definition. Conversely, if the system is nonanticipatory, then there exists an initial nonanticipatory response function from which a dynamical system can be constructed. (Refer to Theorem 1.2.) It is possible that some state transition ϕ_{ot} may not be onto, but in that case as the proof of Theorem 1.1 shows, the state object at t can be restricted to the range of ϕ_{ot} such that every ϕ_{ot} can be onto. Then it follows from Theorem 1.3 that the resultant dynamical system representation is nonanticipatory. Q.E.D.

The following result, which is of some conceptual importance, follows from Theorem 4.1, Chapter II.

Theorem 1.5. If the response function and hence the state-transition functions are accepted as partial functions, every time system has a nonanticipatory dynamical representation.

PROOF. The proof is immediately obtained when Theorem 4.1, Chapter II, is applied to Theorem 1.4. The procedure to define $\bar{\phi}$ is exactly the same as described in Theorem 1.2. Q.E.D.

Nonanticipation of a time system is defined with respect to its initial systems-response function. It is clearly true that even if a system is nonanticipatory, an initial systems-response function for such a system, which in general can be chosen arbitrarily, may not satisfy the nonanticipatory conditions. A general condition for a time system to be nonanticipatory is not known yet. However, a past-determined system is intrinsically nonanticipatory, and moreover, its "natural" initial systems-response function is nonanticipatory too. This fortunate fact is of special importance in the systems theory in general, in view of a widespread application of the past-determined systems. Past-determined systems will be discussed in this light in Chapter V.

2. CANONICAL REPRESENTATION (DECOMPOSITION) OF DYNAMICAL SYSTEMS AND CHARACTERIZATION OF STATES

(a) Canonical Representation of Dynamical Systems

Let $\bar{\rho}$ be an arbitrary response family for a system S and $\bar{C} = \{C_t : t \in T\}$ the family of associated state objects. Since C_t is arbitrary, it might have more states than necessary for consistency with S. An obvious way to eliminate some of the redundant states is to consider any two states c_t and $\hat{c}_t$ as equal if the future behavior of the system starting from c_t and from $\hat{c}_t$ is the same. More precisely, let $E_t \subset C_t \times C_t$ be the relation such that

$$(c_t, \hat{c}_t) \in E_t \leftrightarrow (\forall x_t)[\rho_t(c_t, x_t) = \rho_t(\hat{c}_t, x_t)]$$

Apparently, E_t is an equivalence relation. Starting from arbitrary C_t, we can introduce a reduced state object $\hat{C}_t = C_t/E_t$ whose elements are equivalence classes. The corresponding response family $\bar{\hat{\rho}} = \{\hat{\rho}_t : \hat{C}_t \times X_t \to Y_t\}$ such that

$$\hat{\rho}([c_t], x_t) = y_t \leftrightarrow \rho_t(c_t, x_t) = y_t$$

is termed *reduced*. The following simple fact will be used in the sequel.

Proposition 2.1. Let $\hat{\rho}$ be an arbitrary response family and $\bar{\hat{\rho}}$ the corresponding reduced family. $\bar{\hat{\rho}}$ is consistent with a system S if and only if $\bar{\rho}$ is consistent with S.

PROOF. Since the relation

$$\rho_t(c_t, x_t) = \rho_0(c_0, x^t \cdot x_t) \mid T_t \leftrightarrow \hat{\rho}_t([c_t], x_t) = \hat{\rho}_0([c_0], x^t \cdot x_t) \mid T_t$$

holds, the proposition follows from Theorem 2.1, Chapter II. Q.E.D.

Definition 2.1. Let S be a time system with a given family of output-generating functions $\bar{\mu} = \{\mu_{tt'} : t, t' \in T\}$. A pair $(\bar{\phi}, \bar{\lambda})$, where $\bar{\phi}$ is a state-transition family while $\bar{\lambda}$ is a family of output functions, will be referred to as a canonical (dynamical) representation of S if and only if for any $t, t' \in T$, the diagram

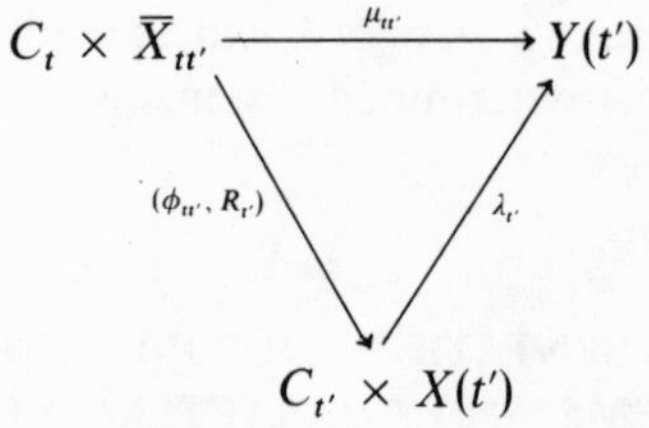

is commutative, where $(\phi_{tt'}, R_{t'})(c_t, \bar{x}_{tt'}) = (\phi_{tt'}(c_t, x_{tt'}), \bar{x}_{tt'}(t'))$.

We have now the following theorem.

Theorem 2.1. A time system has a canonical representation if and only if it is nonanticipatory.

PROOF. Let us consider the *if* part first. When a time system is nonanticipatory, it follows from Theorem 1.4 that it has a nonanticipatory dynamical system representation, $(\bar{\rho}, \bar{\phi})$. Then an output function $\lambda_t : C_t \times X(t) \to Y(t)$ such that $\lambda_t(c_t, x_t(t)) = \rho_t(c_t, x_t)(t)$ and an output-generating function $\mu_{tt'} : C_t \times \overline{X}_{tt'} \to Y(t')$ such that

$$\mu_{tt'}(c_t, \bar{x}_{tt'}) = \rho_t(c_t, x_{tt'} \cdot \hat{x}_{t'})(t')$$

where $\hat{x}_{t'}$ is arbitrary with $\hat{x}_{t'}(t') = \bar{x}_{tt'}(t')$, are properly defined. Furthermore, since

$$\begin{aligned}\rho_t(c_t, x_{tt'} \cdot \hat{x}_{t'})(t') &= \rho_{t'}(\phi_{tt'}(c_t, x_{tt'}), \hat{x}_{t'})(t') = \lambda_{t'}(\phi_{tt'}(c_t, x_{tt'}), \hat{x}_{t'}(t')) \\ &= \mu_{tt'}(c_t, \bar{x}_{tt'})\end{aligned}$$

we have the commutative diagram. Hence, the system has a canonical representation. Conversely, let a time system have a canonical representation. Let $\rho_t : C_t \times X_t \to Y_t$ be such that

$$\rho_t(c_t, x_t) = y_t \leftrightarrow y_t(\tau) = \lambda_\tau(\phi_{t\tau}(c_t, x_{t\tau}), x_t(\tau)) \qquad \text{for} \quad \tau \geq t \quad \text{and} \quad x_{t\tau} = x_t \mid T^\tau$$

Now, we shall show that

$$x \mid \overline{T}^t = x' \mid \overline{T}^t \to \rho_0(c_0, x) \mid \overline{T}^t = \rho_0(c_0, x') \mid \overline{T}^t$$

for any $c_0 \in C_0$. Let $\tau \in \overline{T}^t$ be arbitrary; then

$$\rho_0(c_0, x)(\tau) = \lambda_\tau(\phi_{0\tau}(c_0, x^\tau), x(\tau)),$$

and

$$\rho_0(c_0, x')(\tau) = \lambda_\tau(\phi_{0\tau}(c_0, x'^\tau), x'(\tau))$$

since $x^{\tau} = x'^{\tau}$ and $x(\tau) = x'(\tau)$ for $\tau \in \bar{T}^t$, we have

$$\rho_0(c_0, x)(\tau) = \rho_0(c_0, x')(\tau) \qquad \text{for} \quad \tau \in \bar{T}^t$$

Hence ρ_0 is nonanticipatory. Q.E.D.

The canonical representation essentially means a decomposition of the system into the subsystems in a cascade as shown in Fig. 2.1. The first

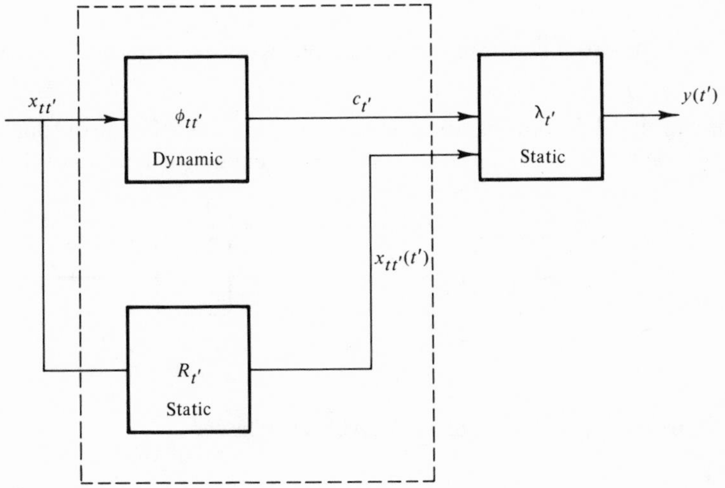

FIG. 2.1

subsystem, denoted by $\phi_{tt'}$ in Fig. 2.1, represents fully the dynamic behavior of the system, while the other two subsystems, $\lambda_{t'}$ and $R_{t'}$, are static and indicate only how the output values are generated from the given states and the current input values. The first subsystem is defined solely in terms of $\bar{\phi}$, and therefore the dynamics of the system is fully represented by this single family of functions. If one is concerned solely with the dynamics, attention can be restricted to $\bar{\phi}$.

A somewhat stronger decomposition is possible if the system satisfies conditions for strong nonanticipation. The following theorem is then valid.

Theorem 2.2. The system is strongly nonanticipatory if and only if there exist a state-transition family $\bar{\phi}$, an output-function family $\bar{\lambda} = \{\lambda_t : C_t \rightarrow$

$Y(t)\}$, and an output-generating-function family $\bar{\mu} = \{\mu_{tt'}: C_t \times X_{tt'} \to Y(t')\}$ such that the diagram

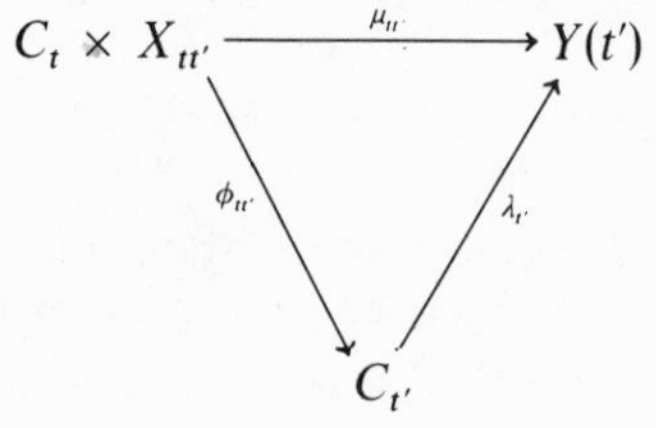

is commutative.

PROOF. The proof is completely analogous with the proof of Theorem 2.1. Q.E.D.

The canonical decomposition is now given as in Fig. 2.2. The output values depend solely on the present state and are not explicitly dependent upon the input.

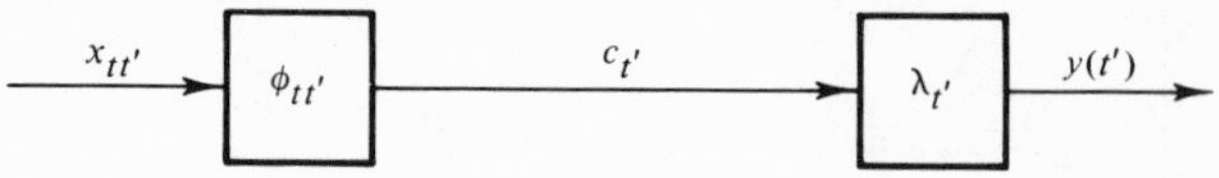

FIG. 2.2

(b) Characterization of State as Equivalence Classes

Conceptually, one of the roles of the concept of state is to reflect the past history of the system's behavior. If two different past behaviors lead to the same state, they are apparently equivalent as far as present and future behavior of the system are concerned. The state at any time can, therefore, be represented as corresponding to an equivalence class generated by an equivalence relation defined on the past of the system. The equivalence that characterizes the states depends upon the particular family of state objects used, and conversely, by introducing an appropriate equivalence relation, one can define, or construct, the corresponding family of state objects and the state space itself.

It should be noticed that we are talking here about the equivalences defined *strictly in terms of the past*. These equivalence relations ought to be distinguished from the equivalences that are defined in reference to the future behavior and are used to eliminate redundant states.

(α) Let ρ_0 be an initial systems-response function of a time system S, i.e., $\rho_0: C_0 \times X \to Y$. The present and the future output y_t of the system is uniquely specified then by the initial-state–initial-input segment pair (c_o, x^t); that is, it is given by the restriction $y_t = \rho_o(c_o, x^t \cdot x_t) \mid T_t$.

This observation suggests the selection of the initial-state–initial-input segment pair (c_o, x^t) as a state at time t, a choice justified by Theorem 1.1. In other words, $C_o \times X^t$ itself can be used as a state object at time t. However, it is possible that the two different elements (c_o, x^t) and $(\hat{c}_o, \hat{x}^t)$ yield completely the same future output. Then, as far as the behavior of the system is concerned, we need not distinguish (c_o, x^t) from $(\hat{c}_o, \hat{x}^t)$. This observation leads to an equivalence relation E^t on $C_o \times X^t$, which is referred to in the literature as the *Nerode equivalence* [2].

Proposition 2.2. Let ρ_o be an initial systems-response function of a time system S. For every $t \in T$, let $E^t \subset (C_o \times X^t) \times (C_o \times X^t)$ be an equivalence relation such that

$$((c_o, x^t), (\hat{c}_o, \hat{x}^t)) \in E^t \leftrightarrow (\forall x_t)[\rho_o(c_o, x^t \cdot x_t) \mid T_t = \rho_o(\hat{c}_o, \hat{x}^t \cdot x_t) \mid T_t]$$

There exists then a family of functions $\bar{\rho} = \{\rho_t : C_t \times X_t \to Y_t\}$ such that $C_t = (C_o \times X^t)/E^t$ and $\bar{\rho}$ is consistent with S and realizable.

PROOF. E^t is apparently an equivalence relation. Let $C_t = (C_o \times X^t)/E^t = \{[c_o, x^t]\}$. Let $\rho_t : C_t \times X_t \to Y_t$ be such that

$$\rho_t([c_o, x^t], x_t) = \rho_o(c_o, x^t \cdot x_t) \mid T_t$$

Since $((c_o, x^t), (\hat{c}_o, \hat{x}^t)) \in E^t$ implies that

$$\rho_o(c_o, x^t \cdot x_t) \mid T_t = \rho_o(\hat{c}_o, \hat{x}^t \cdot x_t) \mid T_t$$

for every x_t, ρ_t is properly defined. Then, since the consistency conditions (P1) and (P2) of Theorem 2.1, Chapter II, are trivially satisfied, $\bar{\rho}$ is consistent with S. In the similar way, the conditions of Theorem 1.1 are shown to be satisfied. Q.E.D.

Proposition 2.3. Let $(\bar{\rho}, \bar{\phi})$ be a dynamical system and $E_t \subset C_t \times C_t$ a relation such that

$$(c_t, \hat{c}_t) \in E_t \leftrightarrow (\forall x_t)[\rho_t(c_t, x_t) = \rho_t(\hat{c}_t, x_t)]$$

There exists then a one-to-one mapping

$$F : (C_o \times X^t)/E^t \to C_t/E_t$$

such that

$$F([c_o, x^t]) = [\phi_{ot}(c_o, x^t)]$$

PROOF. Since

$$\begin{aligned}[c_0, x^t] = [c_0', x'^t] &\leftrightarrow (c_0, x^t)E^t(c_o', x'^t)\\ &\leftrightarrow (\forall x_t)(\rho_o(c_0, x^t \cdot x_t) \mid T_t = \rho_0(c_o', x'^t \cdot x_t) \mid T_t)\\ &\leftrightarrow (\forall x_t)(\rho_t(\phi_{0t}(c_0, x^t), x_t) = \rho_t(\phi_{0t}(c_o', x'^t), x_t))\\ &\leftrightarrow \phi_{ot}(c_0, x^t)E_t\phi_{0t}(c_o', x'^t)\\ &\leftrightarrow [\phi_{0t}(c_0, x^t)] = [\phi_{ot}(c_o', x_o'^t)]\end{aligned}$$

F is well defined and one to one. Q.E.D.

The preceding two propositions suggest a construction procedure for the reduced state objects (where redundant states are eliminated) whenever an initial-response function ρ_0 is given. First, the equivalences $\{E^t : t \in T\}$ are introduced, and the state objects are defined by $C_t = (C_0 \times X^t)/E^t$. The response functions are then given by $\rho_t : C_t \times X_t \to Y_t$ such that

$$\rho_t(c_t, x_t) = \rho_0(c_0, x^t \cdot x_t) \mid T_t \qquad \text{where} \quad c_t = [c_0, x^t]$$

Apparently, ρ_t is well defined and Theorem 1.1 guarantees that $\bar{\rho} = \{\rho_t\}$ is realizable. Furthermore, Proposition 2.3 guarantees that $\bar{C} = \{C_t : t \in T\}$ is the set of smallest state objects in the sense that starting from any other set of state objects $\tilde{C}_t$ and the initial systems-response function ρ_0, after such state objects are reduced by the appropriate equivalence E_t (i.e., after redundant states are eliminated), there exists a one-to-one map from $\bar{C}_t$ into the reduced state object $\tilde{C}_t/E_t$.

(β) The preceding equivalence relation E^t is defined for a given initial response function ρ_0, i.e., in reference to a given initial state object C_0. When the system is past-determined, it is possible to characterize reduced states by an equivalence defined solely on the input–output objects, i.e., *without reference to the initial states.* The states are then defined in terms of the primary concepts X and Y, which are used to define the system S itself. This will be discussed in Chapter V.

3. CONSTRUCTIVE ORIGIN OF STATE-SPACE REPRESENTATION

(a) Construction of a State Space and a Canonical Representation

A state space can be introduced directly by assuming an abstract set that satisfies the necessary requirements or by constructing the state space from the previously introduced family of state objects. The general procedure for such a construction involves the following steps:

(1) All state objects are aggregated, e.g., by the union operation $\tilde{C} = \bigcup\{C_t : t \in T\}$ or by the Cartesian product operation $\tilde{C} = \times_{t \in T} C_t$.

(2) An equivalence relation $E_c \subset \tilde{C} \times \tilde{C}$, which has to satisfy some necessary conditions, is introduced as given below.

(3) The quotient set $\tilde{C}/E_c$ or a set isomorphic to it,

$$C = \tilde{C}/E_c$$

is taken as the state space.

The necessary conditions mentioned in (2) are the following:

(i) $(\forall[c])(\forall x_{tt'})(\exists[c'])(\forall c_t)(c_t \in [c] \rightarrow \phi_{tt'}(c_t, x_{tt'}) \in [c'])$

(ii) $(\forall[c])(\forall c_t)(\forall c_t')(\forall x_t)(c_t \in [c] \;\&\; c_t' \in [c] \rightarrow \rho_t(c_t, x_t) = \rho_t(c_t', x_t))$

or

(ii)′ $(\forall[c])(\forall c_t)(\forall c_{t'})(\forall a)(c_t \in [c] \;\&\; c_{t'} \in [c] \rightarrow \lambda_t(c_t, a) = \lambda_t(c_{t'}, a))$

Condition (i) requires simply that the evolution of the system in time is represented as "uninterrupted" much as in the family of state objects. Condition (ii) is used when the system is defined by its dynamical representation [i.e., the pairs $(\bar{\rho}, \bar{\phi})$], while condition (ii)′ is used when the canonical representation [i.e., the pair $(\bar{\phi}, \bar{\lambda})$] is given. If the state objects have additional algebraic structures (for instance, they are linear algebras as required for the linear system case), the equivalence relation which generates a state space should preserve those structures in the quotient set.

The results derived in terms of the state objects have apparent counterparts when the state space is introduced. Of particular interest is the canonical representation.

Before presenting the form of the canonical representation in the state space, we shall introduce the following notational convention.

Let $T_{\hat{t}\hat{t}'} \subset T_{tt'}$ and $X_{tt'}$ be an input set. A mapping $R_{T_{\hat{t}\hat{t}'}}$ can be then defined:

$$R_{T_{\hat{t}\hat{t}'}} : X_{tt'} \rightarrow X_{\hat{t}\hat{t}'}$$

such that

$$(\forall x_{tt'})[R_{T_{\hat{t}\hat{t}'}}(x_{tt'}) = x_{tt'} \mid T_{\hat{t}\hat{t}'}]$$

where $R_{T_{\hat{t}\hat{t}'}}$ is termed a restriction operator. For simplicity of notation, the index will always denote the codomain of a restriction operator, i.e., the subset on which the original function is restricted, while the domain (i.e., the time set of the original function being restricted) is understood from the context. The same symbol R will be used in a similar way for other objects too, e.g., Y, $Y_{tt'}$, etc. We shall have then, e.g.,

$$R_{T_t} : Y \rightarrow Y_t$$

such that

$$(\forall y)[R_{T_t}(y) = y \mid T_t]$$

Furthermore, we use the convention that $R_{\{t\}}: X \to X(t)$ such that

$$(\forall x)[R_{\{t\}}(x) = x(t)]$$

Let $\phi_{tt'}$ and $\mu_{tt'}$ represent the state-transition and the output-generating functions defined on the state space, while $\hat{\phi}_{tt'}$ and $\hat{\mu}_{tt'}$ represent the respective functions defined on the family of state objects. (How $\phi_{tt'}$ and $\mu_{tt'}$ are defined by means of $\hat{\phi}_{tt'}$ and $\hat{\mu}_{tt'}$ will be shown in reference to specific types of equivalences in the subsequent sections.) The canonical representation is now defined in terms of a family of state-space transitions $\bar{\phi} = \{\phi_{tt'}: C \times X_{tt'} \to C\}$ and maps $\bar{\lambda} = \{\lambda_t: C \times A \to B\}$. The diagram of the decomposition is as given in Figs. 3.1 and 3.2. The pair of mappings $(\phi_{tt'}, R^*_{\{t'\}})$ used in Fig. 3.1 is applied in the following way

$$(\phi_{tt'}, R^*_{\{t'\}})(c, \bar{x}_{tt'}) = (\phi_{tt'}(c, x_{tt'}), \bar{x}_{tt'}(t'))$$

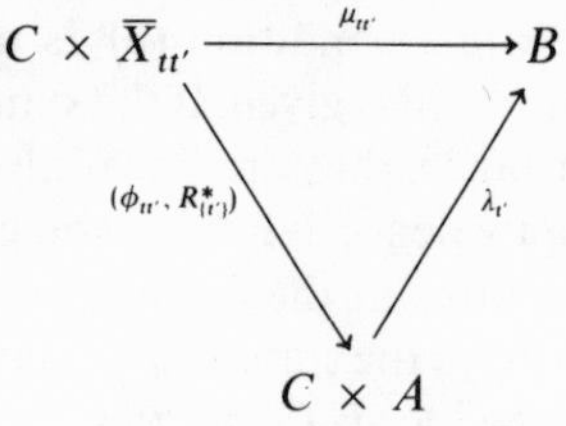

FIG. 3.1

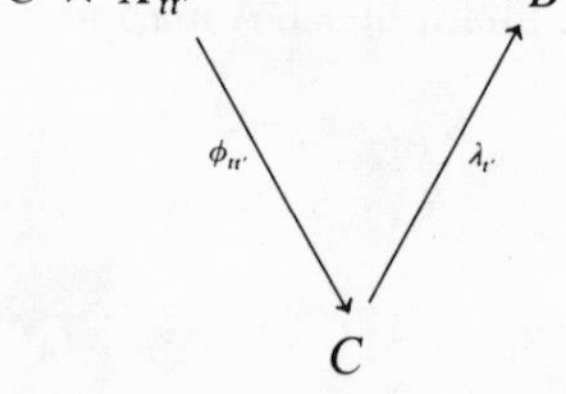

FIG. 3.2

The importance of the canonical representation can now be fully recognized: The dynamics of the system is represented by a family of transformations of the set C and the corresponding input restriction $X_{tt'}$ into C itself. The set C, both in the domain and range of $\phi_{tt'}$, is the same for any $t, t' \in T$; the only difference between the state-transition functions for any two intervals, e.g., $\phi_{tt'}$ and $\phi_{\bar{t}\bar{t}'}$, is in the corresponding input restriction.

A specific connection can be now established with the classical concept of a dynamical system as, e.g., used in topological dynamics.

To any input $x \in X$ of a general dynamical system corresponds a set of transformations

$$\bar{\phi}^x = \{\phi^x_{tt'} : C \to C\}$$

such that

$$\phi^x_{tt'} = \phi_{tt'} \mid \{x \mid T_{tt'}\} \times C$$

If the state space satisfies the necessary additional requirements, e.g., it is an appropriately defined topological space, the family $\bar{\phi}^x$ is referred to as a dynamical system since it fully specifies the time evolution of the system for the given input x. It can be readily shown that $\bar{\phi}^x$ has the required properties for a dynamical system as stated in the classical literature, in particular, the composition (semigroup) property and the consistency property [3].

In conclusion, the usefulness of the state-space approach for the study of time systems is primarily due to the following:

(1) The evolution in time, dynamics, *for any given input*, is described fully by the mappings of the state space C into itself.

(2) The output at any time is obtained by using a static function from the state, if the system is strongly nonanticipatory, or from the current input value also if the system is only nonanticipatory.

(b) State-Space-Generating Equivalences and Dynamical Systems in a State Space

There are many equivalences that satisfy conditions (i) and (ii) from Section 3a; i.e., the quotient set generated by such an equivalence can be used as a state space. In this section, we shall introduce two typical state-space-generating equivalences.

(α) The first way to define an equivalence in $\tilde{C}$ is in terms of an output function. Informally, two states will be considered as equivalent, even if at different times, if for the same value of inputs they have the same value of outputs, and furthermore, the state-transition function will always generate the equivalent states if the inputs are the same. More precisely, we shall introduce a relation

$$E_\lambda^\alpha \subset \tilde{C} \times \tilde{C} \qquad \text{where} \quad \tilde{C} = \bigcup_{t \in T} C_t$$

such that

$$(c_t, \hat{c}_{t'}) \in E_\lambda{}^\alpha \to (\forall a)[\lambda_t(c_t, a) = \lambda_{t'}(\hat{c}_{t'}, a)] \;\&\; \{t = t' \to (\forall x_{tt''})[(\phi_{tt''}(c_t, x_{tt''}), \phi_{tt''}(\hat{c}_{t'}, x_{tt''})) \in E_\lambda{}^\alpha]\} \tag{3.5}$$

Let $\bar{E}$ be the family of all such equivalences that satisfy relation (3.5), $\bar{E} = \{E_\lambda{}^\alpha : \alpha \in A\}$. The trivial equivalence relation $I = \{(c_t, c_t) : c_t \in \tilde{C}\}$ satisfies (3.5). Therefore, $\bar{E}$ is not empty. In general, there are many equivalence relations that satisfy (3.5). The family $\bar{E}$ can be ordered by the set-theoretic inclusion $\subseteq$. For such an ordered set, the following lemma holds.

Lemma 3.1.† The set $\bar{E}$ has a maximal element $E_m{}^\lambda$ with respect to the ordering $\leq$.

PROOF. Let P be an arbitrary nonempty chain in $\bar{E}$; that is, P is a linearly ordered subset of $\bar{E}$, where P is denoted by $P = \{E_\lambda{}^\beta \mid \beta \in B\}$. Let

$$E_0 = \bigcup_{\beta \in B} E_\lambda{}^\beta$$

E_0 will be shown to be an element of $\bar{E}$, i.e., $E_0 \in \bar{E}$. Since for any c_t, $c_{t'}$, and $c_{t''}$, E_0 satisfies

Reflexivity: $(c_t, c_t) \in \tilde{C} \times \tilde{C} \to (c_t, c_t) \in E_\lambda{}^\beta$ for every $\beta \in B \to (c_t, c_t) \in E_0$

Symmetry: $(c_t, c_{t'}) \in E_0 \to (c_t, c_{t'}) \in E_\lambda{}^\beta$ for some $\beta \in B \to (c_{t'}, c_t) \in E_\lambda{}^\beta \to (c_{t'}, c_t) \in E_0$

Transitivity: $(c_t, c_{t'}) \in E_0 \;\&\; (c_t, c_{t''}) \in E_0 \to (c_t, c_{t'}) \in E_\lambda{}^\beta$ and $(c_{t'}, c_{t''}) \in E_\lambda{}^{\beta'}$ for some $\beta, \beta' \in B$. Since P is linearly ordered, $(c_t, c_{t'}) \in E_\lambda{}^{\beta''}$ and $(c_{t'}, c_{t''}) \in E_\lambda{}^{\beta''}$ where $\beta'' = \beta$ or $\beta'' = \beta' \to (c_t, c_{t''}) \in E_\lambda{}^{\beta''} \to (c_t, c_{t''}) \in E_0$

E_0 is an equivalence relation, and furthermore, for any $(c_t, c_{t'})$ the following holds:

$$\begin{aligned}(c_t, c_{t'}) \in E_0 \to (c_t, c_{t'}) \in E_\lambda{}^\beta \quad &\text{for some} \quad \beta \in B \\ \to (\forall a)(\lambda_t(c_t, a) &= \lambda_{t'}(c_{t'}, a)) \\ \& \{t = t' \to (\forall x_{tt''})&[(\phi_{tt''}(c_t, x_{tt''}), \phi_{tt''}(c_{t'}, x_{tt''})) \in E_\lambda{}^\beta]\} \\ \to (\forall a)(\lambda_t(c_t, a) &= \lambda_{t'}(c_{t'}, a)) \\ \& \{t = t' \to (\forall x_{tt''})&[(\phi_{tt''}(c_t, x_{tt''}), \phi_{tt''}(c_{t'}, x_{tt''})) \in E_o]\}\end{aligned}$$

† This result holds under a more general condition. Let $E^\alpha \subset \tilde{C} \times \tilde{C}$ be such that

$$(c_t, \hat{c}_{t'}) \in E^\alpha \to P(c_t, \hat{c}_{t'}) \;\&\; (t = t' \to (\forall a)[\lambda_t(c_t, a) = \lambda_{t'}(\hat{c}_t, a)] \;\&\; (\forall x_{tt''})[(\phi_{tt''}(c_t, x_{tt''}), \phi_{t't''}(c_{t'}, x_{tt''})) \in E^\alpha])$$

where $P(c_t, c_{t'})$ is an arbitrary predicate such that $P(c_t, c_t)$ is true for any c_t. If $\bar{E}$ is the family of all such equivalences as E^α, Lemma 3.1 holds.

Hence, E_0 is an element of $\bar{E}$. Consequently, since an arbitrary chain in $\bar{E}$ has been shown to have an upper bound, $\bar{E}$ has a maximal element $E_m{}^\lambda$ from Zorn's lemma. Q.E.D.

Apparently, it is desirable to use the maximal equivalence $E_m{}^\lambda$ and to define the state space as

$$C = \tilde{C}/E_m{}^\lambda$$

since $E_m{}^\lambda$ yields the most "reduced" state space.

Consider now how the auxiliary functions can be defined in terms of the state space C so as to be consistent with the respective functions defined in terms of the state objects. Let $I_t : \tilde{C}/E_m{}^\lambda \to C_t$ be such that

$$I_t([c]) = \begin{cases} c_t, & \text{if } \ c_t \in [c] \cap C_t \\ c_t{}^* \in C_t, & \text{if } \ [c] \cap C_t = \phi \end{cases}$$

where $c_t{}^*$ is an arbitrary element of C_t and, in general, different for each $[c]$. Then let the state-transition function $\hat{\phi}_{tt'}$ in the state space C be defined by

$$\hat{\phi}_{tt'} : (\tilde{C}/E_m{}^\lambda) \times X_{tt'} \to \tilde{C}/E_m{}^\lambda$$

such that

$$\hat{\phi}_{tt'}([c], x_{tt'}) = [\phi_{tt'}(I_t([c]), x_{tt'})]$$

Apparently, $\hat{\phi}_{tt'}$ is well defined. In order to show that it has the required composition property, notice that

(i) $I_t([c_t]) \in [c_t] \to c_t \equiv I_t([c_t])$

(ii) $c_t \equiv c_t' \to \phi_{tt'}(c_t, x_{tt'}) \equiv \phi_{tt'}(c_t', x_{tt'})$ by relation (3.5),

where $c_t \equiv c_{t'}$ represents $(c_t, c_{t'}) \in E_m{}^\lambda$.

(iii)
$$\begin{aligned} I_t([c_t]) \equiv c_t &\to \phi_{tt'}(c_t, x_{tt'}) \equiv \phi_{tt'}(I_t([c_t]), x_{tt'}) \\ &\to [\phi_{tt'}(c_t, x_{tt'})] = [\phi_{tt'}(I_t([c_t]), x_{tt'})] \\ &\to [\phi_{tt'}(c_t, x_{tt'})] = \hat{\phi}_{tt'}([c_t], x_{tt'}) \end{aligned} \tag{3.6}$$

Now, we can prove the composition property of $\bar{\hat{\phi}}$:

$$\begin{aligned} \hat{\phi}_{tt''}([c], x_{tt''}) &= [\phi_{tt''}(I_t([c]), x_{tt''})] && \text{(by definition)} \\ &= [\phi_{t't''}(\phi_{tt'}(I_t([c]), x_{tt'}), x_{t't''})] && \text{(from the composition property of } \phi_{tt'}) \\ &= \hat{\phi}_{t't''}([\phi_{tt'}(I_t([c]), x_{tt'})], x_{t't''}) && \text{[from (3.6)]} \\ &= \hat{\phi}_{t't''}(\hat{\phi}_{tt'}([c], x_{tt'}), x_{t't''}) \end{aligned}$$

Hence, $\bar{\hat{\phi}}$ satisfies the required composition property.

Consider now the output function $\hat{\lambda}_t$ defined on the state space

$$\hat{\lambda}_t : (\tilde{C}/E_m{}^{\lambda}) \times A \to B$$

such that

$$\hat{\lambda}_t([c], a) = \lambda_t(I_t([c]), a)$$

Since $c_t \equiv I_t[c_t]$, we have

$$\lambda_t(c_t, a) = \lambda_t(I_t([c_t]), a) = \hat{\lambda}_t([c_t], a) \tag{3.7}$$

as required.

In summary, then, from (3.6) and (3.7), respectively, we have

$$\hat{\phi}_{tt'}([c_t], x_{tt'}) = [\phi_{tt'}(c_t, x_{tt'})]$$
$$\hat{\lambda}_t([c_t], a) = \lambda_t(c_t, a)$$

so that the following diagram is commutative:

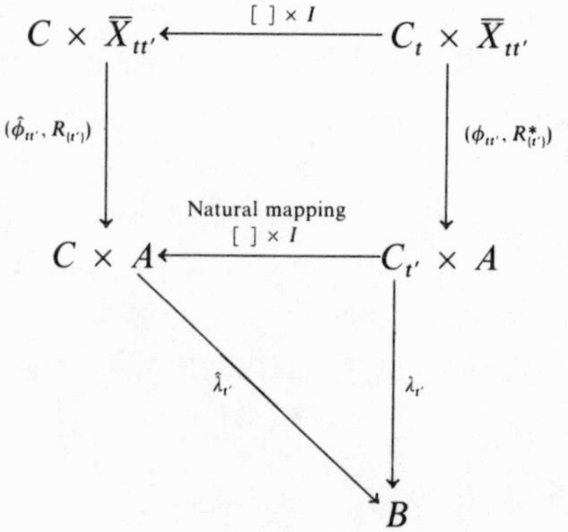

FIG. 3.3

Hence, $C = \tilde{C}/E_m{}^{\lambda}$ is the state space, and the mappings

$$\hat{\phi}_{tt'} : C \times X_{tt'} \to C$$

$$\hat{\lambda}_t : C \times A \to B$$

define a state transition and an output function, respectively.

(β) An alternative way to construct a state space is in terms of an equivalence relation defined by means of the shifting operator. Essentially, the states will be defined as equivalent (regardless at what time they appear) if the respective system's behavior is the same (i.e., the same input–output functions) except for a shift in time.

To introduce this equivalence, the following assumptions will be added:

(iii) The time set T is a stationary time set.
(iv) X is a stationary object, i.e., for every $t \in T$

$$F^t(X) = X_t$$

where F^t is the shift operator defined in Section 3, Chapter II.

Let $\tilde{C} = \bigcup_{t \in T} C_t$. We shall introduce a relation $E_\alpha{}^\rho \subset \tilde{C} \times \tilde{C}$ such that for $t' \geq t$,

$$(c_t, \hat{c}_{t'}) \in E_\alpha{}^\rho \to (\forall x_t)[F^{t'-t}(\rho_t(c_t, x_t)) = \rho_{t'}(\hat{c}_{t'}, F^{t'-t}(x_t))]$$

and

$$\{(t = t') \to (\forall x_{tt''})[(\phi_{tt''}(c_t, x_{tt''}), \phi_{tt''}(c_{t'}, x_{tt''})) \in E_\alpha{}^\rho]\} \tag{3.8}$$

Let $\bar{E}^\rho$ be the family of equivalence relations which satisfies (3.8); the trivial equivalence relation $\{(c_t, c_t) : c_t \in \tilde{C}\}$ again satisfies (3.8). Hence, $\bar{E}^\rho = \{\bar{E}_\alpha{}^\rho : \alpha \in A\}$ is not empty; furthermore, there may be many equivalence relations satisfying (3.8). The existence of a maximal equivalence relation $E_m{}^\rho$ in $\bar{E}^\rho$ can be proven by the same argument as used to show the existence of $E_m{}^\lambda$.

When the maximal equivalence is used, the state space is given by the quotient set

$$C = \tilde{C}/E_m{}^\rho$$

and the system response at t is the function

$$\hat{\rho}_t : C \times X_t \to Y_t$$

such that

$$\hat{\rho}_t([c], x_t) = \begin{cases} \rho_t(c_t, x_t), & \text{if } (\exists c_t)(c_t \in [c]) \\ \text{undefined} & \text{otherwise} \end{cases}$$

while the state-transition function is

$$\hat{\phi}_{tt'} : C \times X_{tt'} \to C$$

such that

$$\hat{\phi}_{tt'}([c], x_{tt'}) = \begin{cases} \phi_{tt'}(c_t, x_{tt'}), & \text{if} \quad (\exists c_t)(c_t \in [c]) \\ \text{undefined} & \text{otherwise} \end{cases}$$

If $\hat{\rho}_t$ and $\hat{\phi}_{tt'}$ are to be total functions, they ought to be extended appropriately; in particular, the extended system ought to preserve basic properties of the original system, e.g., linearity.

Which of the above equivalences $E_\alpha{}^\lambda$ or $E_\alpha{}^\rho$ is used to introduce the state space depends ultimately upon the circumstances in application. However, if both equivalences are equally good candidates, a comparison between the two approaches indicates the following key distinctions:

(1) Method (β) requires additional assumptions, namely, (iii) and (iv).

(2) The state space constructed by $E_m{}^\rho$ is, in general, larger than the space constructed by $E_m{}^\lambda$.

The state space can be constructed by using other equivalence relations. A notable example is when T is a metric space. This case can be shown to be a special case of method (β).

In conclusion, it should be pointed out again that we consider the state space as a secondary concept, which when introduced has to be consistent with the primary information given in the input–output pairs or has to be constructed in a manner shown in this section. Often the state space is given a priori in the definition of the system itself. The discussion in this section explains what the origin of such a concept is and what condition it has to satisfy even if introduced a priori.

(c) Commutative Diagram of Auxiliary Functions

Relationships between auxiliary functions can be represented in a convenient way by the diagram shown in Fig. 3.4. The arrows denote mappings, and each of the loops represents a commutative relationship of the mappings involved.

The mappings in the diagram are: the auxiliary functions ρ_0, ρ_t, $\mu_{tt'}$, λ_t, $\hat{\lambda}_t$; the restrictions $R_{T_{t'}}$, $R_{\{t'\}}$, etc.; the identity operator I; and the appropriate composition of those mappings F_1, F_2, F_3, F_4 defined as follows.

The mapping F_1 is given by the algebraic diagram in Fig. 3.5., where I represents the identity, and $I \times R_T t$ indicates that the mapping is applied componentwise; i.e., the first component is simply mapped into itself, while the second is transformed by the restriction operator.

The second composite mapping F_2 is defined by the diagram in Fig. 3.6. Again, $I \times R_{\overline{T}_{tt'}}$ indicates that the first component is mapped into itself while the second is restricted to $\overline{T}_{tt'}$.

The third mapping F_3 is given by the diagram in Fig. 3.7, where $\phi_{tt'}$ is the state-transition function, while $I \times R_{T_{tt'}}$ has the same interpretation as for F_1 and F_2.

Finally, the mapping F_4 is given by the diagram in Fig. 3.8, where nat $E_m{}^{\lambda}$ is the natural mapping of the equivalence relation $E_m{}^{\lambda}$.

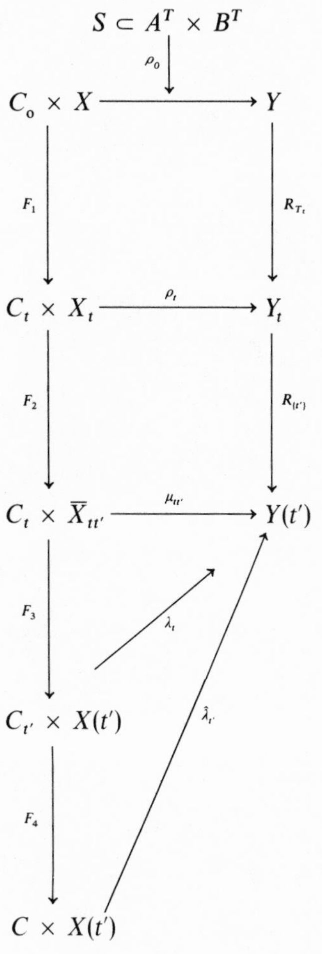

FIG. 3.4

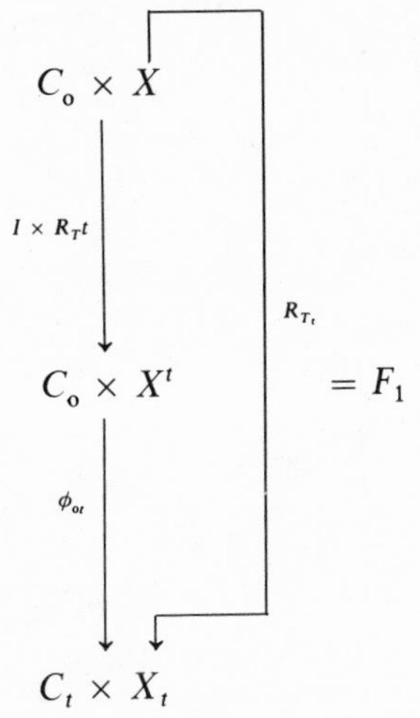

FIG. 3.5

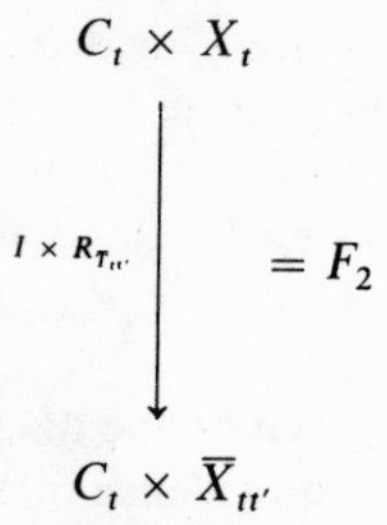

FIG. 3.6

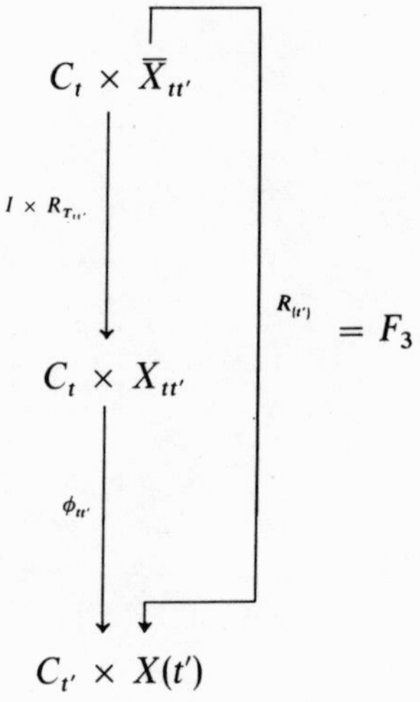

FIG. 3.7

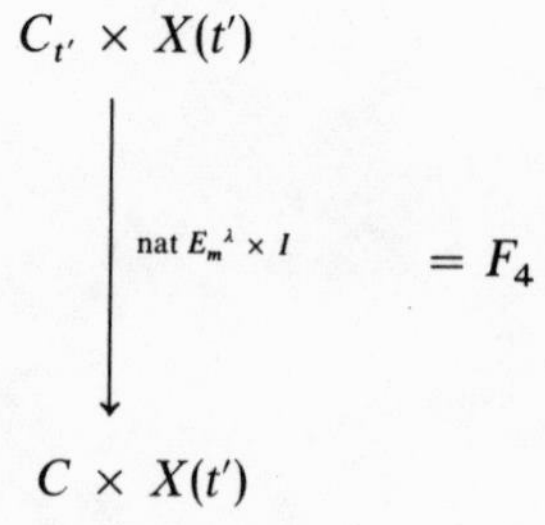

FIG. 3.8

(d) Construction of the State-Space Representation

We can now indicate how the results developed so far are related in providing conditions for the existence of various auxiliary functions and associated system representations and furthermore how they can be used in an orderly procedure to construct all the necessary machinery for a state-space description of a general time system.

Figure 3.9 indicates the relationship between theorems that give the conditions for the existence of different kinds of representations. At each step a new auxiliary function is introduced, and a different representation of the system becomes possible. Some standard terminology usually associated with the conditions given by the corresponding theorems is also indicated in the diagram.

Condition	System representation	Auxiliary function
	$S \subset X^T \times Y^T$ ↓	Time system
Existence of initial-response function	Theorems 1.1 and 4.2, Chapter II ↓ ρ_0 ↓	$\rho_0 : C_0 \times X \to Y$ (initial-response function)
Existence of systems-response family	Theorems 2.1 and 4.2, Chapter II ↓ $\bar{\rho}$ ↓	$\bar{\rho} = \{\rho_t : C_t \times X_t \to Y_t\}$ (systems-response family)
Realizability	Theorems 1.1 and 1.2, Chapter III ↓ $(\bar{\rho}, \bar{\phi})$ ↓	$\bar{\phi} = \{\phi_{tt'} : C_t \times X_{tt'} \to C_{t'}\}$ (state-transition family)
Canonical (decomposition) representation	Theorem 2.1, Chapter III ↓ $(\bar{\phi}, \hat{\lambda})$ ↓	$\hat{\lambda} = \{\lambda_t : C_t \times X(t) \to Y(t)\}$ (output-function family)
State-space representation	Theorem 3.1, Chapter III ↓ $(\bar{\bar{\phi}}, \bar{\bar{\lambda}})$	$\bar{\bar{\phi}} = \{\phi_{tt'} : C \times X_{tt'} \to C\}$ $\bar{\bar{\lambda}} = \{\hat{\lambda}_t : C \times A \to B\}$

FIG. 3.9

Figure 3.10 gives a procedure for the construction of the state space. Starting from given initial state object C_0, there are three steps in the construction process:

(i) First, for any $t \in T$, the set $C_0 \times X^t$ is taken as the state object, i.e., $C_t = C_0 \times X^t$.

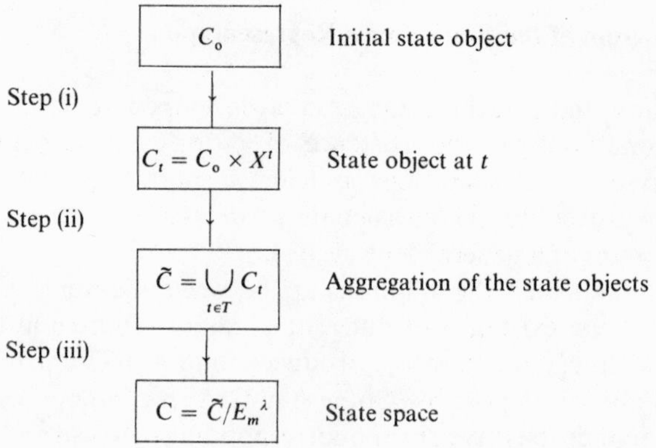

FIG. 3.10

(ii) Second, the union of all state objects is formed

$$\tilde{C} = \bigcup_{t \in T} C_t$$

(iii) Finally, the equivalence E_m^{λ} is defined in $\tilde{C}$, and the quotient set $C = \tilde{C}/E_m^{\lambda}$ is taken as the state space.

Let us show how the auxiliary functions associated with the construction process shown in Fig. 3.9 can be defined.

Given the initial response

$$\rho_o : C_o \times X \to Y$$

the response function at t

$$\rho_t : C_t \times X_t \to Y_t$$

associated with the state $C_t = C_o \times X^t$ is defined in terms of ρ_o so that

$$\rho_t(c_t, x_t) = \rho_o(c_o, x^t \cdot x_t) \mid T_t \tag{3.9}$$

The state-transition function on $T_{tt'}$

$$\phi_{tt'} : C_t \times X_{tt'} \to C_{t'}$$

associated with $C_t = C_o \times X^t$ and $C_{t'} = C_o \times X^{t'}$ is defined by

$$\phi_{tt'}((c_o, x^t), x_{tt'}) = (c_o, x^t \cdot x_{tt'}) \tag{3.10}$$

To show consistency of $\bar{\phi}$ as defined by Eq. (3.10) with the given response family $\bar{\rho}$, observe that for a given $c_t = (c_o, x^t)$

$$\begin{aligned}
\rho_{t'}(\phi_{tt'}(c_t, x_{tt'}), x_{t'}) &= \rho_{t'}((c_o, x^t \cdot x_{tt'}), x_{t'}) \\
&= \rho_o(c_o, x^t \cdot x_{tt'} \cdot x_{t'}) \mid T_{t'} \\
&= (\rho_o(c_o, x^t \cdot x_{tt'} \cdot x_{t'}) \mid T_t) \mid T_{t'} \\
&= \rho_t(\phi_{ot}(c_o, x^t), x_{tt'} \cdot x_{t'}) | T_{t'} = \rho_t(c_t, x_{tt'} \cdot x_{t'}) \mid T_{t'}
\end{aligned}$$

To show that $\bar{\phi}$ has the required composition property, notice that, for a given $c_t = (c_o, x^t)$,

$$\begin{aligned}
\phi_{t't''}(\phi_{tt'}(c_t, x_{tt'}), x_{t't''}) &= \phi_{t't''}((c_o, x^t \cdot x_{tt'}), x_{t't''}) \\
&= (c_o, x^t \cdot x_{tt'} \cdot x_{t't''}) \\
&= \phi_{tt''}((c_o, x^t), x_{tt'} \cdot x_{t't''})
\end{aligned}$$

Therefore, $(\bar{\rho}, \bar{\phi})$ as defined by Eqs. (3.9) and (3.10) is a dynamical system representation of the system S consistent with the given ρ_o.

The output function λ_t is again defined in terms of $\bar{\rho}$

$$\lambda_t(c_t, x_t(t)) = \rho_t(c_t, x_t)(t)$$

Finally, the state-transition and output functions on the state space $C = \tilde{C}/E_m{}^\lambda$ are constructed in a manner described in Section 3b, namely,

$$\hat{\phi}_{tt'}: C \times X_{tt'} \to E$$

such that

$$\hat{\phi}_{tt'}([c], x_{tt'}) = [\phi_{tt'}(I_t([c]), x_{tt'})]$$

and

$$\hat{\lambda}_t: C \times A \to B$$

such that

$$\hat{\lambda}_t([c], a) = \lambda_t(I_t([c]), a)$$

where $I_t: C \to C_t$ such that

$$I_t([c]) = \begin{cases} c_t, & \text{if} \quad c_t \in [c] \cap C_t \neq \phi \\ c_t{}^* \in C_t, & \text{otherwise} \end{cases}$$

and $c_t{}^*$ is an arbitrary element of C_t. As has been shown in Section 3b, $(\hat{\phi}, \hat{\lambda})$ as defined above satisfies the required consistency and composition properties so that $(\hat{\phi}, \hat{\lambda})$ is a canonical state space representation of S.

The construction procedure shown in Figs. 3.9 and 3.10 starts from a given initial systems response, i.e., a given initial state object. For different selections of initial state objects, different state spaces will be constructed. A past-determined system, however, yields a unique natural state space and a unique state-space representation. This will be discussed in Chapter V.

Chapter IV

LINEARITY

The objective of this chapter is to investigate how problems considered so far are affected by the assumption of linearity. The results from this chapter will point out some properties due to linearity and will be used in subsequent chapters when dealing with linear systems.

The separability of the system's response function $\bar{\rho}$ into the input response $\bar{\rho}_1$ and the state response $\bar{\rho}_2$ is established for the case of a linear system. Then the realizability theory for linear systems is developed in reference to these two parts of the system's response. Of particular interest is the realizability of the input-response family $\bar{\rho}_2$. When the system is strongly nonanticipatory, $\bar{\rho}_2$ is realizable if and only if it is equivalent to the consecutive application of two transformations: The first transformation maps inputs into states, while the second maps states into the output values. When a linear system is time invariant, the representation of ρ_2 into two consecutive maps appears in a form which is convenient for the direct application, e.g., to the case of a system specified by differential equations. Actually, the developed results (Theorem 3.3, Chapter IV) give the minimal assumptions needed to represent ρ_2 by the two maps having the required property, which shows that the key assumption is linearity rather than specification of the system by differential equations.

On the basis of the developed results, an orderly procedure for the construction of the state space of a linear system is given. In contrast to the case of a general time system, the construction is not based on an assumed initial state object but rather on a special subset of the total input–output set, referred to as the algebraic core. For any linear system, the state space constructed from such an algebraic core is unique; furthermore, such a state space is minimal in an appropriate sense.

1. LINEAR TIME SYSTEMS

Linearity is an important property for a system because it enables one to arrive at the conclusions valid for the entire class of inputs although considering only a few of them. For a system to be linear, first of all the input object has to have an appropriate structure and, second, the system's transformation has to preserve that structure in a given sense so that an analogous structure is also present in the output set.

Recall the definition of a (complete) linear system given in Chapter II: A system $S \subset X \times Y$ is linear if and only if X and Y are linear algebras over the same field $\mathscr{A}$, and S satisfies the following:

$$(\forall s)(\forall s')[s \in S \;\&\; s' \in S \rightarrow s + s' \in S]$$

$$(\forall s)(\forall \alpha)[\alpha \in \mathscr{A} \;\&\; s \in S \rightarrow \alpha \cdot s \in S]$$

If X and Y are time objects, a usual realization of the algebra operations in X and Y is in terms of the linear structure in the alphabets A and B. For X we have

$$x'' = x + x' \leftrightarrow (\forall t)[x''(t) = x(t) + x'(t)]$$

$$x' = \alpha x \leftrightarrow (\forall t)[x'(t) = \alpha x(t)] \tag{4.1}$$

and completely analogous for Y. In what follows we shall assume that X and Y have algebras defined by Eq. (4.1).

More specifically, then, we have the following definition.

Definition 1.1. Let S be a time system and furthermore

(i) A and B are linear algebras over the same field $\mathscr{A}$;
(ii) X is a linear algebra with the operations $+$ and $\cdot$ such that

$$(\forall t)[(x + x')(t) = x(t) + x'(t)]$$

$$(\forall \alpha)[(\alpha \cdot x)(t) = \alpha \cdot x(t)]$$

Y is a linear algebra similarly defined.

Then S is a (complete) linear time system if and only if

(iii) $(x, y) \in S \;\&\; (\hat{x}, \hat{y}) \in S \rightarrow (x + \hat{x}, y + \hat{y}) \in S$
(iv) $(x, y) \in S \;\&\; \alpha \in \mathscr{A} \rightarrow (\alpha x, \alpha y) \in S$

We shall also assume that the object X is complete relative to concatenation (Definition 2.3, Chapter II) which, due to the linearity of X, can be expressed by the condition

$$(\forall x)(x \in X \to x^t \cdot o \in X)$$

As a matter of fact, suppose X satisfies the above condition and let x, $\hat{x}$ be two arbitrary elements of X; then $x^t \cdot o \in X$ and $\hat{x}^t \cdot o \in X$ by definition. Since X is a linear algebra, $\hat{x} - \hat{x}^t \cdot o = o \cdot \hat{x}_t \in X$ and hence, $x^t \cdot o + o \cdot \hat{x}_t = x^t \cdot \hat{x}_t \in X$, which satisfies the completeness condition of X as given in Definition 2.3, Chapter II.

When a system is linear, some deeper and more specific results can be developed. This is, of course, due to the linear structure that has been added to the system's objects. The development in this chapter will follow the same line as for the general time systems in the preceding chapters. However, the results for the linear systems will not be a direct application of the general results, since additional conditions will have to be satisfied (e.g., the requirement for the linearity of the state objects and the state space). Furthermore, some special facts specific for the linear systems (e.g., the realizability of the strongly nonanticipatory systems) will have to be established.

2. DECOMPOSITION OF SYSTEMS RESPONSE: STATE- AND INPUT-RESPONSE FUNCTIONS

As shown in Chapter II, the systems-response function for a linear system can be decomposed in reference to the algebraic operation in Y, in two functions representing the response of the system due to the state and due to the input separately. The total response is then the sum of such separated responses. Such a separation of responses is, of course, possible for linear time systems also.

Definition 2.1. Let $S \subset X \times Y$ be a linear time system and ρ_o a mapping $\rho_o: C_o \times X \to Y$. ρ_o is termed a linear initial systems-response function if and only if

(i) ρ_o is consistent with S, i.e.,

$$(x, y) \in S \leftrightarrow (\exists c)[y = \rho_o(c, x)]$$

(ii) C_o is a linear algebra over the field $\mathscr{A}$;

(iii) there exist two linear mappings $\rho_{10}: C_o \to Y$ and $\rho_{20}: X \to Y$ such that for all $(c, x) \in C_o \times X$

$$\rho_o(c, x) = \rho_{10}(c) + \rho_{20}(x)$$

C_o is referred to as the linear initial state object. The mapping $\rho_{10}: C_o \to Y$ is termed the initial state response, while $\rho_{20}: X \to Y$ is the initial input response.

Notice the distinction between the initial systems-response function for a general time system and the linear initial systems-response function; the first concept requires only (i), while for the second, conditions (ii) and (iii) have to be satisfied also.

From Theorem 1.2, Chapter II, we have immediately the following proposition.

Proposition 2.1. A time system is linear if and only if it has a linear initial systems-response function.

A response family for a linear system is defined in an obvious manner in the following definition.

Definition 2.2. Let S be a linear time system. Then a family of linear maps $\bar{\rho} = \{\rho_t: C_t \times X_t \to Y_t\}$ is the linear-response family for S if and only if $\bar{\rho}$ is consistent with S; that is, for every $t \in T$, ρ_t is a linear initial systems response for S_t.

In view of Definition 2.2, every linear initial systems response can be decomposed into two maps $\rho_{1t}: C_t \to Y_t$ and $\rho_{2t}: X_t \to Y_t$ such that

$$\rho_t(c_t, x_t) = \rho_{1t}(c_t) + \rho_{2t}(x_t)$$

ρ_{1t} and ρ_{2t} will be referred to as the *state response* and the *input response* at t, respectively.

From Theorem 1.2, Chapter II, we have immediately the following proposition.

Proposition 2.2. Every linear time system has a linear-response family.

Finally, a linear dynamical system is defined in the following manner.

Definition 2.3. Let S be a linear system and $\bar{\rho}$ its linear-response family. S is a linear dynamical system (i.e., has a linear dynamical-system representation) if and only if for all $t, t' \in T$, there exists a pair of linear maps $\phi_{1tt'}: C_t \to C_{t'}$, $\phi_{2tt'}: X_{tt'} \to C_{t'}$ and its sum $\phi_{tt'}(c_t, x_{tt'}) = \phi_{1tt'}(c_t) + \phi_{2tt'}(x_{tt'})$ such that $(\bar{\rho}, \bar{\phi})$ is a dynamical-system representation of S, i.e.,

$$\rho_{t'}(\phi_{tt'}(c_t, x_{tt'}), x_{t'}) = \rho_t(c_t, x_t) \mid T_{t'}$$

$$\phi_{t't''}(\phi_{tt'}(c_t, x_{tt'}), x_{t't''}) = \phi_{tt''}(c_t, x_{tt''})$$

$$\phi_{tt}(c_t, x_{tt}) = c_t$$

3. REALIZATION THEORY

The realization problem for linear systems is analogous to the case of general time systems except that the initially given systems-response function (whose dynamic realizability is investigated) is linear and actually given by its component mappings $\bar{\rho}_1$ and $\bar{\rho}_2$. For the realization by a linear dynamical system, it is required in addition, of course, that $\bar{\phi}$ is linear in the sense that the conditions of Definition 2.3 are satisfied.

The consistency conditions and the realizability conditions for $\bar{\rho}$ are given in Theorem 2.1, Chapter II, and Theorem 1.1, Chapter III, for general time systems. As mentioned above, these theorems cannot cover directly the case of linear time systems. However, some useful information for the linear time systems can be deduced from them.

For the state response, i.e., $\bar{\rho}_1$, the case when the input is zero, i.e., $x_t = \text{o}$, is of special interest. Conditions (P2) and (P3) take now the form

$$(\forall c_t)(\forall x_{tt'})(\exists c_{t'})(\rho_{1t'}(c_{t'}) = \rho_t(c_t, x_{tt'} \cdot \text{o}) \mid T_{t'}) \tag{4.2}$$

$$(\forall c_{t'})(\exists c_t)(\exists x_{tt'})(\rho_{1t'}(c_{t'}) = \rho_t(c_t, x_{tt'} \cdot \text{o}) \mid T_{t'}) \tag{4.3}$$

Conditions (4.2) and (4.3) will be referred to as the *state-response consistency condition*. Their conceptual interpretation is the following: By definition, for every $(c_t, x_{tt'})$ if the state–input pair $(c_t, x_{tt'} \cdot \text{o})$ is applied to the system, the output observed from t' is $y_{t'} = \rho_t(c_t, x_{tt'} \cdot \text{o}) \mid T_{t'}$. The condition (4.2) then requires that for every such output, there exists a state $c_{t'}$ so that if the state–input pair $(c_{t'}, \text{o})$ is applied, $y_{t'}$ will be obtained. Condition (4.3) furthermore requires that only such states are acceptable in $C_{t'}$.

For the input response, i.e., $\bar{\rho}_2$, the case of interest is when $c_t = \text{o}$ and $x_{tt'} = \text{o}$, so that condition (P3) becomes

$$(\exists \hat{c}_{t'})(\forall x_{t'})(\rho_{t'}(\hat{c}_{t'}, x_{t'}) = \rho_t(\text{o}, \text{o} \cdot x_{t'}) \mid T_{t'})$$

Since ρ_t is linear, $\rho_{t'}(\hat{c}_{t'}, \text{o}) = \rho_t(\text{o}, \text{o}) \mid T_{t'} = \text{o}$ holds for $x_{t'} = \text{o}$. Hence we have

$$(\forall x_{t'})(\rho_{2t'}(x_{t'}) = \rho_{2t}(\text{o} \cdot x_{t'}) \mid T_{t'}) \tag{4.4}$$

Condition (4.4) will be referred to as the *input-response consistency condition*. It requires that zero input, when applied to the input-response function over a given time interval, does not affect the future evolution of the input-response function. In other words, when the input is such that its values are zero up to the time t', i.e., $x_t = \text{o} \cdot x_{t'}$, the restriction of the corresponding output from time t' on, $\rho_{2t}(\text{o} \cdot x_{t'}) \mid T_{t'}$, is equal to the output of the input–response function starting from t' and applying $x_{t'}$. In general,

$$\rho_{2t'}(x_{t'}) \neq \rho_{2t}(x_{tt'} \cdot x_{t'}) \mid T_{t'}$$

The consistency of $\bar{\rho}$ for the case of linear time systems can be now characterized in terms of the newly introduced consistency conditions.

Theorem 3.1. Let X and Y be linear time objects $X \subset A^T$, $Y \subset B^T$, $\bar{C} = \{C_t : t \in T\}$ a family of linear spaces, $\bar{\rho}_1, \bar{\rho}_2$, two families of linear mappings $\bar{\rho}_1 = \{\rho_{1t} : C_t \to Y_t\}$, $\bar{\rho}_2 = \{\rho_{2t} : X_t \to Y_t\}$, and

$$\bar{\rho} = \{\rho_t : C_t \times X_t \to Y_t \,\&\, (\forall (c_t, x_t))[\rho_t(c_t, x_t) = \rho_{1t}(c_t) + \rho_{2t}(x_t)]\}$$

Suppose $\bar{\rho}_2$ satisfies the input-response consistency condition: For all $t \leq t'$,

$$(\forall x_{t'})[\rho_{2t'}(x_{t'}) = \rho_{2t}(\text{o} \cdot x_{t'}) \mid T_{t'}]$$

Then there exists a linear system $S \subset X \times Y$ such that $\bar{\rho}$ is a linear-response family for S if and only if $\bar{\rho}$ satisfies the state-response consistency condition: For all $t \leq t'$,

(i) $(\forall (c_t, x_{tt'}))(\exists c_{t'})(\rho_{1t'}(c_{t'}) = \rho_t(c_t, x_{tt'} \cdot \text{o}) \mid T_{t'})$

and

(ii) $(\forall c_{t'})(\exists (c_t, x_{tt'}))(\rho_{1t'}(c_{t'}) = \rho_t(c_t, x_{tt'} \cdot \text{o}) \mid T_{t'})$

PROOF. Let $S = S_{\text{o}}{}^{\rho}$. First, we shall prove the *if* part. Suppose $(x_t, y_t) \mid T_{t'} \in S_t{}^{\rho} \mid T_{t'}$ for $t \leq t'$. Then due to the definition of $S_t{}^{\rho}$, $y_t = \rho_{1t}(c_t) + \rho_{2t}(x_t)$ holds for some $c_t \in C_t$. Hence,

$$\begin{aligned} y_t \mid T_{t'} &= (\rho_{1t}(c_t) + \rho_{2t}(x_t)) \mid T_{t'} = (\rho_{1t}(c_t) + \rho_{2t}(x_{tt'} \cdot \text{o}) + \rho_{2t}(\text{o} \cdot x_{t'})) \mid T_{t'} \\ &= (\rho_{1t}(c_t) + \rho_{2t}(x_{tt'} \cdot \text{o})) \mid T_{t'} + \rho_{2t}(\text{o} \cdot x_{t'}) \mid T_{t'} \end{aligned}$$

It follows from condition (i) that $y_t \mid T_{t'} = \rho_{1t'}(c_{t'}) + \rho_{2t'}(x_{t'})$ for some $c_{t'} \in C_{t'}$. Hence, $(x_t, y_t) \mid T_{t'} \in S_{t'}{}^{\rho}$ or $S_t{}^{\rho} \mid T_{t'} \subseteq S_{t'}{}^{\rho}$. Suppose $(x_{t'}, y_{t'}) \in S_{t'}{}^{\rho}$. Then $y_{t'} = \rho_{1t'}(c_{t'}) + \rho_{2t'}(x_{t'})$ for some $c_{t'} \in C_{t'}$. It follows from condition (ii) that

$$\rho_{1t'}(c_{t'}) = (\rho_{1t}(c_t) + \rho_{2t}(x_{tt'} \cdot \text{o})) \mid T_{t'}$$

for some $c_t \in C_t$ and $x_{tt'} \in X_{tt'}$. Hence,

$$\begin{aligned} y_{t'} &= (\rho_{1t}(c_t) + \rho_{2t}(x_{tt'} \cdot \text{o})) \mid T_{t'} + \rho_{2t}(\text{o} \cdot x_{t'}) \mid T_{t'} \\ &= (\rho_{1t}(c_t) + \rho_{2t}(x_{tt'} \cdot x_{t'})) \mid T_{t'} \end{aligned}$$

Therefore, $(x_{t'}, y_{t'}) \in S_t{}^{\rho} \mid T_{t'}$ or $S_{t'}{}^{\rho} \subseteq S_t{}^{\rho} \mid T_{t'}$. Hence, combining the first part of the proof with the present one, we have $S_{t'}{}^{\rho} = S_t{}^{\rho} \mid T_{t'}$ for $t \leq t'$. Therefore, $S_t{}^{\rho} = S_{\text{o}}{}^{\rho} \mid T_t$.

Next, we shall prove the *only if* part. Notice that $S_t{}^{\rho} = S_{\text{o}}{}^{\rho} \mid T_t$ for every $t \in T$ implies and is implied by $S_{t'}{}^{\rho} = S_t{}^{\rho} \mid T_{t'}$ for any $t \leq t'$, because $S_t{}^{\rho} = S_{\text{o}}{}^{\rho} \mid T_t$ and $S_{t'}{}^{\rho} = S_{\text{o}}{}^{\rho} \mid T_{t'}$ implies

$$S_t{}^{\rho} \mid T_{t'} = (S_{\text{o}}{}^{\rho} \mid T_t) \mid T_{t'} = S_{\text{o}}{}^{\rho} \mid T_{t'} = S_{t'}{}^{\rho}$$

Suppose $S_{t'}{}^{\rho} = S_t{}^{\rho} \mid T_{t'}$ holds for every $t \leq t'$. Then,

$$(\forall x_t)(\forall c_t)(\exists c_{t'})(\rho_{1t'}(c_{t'}) + \rho_{2t'}(x_{t'}) = (\rho_{1t}(c_t) + \rho_{2t}(x_t)) \mid T_{t'})$$

where $x_{t'} = x_t \mid T_{t'}$. Hence,

$$(\forall x_{tt'})(\forall c_t)(\exists c_{t'})(\rho_{1t'}(c_{t'}) = (\rho_{1t}(c_t) + \rho_{2t}(x_{tt'} \cdot o)) \mid T_{t'})$$

Furthermore, due to $S_{t'}{}^{\rho} = S_t{}^{\rho} \mid T_{t'}$ we have

$$(\forall x_{t'})(\forall c_{t'})(\exists x_{tt'})(\exists c_t)(\rho_{1t'}(c_{t'}) + \rho_{2t'}(x_{t'}) = (\rho_{1t}(c_t) + \rho_{2t}(x_{tt'} \cdot x_{t'})) \mid T_{t'})$$

Hence,

$$(\forall c_{t'})(\exists x_{tt'})(\exists c_t)(\rho_{1t'}(c_{t'}) = (\rho_{1t}(c_t) + \rho_{2t}(x_{tt'} \cdot o)) \mid T_{t'}) \qquad \text{Q.E.D.}$$

We can now present the main result in the realization theory for linear systems.

Theorem 3.2. Let X and Y be linear time objects $X \subset A^T$, $Y \subset B^T$, $\bar{C} = \{C_t : t \in T\}$ a family of linear spaces, $\bar{\rho}_1, \bar{\rho}_2$, two families of linear mappings $\bar{\rho}_1 = \{\rho_{1t} : C_t \to Y_t\}$, $\bar{\rho}_2 = \{\rho_{2t} : X_t \to Y_t\}$, and

$$\bar{\rho} = \{\rho_t : C_t \times X_t \to Y_t \,\&\, (\forall (c_t, x_t))[\rho(c_t, x_t) = \rho_{1t}(c_t) + \rho_{2t}(x_t)]\}$$

$\bar{\rho}$ is then realizable by a linear dynamical system (i.e., there exists a family of linear state-transition functions $\bar{\phi}$ such that $(\bar{\rho}, \bar{\phi})$ is a linear dynamical system) if and only if $\bar{\rho}$ satisfies both the conditions of the state-response consistency and the input-response consistency, i.e., for all $t, t' \in T$,

(i) $\{(\forall (c_t, x_{tt'}))(\exists c_{t'})[\rho_{1t'}(c_{t'}) = \rho_t(c_t, x_{tt'} \cdot o) \mid T_{t'}]$
and $(\forall c_{t'})(\exists (c_t, x_{tt'}))[\rho_{1t'}(c_{t'}) = \rho_t(c_t, x_{tt'} \cdot o) \mid T_{t'}]\}$

(ii) $(\forall x_{t'})[\rho_{2t'}(x_{t'}) = \rho_{2t}(o \cdot x_{t'}) \mid T_{t'}]$

PROOF. Let us consider the *only if* part first. Let $(\bar{\rho}, \bar{\phi})$ be a linear dynamical system where $\bar{\rho} = \{\rho_t\}$ is the family of linear mappings described in Theorem 3.2. Then we have

$$\begin{aligned}\rho_{t'}(\phi_{tt'}(c_t, x_{tt'}), x_{t'}) &= \rho_{1t'}(\phi_{tt'}(c_t, x_{tt'})) + \rho_{2t'}(x_{t'}) \\ &= \rho_{1t'}(\phi_{1tt'}(c_t)) + \rho_{1t'}(\phi_{2tt'}(x_{tt'})) + \rho_{2t'}(x_{t'})\end{aligned}$$

and

$$\begin{aligned}\rho_t(c_t, x_t) \mid T_{t'} &= \rho_{1t}(c_t) \mid T_{t'} + \rho_{2t}(x_t) \mid T_{t'} \\ &= \rho_{1t}(c_t) \mid T_{t'} + \rho_{2t}(x_{tt'} \cdot o) \mid T_{t'} + \rho_{2t}(o \cdot x_{t'}) \mid T_{t'}\end{aligned}$$

Since

$$\rho_t(c_t, x_t) \mid T_{t'} = \rho_{t'}(\phi_{tt'}(c_t, x_{tt'}), x_{t'})$$

holds for every $c_t, x_{tt'}$, and $x_{t'}$ we have

$$\rho_{2t'}(x_{t'}) = \rho_{2t}(\mathrm{o} \cdot x_{t'}) \mid T_{t'} \quad \text{when} \quad c_t = \mathrm{o} \quad \text{and} \quad x_{tt'} = \mathrm{o}$$

Hence, $\{\rho_t\}$ satisfies the input-response consistency. Then it follows from Theorem 3.1 that the state-response consistency is also satisfied.

Next, let us consider the *if* part. First, let us consider the case where ρ_{1t} is a one-to-one mapping for every $t \in T$ where $\rho_{1t}: C_t \to \rho_{1t}(C_t)$ has the inverse ρ_{1t}^{-1}. To prove the realizability, we have to (i) show that $\bar{\rho} = \{\rho_t\}$ is a response family; (ii) construct a linear state-transition family $\{\phi_{tt'}\}$; (iii) show that $\{\phi_{tt'}\}$ is consistent with $\{\rho_t\}$ and satisfies the composition property.

(i) Since $\{\rho_t\}$ satisfies the state-response consistency, the input-response consistency $\{\rho_t\}$ is a response family by Theorem 3.1.

(ii) Let $\phi_{1tt'}: C_t \to C_{t'}$ be such that

$$\phi_{1tt'}(c_t) = \rho_{1t'}^{-1}(\rho_{1t}(c_t) \mid T_{t'})$$

$\phi_{1tt'}$ is properly defined because the state-response consistency implies that $\rho_{1t}(c_t) \mid T_{t'} = \rho_{1t'}(c_{t'})$ for some $c_{t'}$ or $\rho_{1t}(c_t) \mid T_{t'} \in \rho_{1t'}(C_{t'})$ for every c_t, and $\rho_{1t'}$ is a one-to-one mapping.

Let $\phi_{2tt'}: X_{tt'} \to C_{t'}$ be such that

$$\phi_{2tt'}(x_{tt'}) = \rho_{1t'}^{-1}(\rho_{2t}(x_{tt'} \cdot \mathrm{o}) \mid T_{t'})$$

$\phi_{2tt'}$ is also properly defined because the state-response consistency implies that $\rho_{2t}(x_{tt'} \cdot \mathrm{o}) \mid T_{t'} \in \rho_{1t'}(C_{t'})$. Since the restriction is a linear operation and since the inverse of a linear operator is linear, $\phi_{1tt'}$ and $\phi_{2tt'}$ are linear and hence so is $\phi_{tt'}$, where

$$\phi_{tt'}(c_t, x_{tt'}) = \phi_{1tt'}(c_t) + \phi_{2tt'}(x_{tt'})$$

(iii) From the construction of $\{\rho_t\}$ and $\{\phi_{tt'}\}$ and the input-response consistency, it follows that

$$\begin{aligned}
\rho_{t'}(\phi_{tt'}(c_t, x_{tt'}), x_{t'}) &= \rho_{1t'}(\phi_{tt'}(c_t, x_{tt'})) + \rho_{2t'}(x_{t'}) \\
&= \rho_{1t'}(\phi_{1tt'}(c_t)) + \rho_{1t'}(\phi_{2tt'}(x_{tt'})) + \rho_{2t'}(x_{t'}) \\
&= \rho_{1t}(c_t) \mid T_{t'} + \rho_{2t}(x_{tt'} \cdot \mathrm{o}) \mid T_{t'} + \rho_{2t}(\mathrm{o} \cdot x_{t'}) \mid T_{t'} \\
&= \rho_t(c_t, x_t) \mid T_{t'}
\end{aligned}$$

Hence, $\{\phi_{tt'}\}$ is consistent with $\{\rho_t\}$.

Since $\bar{\rho}$ is reduced (or ρ_{1t} is a one-to-one mapping), Theorem 2.2, Chapter II, is applicable to the present case, and hence $\{\phi_{tt'}\}$ satisfies the composition property. Hence, $\{\phi_{tt'}\}$ is a desired state-transition family.

The general case will be considered now where ρ_{1t} is not necessarily a one-to-one mapping. Let $E_t \subset C_t \times C_t$ be such that

$$(c_t, c_t') \in E_t \leftrightarrow \rho_{1t}(c_t) = \rho_{1t}(c_t')$$

Then it is easily verified that E_t is a congruence relation. Let $C_t/E_t = \{[c_t]\}$ be the quotient set, which is a linear space in the obvious sense, and for every $t \in T$, a linear map $\tilde{\rho}_{1t}: C_t/E_t \to Y_t$ such that $\tilde{\rho}_{1t}([c_t]) = \rho_{1t}(c_t)$ is properly defined because E_t is a congruence relation. It is easy to show that $\{\tilde{\rho}_t = \langle \tilde{\rho}_{1t}, \rho_{2t} \rangle\}$ satisfies the state-response consistency and the input-response consistency when $\{\rho_t = \langle \rho_{1t}, \rho_{2t} \rangle\}$ does. Therefore, we have a family of state-transition functions $\{\tilde{\phi}_{tt'}\}$ which is consistent with $\{\tilde{\rho}_t\}$. We shall proceed in two steps:

(i) Define a family of state transitions $\{\phi_{tt'}\}$ corresponding to $\{\rho_t\}$.
(ii) Show that $\{\langle \rho_t, \phi_{tt'} \rangle\}$ is a linear dynamical system.

The first step: let $\mu_t: C_t/E_t \to C_t$ be a linear mapping such that $\mu_t([c_t]) \in [c_t]$ which can be constructed by using the same technique as used in the proof of Theorem 1.2, Chapter II. We shall write μ for μ_t when the meaning is apparent from the context. The state-transition family is then defined as

$$\phi_{1tt'}: C_t \to C_{t'} \qquad \text{such that} \quad \phi_{1tt'}(c_t) = \mu(\tilde{\phi}_{1tt'}([c_t]))$$

$$\phi_{2tt'}: X_{tt'} \to C_{t'} \qquad \text{such that} \quad \phi_{2tt'}(x_{tt'}) = \mu(\tilde{\phi}_{2tt'}(x_{tt'}))$$

The second step: we have only to show that $\{\phi_{tt'}\}$ defined above satisfies the composition property and is consistent with $\{\rho_t\}$. First, we should notice that from the definitions of $\tilde{\rho}_{1t}$ and $[c_t]$

$$\rho_{1t}(\mu([c_t])) = \tilde{\rho}_{1t}([c_t]) \tag{4.5}$$

$$[\mu([c_t])] = [c_t] \tag{4.6}$$

It can be readily shown that

$$\rho_t(c_t, x_t) = \tilde{\rho}_t([c_t], x_t) \tag{4.7}$$

$$\phi_{tt'}(c_t, x_{tt'}) = \mu(\tilde{\phi}_{tt'}([c_t], x_{tt'})) \tag{4.8}$$

We can now prove the validity of the consistency

$$\begin{aligned}
\rho_{t'}(\phi_{tt'}(c_t, x_{tt'}), x_{t'}) &= \rho_{t'}(\mu(\tilde{\phi}_{tt'}([c_t], x_{tt'})), x_{t'}) && \text{[from (4.8)]} \\
&= \rho_{1t'}(\mu(\tilde{\phi}_{tt'}([c_t], x_{tt'}))) + \rho_{2t'}(x_{t'}) \\
&= \tilde{\rho}_{1t'}(\tilde{\phi}_{tt'}([c_t], x_{tt'})) + \rho_{2t'}(x_{t'}) && \text{[from (4.5)]} \\
&= \tilde{\rho}_t([c_t], x_t) \mid T_{t'} \\
&= \rho_t(c_t, x_t) \mid T_{t'} && \text{[from (4.7)]}
\end{aligned}$$

and the validity of the composition property

$$\begin{aligned}\phi_{t't''}(\phi_{tt'}(c_t, x_{tt'}), x_{t't''}) &= \phi_{t't''}(\mu(\tilde{\phi}_{tt'}([c_t], x_{tt'})), x_{t't''}) && \text{[from (4.8)]}\\ &= \mu(\tilde{\phi}_{t't''}([\mu(\tilde{\phi}_{tt'}([c_t], x_{tt'}))], x_{t't''})) && \text{[from (4.8)]}\\ &= \mu(\tilde{\phi}_{t't''}(\tilde{\phi}_{tt'}([c_t], x_{tt'}), x_{t't''})) && \text{[from (4.6)]}\\ &= \mu(\tilde{\phi}_{tt''}([c_t], x_{tt''})) = \phi_{tt''}(c_t, x_{tt''}) && \text{[from (4.8)]}\end{aligned}$$

Q.E.D.

Let us briefly compare the reálizability conditions for a general time system and for a linear time system.

Regarding consistency of $\bar{\rho}$ with a time system S, Theorems 2.1 (Chapter II) and 3.1 ought to be compared. Suppose condition (i) from Theorem 3.1 holds, i.e.,

$$(\forall c_o)(\forall x^t)(\exists c_t)(\rho_{1t}(c_t) = (\rho_{10}(c_o) + \rho_{20}(x^t \cdot o)) \mid T_t) \tag{4.9}$$

Then combining Eq. (4.9) with the input-response consistency, we have

$$(\forall c_o)(\forall x^t)(\forall x_t)(\exists c_t)(\rho_{1t}(c_t) + \rho_{2t}(x_t) = (\rho_{10}(c_o) + \rho_{20}(x^t \cdot x_t)) \mid T_t)$$

which is exactly (P1) stated in Theorem 2.1, Chapter II. Property (P2) can be, similarly, derived from the state-response consistency.

Regarding the realizability of $\bar{\rho}$, Theorems 1.1 (Chapter III) and 3.2 ought to be compared. If $\bar{\rho}$ satisfies the input-response consistency, then condition (i) from Theorem 3.2 implies (P3). As a matter of fact, it follows from condition (i) in Theorem 3.2 that

$$(\forall c_t)(\forall x_{tt'})(\exists c_{t'})(\rho_{t'}(c_{t'}, o) = \rho_t(c_t, x_{tt'} \cdot o) \mid T_{t'})$$

Since $(\forall x_{t'})(\rho_{2t'}(x_{t'}) = \rho_{2t}(o \cdot x_{t'}) \mid T_{t'})$, we have

$$(\forall c_t)(\forall x_{tt'})(\exists c_{t'})(\forall x_{t'})(\rho_{t'}(c_{t'}, x_{t'}) = \rho_t(c_t, x_{tt'} \cdot x_{t'}) \mid T_{t'})$$

which is (P3).

We can conclude, then, that the realizability conditions for a linear system are almost the same as for the case of a general time system. However, the present form is more suitable for the application in the case of a linear system, since the conditions are separated into two parts of importance in linear systems analysis: one related with ρ_{1t} and the other with ρ_{2t}. This fact will be used later when considering the problem of a natural realization of a linear time system.

The preceding theorem can be used also for the question of realizability of the component functions $\bar{\rho}_1$ and $\bar{\rho}_2$. For example, let a family of input-response functions $\bar{\rho}_2$ be given; then $\bar{\rho}_2$ is considered as realizable if and only if there exists a state-response function $\bar{\rho}_1$ so that $(\bar{\rho}_1, \bar{\rho}_2)$ is realizable.

Theorem 3.2 then implies that $\bar{\rho}_2$ is realizable if and only if it satisfies the consistency condition (ii), and furthermore there exists a mapping $\bar{\rho}_1$ such that condition (i) (given in terms of both $\bar{\rho}_1$ and $\bar{\rho}_2$) is satisfied.

When the system satisfies the conditions for strong nonanticipation, more specific and stronger conditions for realizability can be derived. In this respect, we have the following theorem.

Theorem 3.3. Let $\bar{\rho}_2 = \{\rho_{2t}: X_t \to Y_t\}$ be a family of linear maps. $\bar{\rho}_2$ is realizable by a strongly nonanticipatory linear system (i.e., it is an input response of a strongly nonanticipatory linear dynamical system) if and only if

(i) $\rho_{2t}(x_t)(\tau) = \rho_{2t}(x_{t\tau} \cdot \text{o})(\tau)$ for $\tau \geq t$ (strong nonanticipation condition);

(ii) there exist two families of linear mappings $\{F_{tt'}: \tilde{C}_t \to B\}$ and $\{G_{tt'}: X_{tt'} \to \tilde{C}_{t'}\}$ such that for any $\hat{t} \in T, \hat{t} \geq t' \geq t$

$$\rho_{2t}(x_{tt'} \cdot \text{o})(\hat{t}) = F_{t'\hat{t}}[G_{tt'}(x_{tt'})]$$

where $G_{tt'}$ satisfies the condition

(iii) $G_{tt''}(0 \cdot x_{t't''}) = G_{t't''}(x_{t't''})$ for $t \leq t' \leq t''$.

PROOF. First, we consider the *if* part. We shall check the input-response consistency and the state-response consistency. Notice that conditions (i) and (ii) imply that $\rho_{2t}(x_t)(\tau) = \rho_{2t}(x_{t\tau} \cdot \text{o})(\tau) = F_{\tau\tau} \cdot G_{t\tau}(x_{t\tau})$. Hence, we have that for $\tau \geq t'$

$$\begin{aligned}(\rho_{2t}(\text{o} \cdot x_{t'}) \mid T_{t'})(\tau) &= \rho_{2t}(\text{o} \cdot x_{t'})(\tau) \\ &= F_{\tau\tau} \cdot G_{t\tau}(\text{o} \cdot x_{t'\tau}) \\ &= F_{\tau\tau} \cdot G_{t'\tau}(x_{t'\tau}) \\ &= \rho_{2t'}(x_{t'})(\tau)\end{aligned}$$

Hence, the input consistency is satisfied.

Now, let $\rho_{20} \mid T_t: X_{\text{o}t} \to Y_t$ such that

$$(\rho_{20} \mid T_t)(x_{\text{o}t}) = \rho_{20}(x_{\text{o}t} \cdot \text{o}) \mid T_t$$

Then we have the following diagram, where $C_t = G_{\text{o}t}(X_{\text{o}t})$:

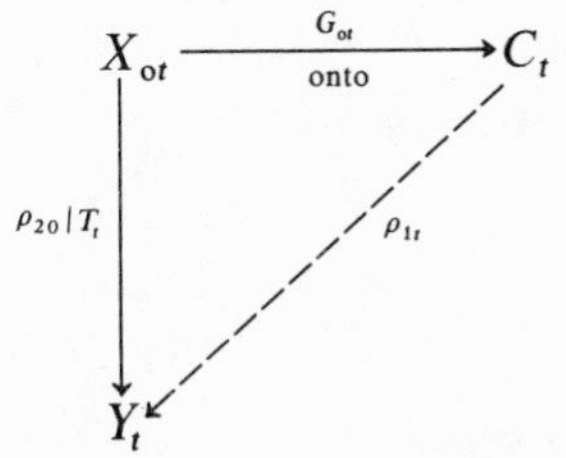

We shall show that there exists a linear mapping $\rho_{1t}: C_t \to Y_t$ such that the above diagram is commutative. As a matter of fact,

$$G_{ot}(x_{ot}) = G_{ot}(x'_{ot}) \to F_{t\tau} \cdot G_{ot}(x_{ot}) = F_{t\tau} \cdot G_{ot}(x'_{ot})$$
$$\to \rho_{20}(x_{ot} \cdot o)(\tau) = \rho_{20}(x'_{ot} \cdot o)(\tau)$$

for every $\tau \geq t$. Hence,

$$G_{ot}(x_{ot}) = G_{ot}(x'_{ot}) \to \rho_{20}(x_{ot} \cdot o) \mid T_t = \rho_{20}(x'_{ot} \cdot o) \mid T_t$$
$$\to (\rho_{20} \mid T_t)(x_{ot}) = (\rho_{20} \mid T_t)(x'_{ot})$$

Furthermore, G_{ot} is an onto mapping by assumption. Hence, $\rho_{1t}: C_t \to Y_t$ can be defined as

$$\rho_{1t}(c_t) = (\rho_{20} \mid T_t)(x_{ot}) \qquad \text{where} \quad c_t = G_{ot}(x_{ot})$$

Let $c_t \in C_t$ and $x_{tt'} \in X_{tt'}$ be arbitrary elements. Then

$$(\rho_{1t}(c_t) + \rho_{2t}(x_{tt'} \cdot o)) \mid T_{t'} = (\rho_{20}(x_{ot} \cdot o) \mid T_t + \rho_{20}(o \cdot x_{tt'} \cdot o) \mid T_t) \mid T_{t'}$$
$$= \rho_{20}(x_{ot} \cdot x_{tt'} \cdot o) \mid T_{t'} = \rho_{1t'}(c_{t'})$$

where $c_{t'} = G_{ot'}(x_{ot} \cdot x_{tt'})$. Let $c_{t'}$ be an arbitrary element of $C_{t'}$. Then

$$\rho_{1t'}(c_{t'}) = \rho_{20}(x_{ot'} \cdot o) \mid T_{t'} \qquad [\text{where } c_{t'} = G_{ot'}(x_{ot'})]$$
$$= ((\rho_{20}(x_{ot} \cdot o) + \rho_{20}(o \cdot x_{tt'} \cdot o)) \mid T_t) \mid T_{t'}$$
$$= (\rho_{1t}(c_t) + \rho_{2t}(x_{tt'} \cdot o)) \mid T_{t'}$$

where $c_t = G_{ot}(x_{ot})$. Hence, the state-response consistency is satisfied. The strong nonanticipation is a direct consequence of condition (i) from Theorem 3.3.

Let us consider the *only if* part. Let $\{\langle \rho_t, \phi_{tt'} \rangle\}$ be a strongly nonanticipatory dynamical system where the input-response family is equal to $\{\rho_{2t}\}$ given in the theorem.

Then the realizability conditions require that

$$\rho_{2t}(x_{tt'} \cdot o) \mid T_{t'} = \rho_{1t'}(\phi_{2tt'}(x_{tt'})) \qquad \text{for every} \quad x_{tt'} \in X_{tt'}$$

Hence, let $G_{tt'}(x_{tt'}) = \phi_{2tt'}(x_{tt'})$, and $F_{t\tau}(c_t) = \rho_{1t}(c_t)(\tau)$. Then

$$\rho_{2t}(x_{tt'} \cdot o)(\tau) = F_{t'\tau} \cdot G_{tt'}(x_{tt'}) \qquad \text{for} \quad \tau \geq t'$$

It follows from the composition property of $\phi_{tt'}$ that

$$\phi_{2tt''}(o \cdot x_{t't''}) = \phi_{t't''}(\phi_{2tt'}(o), x_{t't''})$$
$$= \phi_{2t't''}(x_{t't''})$$

or $G_{tt''}(o \cdot x_{t't''}) = G_{t't''}(x_{t't''})$. Finally, the strong causality implies

$$\rho_{2t}(x_t)(\tau) = \rho_{2t}(x_{t\tau} \cdot o)(\tau) \qquad \text{for} \quad \tau \geq t$$

Q.E.D.

Condition (ii) from the preceding theorem shows that the effect of an input applied during an interval $[t, t')$ can be separated in two parts: (1) changes of the state during the interval $[t, t')$ under direct influence of $x_{tt'}$, (2) free evolution of the state after that interval and its mapping into the output values.

For the special case when the system is specified as a linear differential equation system, it is easily seen that conditions (i) and (iii) are satisfied. For instance, suppose $G_{tt'}$ is represented by

$$c_{t''} = G_{tt''}(x_{tt''}) = \int_t^{t''} w(t'', \sigma)x(\sigma)\, d\sigma$$

If $x(\sigma) = \mathrm{o}$ for $t \leq \sigma \leq t'$, then

$$G_{tt''}(\mathrm{o} \cdot x_{t't''}) = \int_{t'}^{t''} w(t'', \sigma)x(\sigma)\, d\sigma = G_{t't''}(x_{t't''})$$

which is condition (iii).

When $\bar{\rho}$ is a linear systems-response family which is also realizable, there exists a convenient way to construct its state-transition family. Recall that a response family $\bar{\rho}$ is reduced if and only if

$$(\forall x_t)[\rho_t(c_t, x_t) = \rho_t(c_t', x_t)] \rightarrow c_t = c_t'$$

In the linear case, the response family $\bar{\rho}$ is reduced if and only if ρ_{1t} is a one-to-one mapping. We have now the following theorem.

Theorem 3.4. Let $\bar{\rho}$ be a family of linear reduced-response functions which is realizable. Its associated linear state-transition family is then uniquely determined by

$$\phi_{1tt'}(c_t) = \rho_{1t'}^{-1}(\rho_{1t}(c_t) \mid T_{t'})$$

$$\phi_{2tt'}(x_{tt'}) = \rho_{1t'}^{-1}(\rho_{2t}(x_{tt'} \cdot \mathrm{o}) \mid T_{t'})$$

where $\rho_{1t}^{-1} : \rho_{1t}(C_t) \rightarrow C_t$ such that $\rho_{1t}^{-1}(\rho_{1t}(c_t)) = c_t$.

PROOF. Let $(\bar{\rho}, \bar{\phi})$ be a linear dynamical system. By decomposability we have

$$\begin{aligned}\rho_{t'}(\phi_{tt'}(c_t, x_{tt'}), x_{t'}) &= \rho_{1t'}(\phi_{tt'}(c_t, x_{tt'})) + \rho_{2t'}(x_{t'})\\ &= \rho_{1t'}(\phi_{1tt'}(c_t)) + \rho_{1t'}(\phi_{2tt'}(x_{tt'})) + \rho_{2t'}(x_{t'}) \qquad (4.10)\end{aligned}$$

and

$$\begin{aligned}\rho_t(c_t, x_t) \mid T_{t'} &= \rho_{1t}(c_t) \mid T_{t'} + \rho_{2t}(x_t) \mid T_{t'}\\ &= \rho_{1t}(c_t) \mid T_{t'} + \rho_{2t}(x_{tt'} \cdot \mathrm{o}) \mid T_{t'} + \rho_{2t}(\mathrm{o} \cdot x_{t'}) \mid T_{t'} \qquad (4.11)\end{aligned}$$

By Definition 2.3, the left sides of Eqs. (4.10) and (4.11) are equal, and it follows then from the right side that

$$\rho_{1t'}(\phi_{1tt'}(c_t)) = \rho_{1t}(c_t) \mid T_{t'} \tag{4.12}$$

$$\rho_{1t'}(\phi_{2tt'}(x_{tt'})) = \rho_{2t}(x_{tt'} \cdot \text{o}) \mid T_{t'} \tag{4.13}$$

Since $\bar{\rho}$ is reduced, Eqs. (4.12) and (4.13) imply that

$$\phi_{1tt'}(c_t) = \rho_{1t'}^{-1}(\rho_{1t}(c_t) \mid T_{t'})$$

$$\phi_{2tt'}(x_{tt'}) = \rho_{1t'}^{-1}(\rho_{2t}(x_{tt'} \cdot \text{o}) \mid T_{t'})$$

In other words, $\phi_{1tt'}$ and $\phi_{2tt'}$ are uniquely determined by $\bar{\rho}$. Q.E.D.

The preceding theorem provides a way to construct the state-transition family for a linear system starting from a given response family $\bar{\rho}$.

4. CONSTRUCTION OF THE STATE SPACE FOR A LINEAR SYSTEM

The algebraic diagram indicating the commutative relationship of various auxiliary functions for a linear time system has the same format as for a general time system. However, the state objects and the associated auxiliary functions can be constructed more effectively when the special properties due to linearity of the system are used.

Given a (complete) linear time system $S \subset X \times Y$, for every $t \in T$, let $S_t^{\text{o}} \subset S_t$ denote the set

$$S_t^{\text{o}} = \{(\text{o}, y_t) : (\text{o}, y_t) \in S_t\}$$

Notice that S_t^{o} is a linear space for any $t \in T$.

We have now the following theorem.

Theorem 4.1. Let S be a linear time system. Let $C_t = S_t^{\text{o}}$ for every $t \in T$. Then there exists an initial linear response ρ_{o} of S

$$\rho_{\text{o}} : C_{\text{o}} \times X \to Y$$

such that

$$\rho_{\text{o}}(c_{\text{o}}, x) = \rho_{10}(c_{\text{o}}) + \rho_{20}(x)$$

where $\rho_{10}((\text{o}, y)) = y$ and ρ_{20} is linear. Furthermore, if $\bar{\rho} = \{\rho_t\}$ is defined as

$$\rho_{1t} : C_t \to Y_t \quad \text{such that} \quad \rho_{1t}((\text{o}, y_t)) = y_t$$

$$\rho_{2t} : X_t \to Y_t \quad \text{such that} \quad \rho_{2t}(x_t) = \rho_{20}(\text{o} \cdot x_t) \mid T_t$$

and

$$\rho_t(c_t, x_t) = \rho_{1t}(c_t) + \rho_{2t}(x_t)$$

then $\bar{\rho}$ is realizable, and since $\bar{\rho}$ is reduced, a dynamical system representation associated with $\bar{\rho}$ is uniquely determined.

PROOF. The first part of the proposition is a direct consequence of Theorem 1.2, Chapter II. We have only to check, therefore, whether $\{\rho_t\}$ satisfies the input-response consistency and the state-response consistency.

(i) Input-response consistency: It follows from the definition that

$$\rho_{2t'}(x_{t'}) = \rho_{20}(\mathrm{o} \cdot x_{t'}) \mid T_{t'}$$

and for $t \le t'$,

$$\rho_{2t}(\mathrm{o} \cdot x_{t'}) = \rho_{20}(\mathrm{o} \cdot x_{t'}) \mid T_t$$

Hence,

$$\rho_{2t}(\mathrm{o} \cdot x_{t'}) \mid T_{t'} = \rho_{20}(\mathrm{o} \cdot x_{t'}) \mid T_{t'} = \rho_{2t'}(x_{t'})$$

The input-response consistency is therefore satisfied.

(ii) State-response consistency: Let $c_t = (\mathrm{o}, y_t) \in S_t$ and $x_{tt'}$ be arbitrary. Since $(\mathrm{o} \cdot x_{tt'} \cdot \mathrm{o}, \rho_{20}(\mathrm{o} \cdot x_{tt'} \cdot \mathrm{o})) \in S$, we have

$$(x_{tt'} \cdot \mathrm{o}, \rho_{20}(\mathrm{o} \cdot x_{tt'} \cdot \mathrm{o}) \mid T_t) \in S_t$$

Since S_t is linear, we have

$$(x_{tt'} \cdot \mathrm{o}, y_t + \rho_{20}(\mathrm{o} \cdot x_{tt'} \cdot \mathrm{o}) \mid T_t) \in S_t$$

Hence

$$(\mathrm{o}, y_t \mid T_{t'} + \rho_{20}(\mathrm{o} \cdot x_{tt'} \cdot \mathrm{o}) \mid T_{t'}) \in S_{t'}$$

or

$$\rho_{1t'}(c_{t'}) = \rho_{1t}(c_t) \mid T_{t'} + \rho_{2t}(x_{tt'} \cdot \mathrm{o}) \mid T_{t'} \qquad \text{for some} \quad c_{t'} \in C_{t'}$$

(Notice that we used the input-response consistency.) Let $c_{t'} = (\mathrm{o}, y_{t'}) \in S_{t'}$ be arbitrary. Then there exists $(x^{t'}, y^{t'})$ such that $(x^{t'} \cdot \mathrm{o}, y^{t'} \cdot y_{t'}) \in S$. Hence, for some $c_\mathrm{o} \in C_\mathrm{o}$,

$$y^{t'} \cdot y_{t'} = \rho_{10}(c_\mathrm{o}) + \rho_{20}(x^{t'} \cdot \mathrm{o})$$

Therefore,

$$\begin{aligned} y_{t'} &= (\rho_{10}(c_\mathrm{o}) + \rho_{20}(x^{t'} \cdot \mathrm{o})) \mid T_{t'} \\ &= ((\rho_{10}(c_\mathrm{o}) + \rho_{20}(x^t \cdot \mathrm{o}) + \rho_{20}(\mathrm{o} \cdot x_{tt'} \cdot \mathrm{o})) \mid T_t) \mid T_{t'} \\ &= (\rho_{1t}(c_t) + \rho_{2t}(x_{tt'} \cdot \mathrm{o})) \mid T_{t'} \end{aligned}$$

for some $c_t \in C_t$. Notice that the existence of c_t is guaranteed by the first part of the proof of the state-response consistency. $\bar{\rho}$ is apparently reduced. The final result comes from Theorem 3.4. Q.E.D.

Definition 4.1. Let S be a linear system. The set $S^0 = \{(o, y):(o, y) \in S\}$ will be referred to as the (algebraic) core of S.

On the basis of Theorem 4.1, we can now give a specific procedure for the construction of the auxiliary functions and the associated state space for a linear system based on the selection of the algebraic core as the state set. The steps in the construction process are shown in Fig. 4.1:

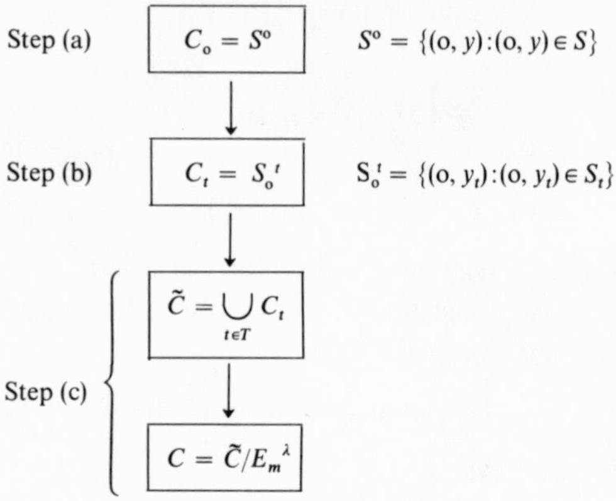

FIG. 4.1

(a) First, the algebraic core S^0 is selected as the initial state object.

(b) Second, the state object at any time t is selected as the core of the respective restriction of the system, i.e., $C_t \doteq S_t^0$.

(c) Third, an appropriate equivalence relation is introduced, and the state space is defined as the quotient set in the usual manner. The conditions that enable such a construction procedure are indicated in Fig. 4.2.

The major distinction between the state-space construction procedure for the case of a general dynamical system and for a linear system is in the selection of the state objects. For the general case, the initial state object is defined a priori; it represents the necessary initial information. For the linear system, the initial state object is defined uniquely by the system itself. The same is true for the selection of the state objects at other times. Another difference between the general and linear case is that in the former the states are defined in terms of the past, while for the latter the states are defined in terms of the future. The present construction procedure will be proven convenient, in particular for a stationary linear time system (as considered in Chapter VI), since it yields a state space in a natural way.

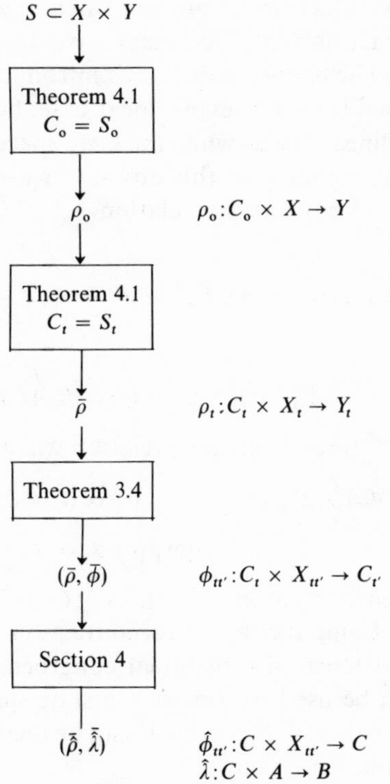

FIG. 4.2

Theorem 4.1 also indicates what the restrictions on the selection of the response functions are. In principle, unless nonanticipation is considered, any function $f : X \to Y$ that satisfies the relation $(x, f(x)) \in S$ for every x and the linearity requirements can be chosen for ρ_{20} and independently from ρ_{10}. Such a selection of ρ_{20} is essentially the only freedom in the construction procedure implied by Theorem 4.1; everything else is defined by the given system S itself. It should be noticed that $\bar{\rho}$ in Theorem 4.1 is already a reduced realizable response family, and hence the state-transition family $\bar{\phi}$ associated with $\bar{\rho}$ is uniquely determined by Theorem 3.4, that is,

$$\phi_{1tt'}(c_t) = \rho_{t'}^{-1}(\rho_{1t}(c_t) \mid T_{t'})$$

and

$$\phi_{2tt'}(x_{tt'}) = \rho_{t'}^{-1}(\rho_{2t}(x_{tt'} \cdot 0) \mid T_{t'})$$

Furthermore, if nonanticipation is satisfied, the system can also be represented by $\bar{\phi}$ and $\bar{\lambda}$. This part of the procedure is completely the same as for the general time systems case except that $\bar{\lambda}$ is automatically linear for the present case.

Consider now the selection of an equivalence which specifies a state space and the construction of the necessary auxiliary functions.

Let $\tilde{C} = \bigcup_{t \in T} C_t$. The equivalence $E_m{}^\lambda$ as introduced for the general time system cannot be used for the present linear case, because the quotient set $\tilde{C}/E_m{}^\lambda$ may not be a linear space, while the state space of a linear system has to be linear. In order to overcome this obstacle, we start with a congruence relation. Recall that an equivalence relation $E_\alpha{}^\lambda \subset \tilde{C} \times \tilde{C}$ is a congruence relation if and only if

$$(c_t, c_{t'}) \in E_\alpha{}^\lambda \;\&\; (c_t', c_{t'}') \in E_\alpha{}^\lambda \rightarrow (c_t + c_t', c_{t'} + c_{t'}') \in E_\alpha{}^\lambda$$

and

$$(c_t, c_t') \in E_\alpha{}^\lambda \;\&\; \beta \in \mathscr{A} \rightarrow (\beta c_t, \beta c_t') \in E_\alpha{}^\lambda$$

Now, let $E_\alpha{}^\lambda \subset \tilde{C} \times \tilde{C}$ be a congruence relation which satisfies the condition

$$(c_t, c_{t'}) \in E_\alpha{}^\lambda \rightarrow (\forall a)(\lambda_t(c_t, a) = \lambda_{t'}(c_{t'}, a)) \;\&\; (t = t' \rightarrow (\forall x_{tt''}) \cdot ((\phi_{tt''}(c_t, x_{tt''}), \phi_{tt''}(c_{t'}, x_{tt''})) \in E_\alpha{}^\lambda))$$

The trivial equivalence relation $I = \{(c_t, c_t) : C_t \in \tilde{C}\} \subset \tilde{C} \times \tilde{C}$ is such a congruence relation. Using the same procedure as in Section 3, Chapter III, we can prove the existence of a maximal congruence relation $E_m{}^\lambda$, which equivalence can then be used to construct a state space. It is easy to show that such a state space $C = \tilde{C}/E_m{}^\lambda = \{[c]\}$ can be linear. Indeed, assume that the system is *full* in the sense that

$$(\forall t)(\forall [c])[C_t \cap [c] \neq \phi]$$

i.e., at every time t there exists at least one state for each of the equivalence classes of $E_m{}^\lambda$. Then C is a linear space in the sense that

$$[c] + [\hat{c}] = [c_0 + \hat{c}_0] \quad \text{where} \quad c_0 \in [c] \quad \text{and} \quad \hat{c}_0 \in [\hat{c}]$$

$$\alpha[c] = [\alpha c_0] \qquad \text{where } c_0 \in [c]$$

Since $E_m{}^\lambda$ is the congruence relation, the above definitions are proper, and furthermore for every $t \in T$, if $c_t \in [c]$ and $\hat{c}_t \in [\hat{c}]$,

$$[c] + [\hat{c}] = [c_t + \hat{c}_t] \quad \text{and} \quad \alpha[c] = [\alpha c_t]$$

The state-transition and output functions can be defined now by

$$\hat{\phi}_{tt'} : C \times X_{tt'} \rightarrow C$$

such that

$$\hat{\phi}_{tt'}([c_t], x_{tt'}) = [\phi_{tt'}(c_t, x_{tt'})]$$

and

$$\hat{\lambda}: C \times A \to B$$

such that

$$\hat{\lambda}([c], a) = \lambda_t(c_t, a) \qquad \text{where} \quad c_t \in [c]$$

Notice that $\hat{\lambda}$ is time invariant (because the system is assumed full).

Because the state objects constructed above are defined directly in terms of the primary systems concepts (i.e., inputs and outputs), the dynamical representation for a *linear system* based on the construction process as illustrated in Figs. 4.1 and 4.2 will be referred to as *natural realization.*

Chapter V

PAST-DETERMINACY

In this chapter, we shall investigate in some more detail the class of past-determined systems. For the sake of illustration, some examples of specific past-determined systems are given in order to show how widespread these systems really are. It is shown then how a well-behaved response function can be constructed directly from the input–output pairs, i.e., without explicit reference to the initial state object. Furthermore, such a response function is nonanticipatory, realizable, and the associated state objects themselves can be generated from the past input–output pairs. Such a response family and the state objects are termed natural since their construction is fully determined by the given input–output pairs, i.e., system itself. This allows then the construction of state space, which is also termed natural for the same reason.

In order to characterize some past-determined systems, the notion of a finite-memory system is introduced. It is shown then that a past-determined and stationary system is of a finite memory, and conversely, if the time set of a finite-memory and stationary system is well ordered, the system is past-determined.

1. ON THE CLASS OF PAST-DETERMINED SYSTEMS

A central idea in our approach via formalization is to stay as close as possible to the observations in the process of developing an explanation for a phenomenon and introduce additional constructs only when absolutely necessary and with minimal additional assumptions. Of special interest, therefore, are the systems in which additional constructs can be introduced with no extra assumptions, directly from observations, i.e., input–output pairs. If one is interested in the state-type representation, a class of systems that allows such a natural treatment is the class of past-determined systems.

The concept of past-determinacy has been already introduced in Chapter II as a causality-type concept. Conceptually, for a past-determined system if one observes the input–output pair long enough, one can deduce everything he needs to construct the state-transition machinery and therefore study the dynamic aspects of the system's behavior. Specifically, we shall use in this chapter the following concept.

Definition 1.1. A time system $S \subset A^T \times B^T$ is a past-determined system if and only if there exist $\hat{t} \in T$ such that

(i) $(x^{\hat{t}}, y^{\hat{t}}) = (x'^{\hat{t}}, y'^{\hat{t}})$ and $x^t = x'^t \rightarrow \bar{y}^t = \bar{y}'^t$ for all $x', x \in X$ and $t \geq \hat{t}$

(ii) $(\forall (x^{\hat{t}}, y^{\hat{t}}) \in S^{\hat{t}})(\forall x_{\hat{t}})$

$$(\exists y_{\hat{t}})(x^{\hat{t}} \cdot x_{\hat{t}}, y^{\hat{t}} \cdot y_{\hat{t}}) \in S$$

Recall that condition (ii) is called the completeness property. The completeness property allows the existence of the natural systems-response family.

The importance of the past-determined systems, in reference to the state-type representation, is, among others, due to the following factors:

(i) Past-determined systems are intrinsically nonanticipatory as will be shown subsequently.

(ii) It is possible to construct a response function for the system in a "natural" way that preserves essential properties such as linearity, nonanticipation, stationarity, and others. Most of the results in the representation theory of general systems depend on the existence of a well-behaved response function. Unfortunately, in general it is not known how to construct such a function; however, for the past-determined systems such a procedure is possible as we shall show in this chapter.

(iii) The class of past-determined systems is rather large, covering a number of different types of systems of interest both in application and for theoretical investigation.

(iv) A conceptual argument can be presented that a well-defined system, in principle, ought to be past-determined and that the absence of past-determinacy results from the lack of information on the system itself.

To illustrate (iii) we shall consider several examples.

(a) Linear Constant-Coefficient Ordinary Differential Equation Systems

Let us consider the following differential equation system:

$$d\mathbf{z}/dt = F\mathbf{z} + G\mathbf{x} \tag{5.1}$$

$$\mathbf{y} = H\mathbf{z} \tag{5.2}$$

where F, G, and H are constant matrices, while $\mathbf{x}$, $\mathbf{y}$, and $\mathbf{z}$ are vectors. The dimension of $\mathbf{z}$ is assumed to be n. Let $T = [0, \infty)$. Then we have the following proposition.

Proposition 1.1. The class of linear time-invariant ordinary differential equation systems described by Eqs. (5.1) and (5.2) is past-determined from $\hat{t} > 0$, where $\hat{t}$ is an arbitrary small positive number.

PROOF. It follows from Eqs. (5.1) and (5.2) that

$$\mathbf{y}(t) = H\, e^{Ft}\, \mathbf{z}(0) + H \int_0^t e^{F(t-\tau)} G\mathbf{x}(\tau)\, d\tau \tag{5.3}$$

Let $\hat{t}$ be an arbitrary positive number. Let $(\hat{\mathbf{x}}, \hat{\mathbf{y}})$ be

$$\hat{\mathbf{y}}(t) = He^{Ft}\hat{\mathbf{z}}(0) + H \int_0^t e^{F(t-\tau)} G\hat{\mathbf{x}}(\tau)\, d\tau \tag{5.4}$$

Then $(\mathbf{x}^{\hat{t}}, \mathbf{y}^{\hat{t}}) = (\hat{\mathbf{x}}^{\hat{t}}, \hat{\mathbf{y}}^{\hat{t}})$ implies that $He^{F\sigma}\mathbf{z}(0) = He^{F\sigma}\hat{\mathbf{z}}(0)$ holds for $0 \leq \sigma < \hat{t}$. Consequently, we have from an elementary calculation that $H\mathbf{z}(0) = H\hat{\mathbf{z}}(0)$, $HF\mathbf{z}(0) = HF\hat{\mathbf{z}}(0), \ldots, HF^{n-1}\mathbf{z}(0) = HF^{n-1}\hat{\mathbf{z}}(0)$. Therefore,

$$He^{F\sigma}\mathbf{z}(0) = He^{F\sigma}\hat{\mathbf{z}}(0)$$

holds for any $\sigma \geq 0$, which implies that if $\mathbf{x}^t = \hat{\mathbf{x}}^t$ $(t \geq \hat{t})$, $\bar{\mathbf{y}}^t = \bar{\hat{\mathbf{y}}}^t$ holds. The completeness property is clearly satisfied. Q.E.D.

(b) Linear Constant-Coefficient Difference Equation System

Let us consider the following difference equation system:

$$\mathbf{z}_{k+1} = F\mathbf{z}_k + G\mathbf{x}_k \tag{5.5}$$

$$\mathbf{y}_k = H\mathbf{z}_k \tag{5.6}$$

where F, G, and H are constant matrices, while $\mathbf{x}_k$, $\mathbf{y}_k$, and $\mathbf{z}_k$ are vectors. The dimension of $\mathbf{z}_k$ is assumed to be n. Let $T = \{0, 1, 2, \ldots\}$. Then we can introduce the following proposition.

Proposition 1.2. The class of linear constant-coefficient difference equation systems as described by Eqs. (5.5) and (5.6) is past-determined from $\hat{t} = n$, where n is the dimension of the state space.

PROOF. We have from Eqs. (5.5) and (5.6) that

$$\mathbf{y}_k = HF^k\mathbf{z}_0 + H\sum_{i=0}^{k-1} F^{k-1-i}G\mathbf{x}_i$$

Let

$$\hat{\mathbf{y}}_k = HF^k\hat{\mathbf{z}}_0 + H\sum_{i=0}^{k-1} F^{k-1-i}G\hat{\mathbf{x}}_i$$

Suppose $(\mathbf{x}^{\hat{t}}, \mathbf{y}^{\hat{t}}) = (\hat{\mathbf{x}}^{\hat{t}}, \hat{\mathbf{y}}^{\hat{t}})$ for $\hat{t} = n$. Then we have that $H\mathbf{z}_0 = H\hat{\mathbf{z}}_0$, $HF\mathbf{z}_0 = HF\hat{\mathbf{z}}_0, \ldots,$ and $HF^{n-1}\mathbf{z}_0 = HF^{n-1}\hat{\mathbf{z}}_0$. Consequently, $HF^k\mathbf{z}_0 = HF^k\hat{\mathbf{z}}_0$ holds for any $k \geq$ o. This result shows that if $\mathbf{x}^t = \hat{\mathbf{x}}^t$ holds, then $\bar{\mathbf{y}}^t = \bar{\hat{\mathbf{y}}}^t$ for $t \geq \hat{t} = n$. The completeness property is clearly satisfied. Q.E.D.

The proof of Proposition 1.2 shows that, in general, $\hat{t}$ can be smaller than n.

(c) Automata Systems

As shown below, a general automaton is not a past-determined system. However, every finite automaton has a subsystem which is past-determined. Let $S = \langle A, B, C, \delta, \lambda \rangle$ be a finite automaton whose A is a finite input alphabet, B a finite output alphabet, C a finite state space, δ a next-state function such that $\delta: C \times A \to C$, and λ a output function such that $\lambda: C \times A \to B$. Let the cardinality of C be n. Let A^* and B^* be the free monoids of A and B, respectively. Let $\delta: C \times A^* \to C$ and $\lambda: C \times A^* \to B^*$ be the extensions of δ and λ as usually defined. Let $\bar{a}$ be an infinite string of a, i.e., $\bar{a} = a \cdot a \cdots$. Let us define a subsystem $\bar{S}$ of S as below. Let S be a discrete time system with the time set $T = \{0, 1, 2, \ldots\}$ and $\bar{S}$ a subset of S such that

$$\bar{S} = \{(\bar{a}, \lambda(c, \bar{a})) : c \in C\} \tag{5.7}$$

where $a \in A$. Then we have the following proposition.

Proposition 1.3. Let S be a finite automaton, $S = \langle A, B, C, \delta, \lambda \rangle$ and n the cardinality of C. S then has a subsystem $\bar{S}$ as defined by Eq. (5.7), which is past-determined from n.

PROOF. Let us consider a finite automaton

$$S_a = \langle \{a\}, B, C, \delta \mid C \times \{a\}, \lambda \mid C \times \{a\} \rangle$$

Notice that only an input string of S_a is a^k (a string of k a's), where $k = 0, 1, 2, \ldots$. Then it follows from the well-known property of a finite automaton that $\lambda(c, a^n) = \lambda(\hat{c}, a^n)$ implies $\lambda(c, a^k) = \lambda(\hat{c}, a^k)$ for any k $(k = 0, 1, 2, \ldots)$. Consequently, $\bar{S}$ is past-determined from $\hat{t} = n$. The completeness property is clearly satisfied. Q.E.D.

(d) Non-Past-Determined Automata Systems

As we mentioned above, an automaton, in general, is not past-determined. Let us give a simple example to show that is the case. Let a finite automaton $S = \langle A, B, C, \delta, \lambda \rangle$ be defined by

$$A = B = C = \{0, 1\}$$

$$\delta(0, 0) = \delta(0, 1) = 0, \ \delta(1, 0) = \delta(1, 1) = 1$$

$$\lambda(0, 0) = \lambda(0, 1) = 0, \ \lambda(1, 0) = 0, \ \lambda(1, 1) = 1$$

The state-transition diagram is given in Fig. 1.1.

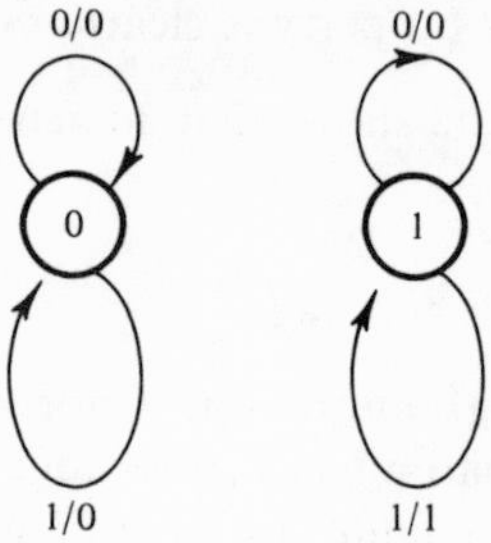

FIG. 1.1

It is clear from the definition that for any large n, $\lambda(0, 0^n) = \lambda(1, 0^n)$ holds. However, $\lambda(0, 0^n \cdot 1) \neq \lambda(1, 0^n \cdot 1)$. Consequently, such an automaton is not past-determined.

2. STATE-SPACE REPRESENTATION

In the previous two chapters, state-space representations of time systems and general dynamical systems have been considered. The results, in essence, depend on showing the existence of a nonanticipatory systems-response function.

No general procedure is presently known to find such a systems-response function for a given general system except when the system is past-determined.

(a) Past-Determined General Time Systems

Most of the results in the investigation of the past-determined systems are based on the following proposition.

Proposition 2.1. Let S be a past-determined system from $\hat{t}$. Then

(i) there exists a strongly nonanticipatory initial systems response $\rho_{\hat{t}}$ at $\hat{t}$ whose state object is $S^{\hat{t}} = S \mid T^{\hat{t}}$;

(ii) there exists a one-to-one correspondence between S^t $(t \geq \hat{t})$ and $S^{\hat{t}} \times X_{\hat{t}t}$, i.e., $S^t \leftrightarrow S^{\hat{t}} \times X_{\hat{t}t}$, and hence S^t can be used as a state object for a realizable strongly nonanticipatory response family of $S_{\hat{t}}$.

PROOF. It follows from Proposition 4.4, Chapter II, that

$$S(x^t, y^t) = \{(x_t, y_t) : (x^t \cdot x_t, y^t \cdot y_t) \in S\}$$

is a strongly nonanticipatory function. Then let $C_{\hat{t}} = S^{\hat{t}}$ and let $\rho_{\hat{t}} : C_{\hat{t}} \times X_{\hat{t}} \to Y_{\hat{t}}$ be such that $\rho_{\hat{t}}(c_{\hat{t}}, x_{\hat{t}}) = S(c_{\hat{t}})(x_{\hat{t}})$. Then $\rho_{\hat{t}}$ is a strongly nonanticipatory initial system response.

Let $F : S^{\hat{t}} \times X_{\hat{t}t} \to S^t$ be such that for $c_{\hat{t}} = (x^{\hat{t}}, y^{\hat{t}})$ and for $x_{\hat{t}} \mid T_{\hat{t}t} = x_{\hat{t}t}$

$$F((x^{\hat{t}}, y^{\hat{t}}), x_{\hat{t}t}) = (x^{\hat{t}} \cdot x_{\hat{t}t}, y^{\hat{t}} \cdot \rho_{\hat{t}}(c_{\hat{t}}, x_{\hat{t}}) \mid T_{\hat{t}t}) \in S^t$$

F is well defined because $\rho_{\hat{t}}$ is nonanticipatory. If $F((x^{\hat{t}}, y^{\hat{t}}), x_{\hat{t}t}) = F((\hat{x}^{\hat{t}}, \hat{y}^{\hat{t}}), \hat{x}_{\hat{t}t})$,

$$(x^{\hat{t}} \cdot x_{\hat{t}t}, y^{\hat{t}} \cdot \rho_{\hat{t}}(c_{\hat{t}}, x_{\hat{t}}) \mid T_{\hat{t}t}) = (\hat{x}^{\hat{t}} \cdot \hat{x}_{\hat{t}t}, \hat{y}^{\hat{t}} \cdot \rho_{\hat{t}}(\hat{c}_{\hat{t}}, \hat{x}_{\hat{t}}) \mid T_{\hat{t}t})$$

holds, where $c_{\hat{t}} = (x^{\hat{t}}, y^{\hat{t}})$ and $\hat{c}_{\hat{t}} = (\hat{x}^{\hat{t}}, \hat{y}^{\hat{t}})$. Hence, $x^t \cdot x_{\hat{t}t} = \hat{x}^{\hat{t}} \cdot \hat{x}_{\hat{t}t}$ and $y^{\hat{t}} = \hat{y}^{\hat{t}}$ hold, which implies that F is a one-to-one mapping. Let $(x^t \cdot x_{\hat{t}t}, y^t \cdot y_{\hat{t}t}) \in S^t$. Then

$$F((x^{\hat{t}}, y^{\hat{t}}), x_{\hat{t}t}) = (x^{\hat{t}} \cdot x_{\hat{t}t}, y^{\hat{t}} \cdot \rho_{\hat{t}}(c_{\hat{t}}, x_{\hat{t}}) \mid T_{\hat{t}t}) \in S^t$$

where $x_{\hat{t}} \mid T_{\hat{t}t} = x_{\hat{t}t}$. Since S is past-determined, $\rho_{\hat{t}}(c_{\hat{t}}, x_{\hat{t}}) \mid T_{\hat{t}t} = y_{\hat{t}t}$. Consequently,

$$F((x^{\hat{t}}, y^{\hat{t}}), x_{\hat{t}t}) = (x^{\hat{t}} \cdot x_{\hat{t}t}, y^{\hat{t}} \cdot y_{\hat{t}t})$$

Hence, F is a one-to-one correspondence.

Let $C_t = S^t$. Let $\rho_t : C_t \times X_t \to Y_t$ such that when $F^{-1}(c_t) = (c_{\hat{t}}, x_{\hat{t}t})$,

$$\rho_t(c_t, x_t) = \rho_{\hat{t}}(c_{\hat{t}}, x_{\hat{t}t} \cdot x_t) \mid T_t$$

Then Theorem 1.1, Chapter III, shows that $\{\rho_t : t \geq \hat{t}\}$ is a realizable response family of $S_{\hat{t}}$. Q.E.D.

Proposition 2.1 shows explicitly that a past-determined system has a nonanticipatory realizable response family and state objects which are determined from the input–output pairs of the system without introducing an auxiliary object, i.e., initial object. (Refer to the state-generating function of Proposition 4.5, Chapter II.) Hence, those response family and state objects will be referred to as *natural*.

The natural state objects are, in general, not reduced. As an illustration, let us consider Eq. (5.1). Let $\mathbf{z}(\mathrm{o}) \neq \hat{\mathbf{z}}(\mathrm{o})$, $\mathbf{x}^t \neq \hat{\mathbf{x}}^t$, and $\mathbf{y}^t \neq \hat{\mathbf{y}}^t$ be such that

$$e^{Ft}\mathbf{z}(\mathrm{o}) + \int_{\mathrm{o}}^{t} e^{F(t-\sigma)}G\mathbf{x}^t(\sigma)\, d\sigma = e^{Ft}\hat{\mathbf{z}}(\mathrm{o}) + \int_{\mathrm{o}}^{t} e^{F(t-\sigma)}G\hat{\mathbf{x}}^t(\sigma)\, d\sigma \tag{5.8}$$

and

$$\mathbf{y}^t = \rho_{\mathrm{o}}(\mathbf{z}(\mathrm{o}), \mathbf{x}^t \cdot \mathbf{x}_t) \mid T^t$$

and

$$\hat{\mathbf{y}}^t = \rho_{\mathrm{o}}(\hat{\mathbf{z}}(\mathrm{o}), \hat{\mathbf{x}}^t \cdot \hat{\mathbf{x}}_t) \mid T^t$$

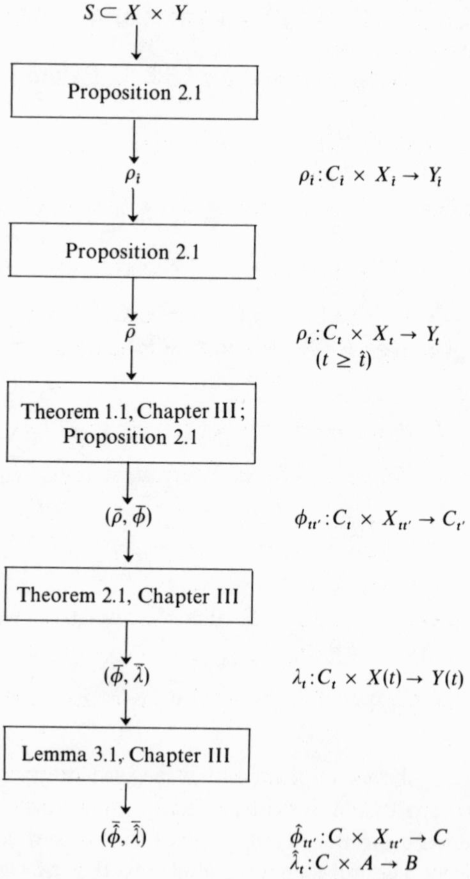

FIG. 2.1 Past-determined system from $\hat{t}$.

Apparently, $(\mathbf{x}^t, \mathbf{y}^t) \neq (\hat{\mathbf{x}}^t, \hat{\mathbf{y}}^t)$, but Eq. (5.8) implies that $\rho_t((\mathbf{x}^t, \mathbf{y}^t), \mathbf{x}_t) = \rho_t((\hat{\mathbf{x}}^t, \hat{\mathbf{y}}^t), \mathbf{x}_t)$ hold for every $\mathbf{x}_t$, where ρ_t is the natural systems-response function.

Once a nonanticipatory response family is given, a state space and other auxiliary functions can be constructed in the way analogous to the case of general time systems. Such a construction is shown in Figs. 2.1 and 2.2.

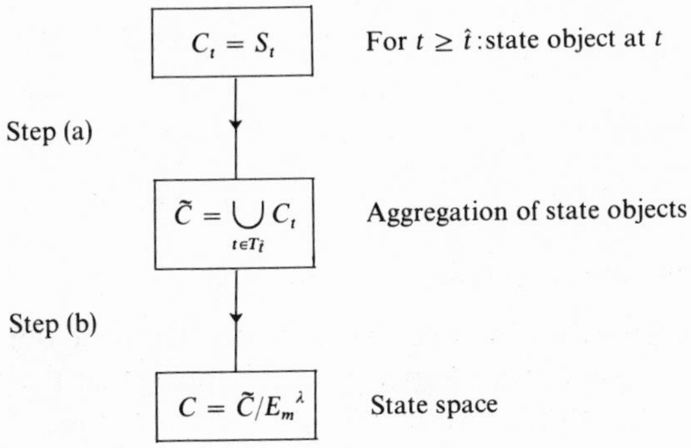

FIG. 2.2

Comparison between the present construction method and the previous one given in Chapter III indicates the following:

(i) The method given in Fig. 3.9, Chapter III, is more general since the system does not have to be past-determined.

(ii) The method given in Fig. 2.2 here yields a unique state-space representation defined solely in terms of the primary information (input–output restrictions), i.e., without introducing a secondary concept, such as the initial state object required to initiate construction procedure given in Fig. 3.10, Chapter III.

(iii) The method given in Fig. 2.1 here yields the representation of $S_{\hat{t}}$ rather than S itself. This may be viewed as a shortcoming, but as Proposition 1.1 shows, for certain classes of systems, $\hat{t}$ can be taken as small as we wish.

(b) Past-Determined Linear Time Systems

The procedure to determine the state space representation of a linear past-determined system is completely analogous to the general, nonlinear case.

Proposition 2.2. Let $\rho_{\hat{t}}:S^{\hat{t}} \times X_{\hat{t}} \to Y_{\hat{t}}$ be the natural systems-response function of a past-determined linear system $S \subset X \times Y$ from $\hat{t}$. Then

(i) $S^{\hat{t}}$ is a linear algebra;

(ii) $\rho_{\hat{t}}$ is a nonanticipatory linear mapping;

(iii) the state-response function $\rho_{1\hat{t}}:S^{\hat{t}} \to Y^{\hat{t}}$ and the input-response function $\rho_{2\hat{t}}:X_{\hat{t}} \to Y_{\hat{t}}$ are given by

$$\rho_{1\hat{t}}(x^{\hat{t}}, y^{\hat{t}}) = y_{\hat{t}} \leftrightarrow (x^{\hat{t}} \cdot o, y^{\hat{t}} \cdot y_{\hat{t}}) \in S$$

$$\rho_{2\hat{t}}(x_{\hat{t}}) = y_{\hat{t}} \leftrightarrow (o \cdot x_{\hat{t}}, o \cdot y_{\hat{t}}) \in S$$

PROOF. Notice that

$$\rho_{\hat{t}}((x^{\hat{t}}, y^{\hat{t}}), x_{\hat{t}}) = y_{\hat{t}} \leftrightarrow (x^{\hat{t}} \cdot x_{\hat{t}}, y^{\hat{t}} \cdot y_{\hat{t}}) \in S$$

The result follows, then, from direct computations. Q.E.D.

For the natural realization of a linear system, the state object $C_{\hat{t}}$ is given by $C_{\hat{t}} = S_{\hat{t}}^{o} = \{y_{\hat{t}} \mid (o, y_{\hat{t}}) \in S\}$. Proposition 2.2 shows that the state-response function $\rho_{1\hat{t}}$ of the natural response function is given by

$$\rho_{1\hat{t}}(x^{\hat{t}}, y^{\hat{t}}) = y_{\hat{t}} \leftrightarrow (x^{\hat{t}} \cdot o, y^{\hat{t}} \cdot y_{\hat{t}}) \in S$$

Therefore, if $\rho_{1\hat{t}}$ is considered a mapping from $S^{\hat{t}}$ into $S_{\hat{t}}^{o}$, $\rho_{1\hat{t}}:S^{\hat{t}} \to S_{\hat{t}}^{o}$ is an onto function, and furthermore, the reduced natural state is isomorphic to $S_{\hat{t}}^{o}$.

Since the natural realization method is applicable to the present case, a state space and auxiliary functions are constructed in the completely same way as in the previous chapter.

3. CHARACTERIZATION OF PAST-DETERMINED SYSTEMS

A general characterization of the past-determined systems is not known at present. However, the linear past-determined system can be characterized by nonanticipation and some continuity property. For this we need the following concept.

Definition 3.1. Let $\rho_{10}:C \times X \to Y$ be a linear systems-response function of a linear time system. For some $\hat{t}$, if ρ_0 satisfies the condition

$$(\forall c)(\rho_{10}(c) \mid T^{\hat{t}} = 0 \mid T^{\hat{t}} \to \rho_{10}(c) = 0)$$

ρ_0 is called analytic from left at $\hat{t}$.

Proposition 3.1. If a linear time system $S \subset X \times Y$ has a systems-response function $\rho_o : C_o \times X \to Y$ which is strongly nonanticipatory and analytic from left at $\hat{t}$, then S is past-determined from $\hat{t}$.

PROOF. Let $\rho_o : C_o \times X \to Y$ be a linear systems-response function which is strongly nonanticipatory and analytic from left at $\hat{t}$. Suppose $(x^{\hat{t}}, y^{\hat{t}}) = (\hat{x}^{\hat{t}}, \hat{y}^{\hat{t}})$, where $y = \rho_{10}(c) + \rho_{20}(x)$ and $\hat{y} = \rho_{10}(\hat{c}) + \rho_{20}(\hat{x})$. Since ρ_o is strongly nonanticipatory, $x^{\hat{t}} = \hat{x}^{\hat{t}}$ implies that $\rho_{20}(x) \mid \bar{T}^{\hat{t}} = \rho_{20}(\hat{x}) \mid \bar{T}^{\hat{t}}$. Consequently, we have $\rho_{10}(c) \mid T^{\hat{t}} = \rho_{10}(\hat{c}) \mid T^{\hat{t}}$ because of $y \mid T^{\hat{t}} = \hat{y} \mid T^{\hat{t}}$. Furthermore, since ρ_o is analytic, we have that

$$\rho_{10}(c - \hat{c}) \mid T^{\hat{t}} = 0 \mid T^{\hat{t}} \to \rho_{10}(c - \hat{c}) = 0 \to \rho_{10}(c) = \rho_{10}(\hat{c})$$

Therefore, $x \mid T^t = \hat{x} \mid T^t$ implies $y \mid \bar{T}^t = \hat{y} \mid \bar{T}^t$, where $t \geq \hat{t}$. Hence, S is past-determined from $\hat{t}$. The completeness property is clearly satisfied. Q.E.D.

Notice that a system described by Eqs. (1.1) and (1.2) is strongly non-anticipatory, and ρ_{10} is an analytic function because ρ_{10} is the integral of $(d\mathbf{z}/dt) = F\mathbf{z}$. Consequently, Proposition 1.1 is a direct result of the above proposition.

The converse of Proposition 3.1 is given by Proposition 3.2 in the following.

Proposition 3.2. Suppose a time system $S \subset X \times Y$ is past-determined from $\hat{t}$. Then $S_{\hat{t}}$ is strongly nonanticipatory, and furthermore, if S is linear, every linear systems-response function of S is analytic from left at $\hat{t}$.

PROOF. The first part is proved in Proposition 2.1. Let $\rho_o : C_o \times X \to Y$ be a linear systems-response function. Let $c \in C_o$ be arbitrary such that $\rho_o(c) \mid T^{\hat{t}} = 0 \mid T^{\hat{t}}$. Let $x \in X$ be arbitrary. Then $(x, \rho_{20}(x)) \in S$ and $(x, \rho_{10}(c) + \rho_{20}(x)) \in S$ hold. Hence, $\rho_{10}(c) \mid T^{\hat{t}} = 0 \mid T^{\hat{t}}$ implies that

$$(x \mid T^{\hat{t}}, \rho_{20}(x) \mid T^{\hat{t}}) = (x \mid T^{\hat{t}}, (\rho_{10}(c) + \rho_{20}(x)) \mid T^{\hat{t}})$$

Consequently, past-determinacy implies that $\rho_{20}(x) = \rho_{10}(c) + \rho_{20}(x)$, that is, $\rho_{10}(c) = 0$. Q.E.D.

The concept of a finite-memory machine defined in the automata theory [4] is intimately related with that of a past-determined system. Let us explore that relationship.

Definition 3.2. Let $S \subset A^T \times B^T$ be a time system, where T is a stationary time set. Let $\hat{t} \in T$ be fixed. Suppose there exists a mapping $h : X^{\hat{t}} \times Y^{\hat{t}} \to B$, such that $y(t') = (F^{-t}(x_{tt'}, y_{tt'}))$ for any $(x, y) \in S$ and $t' = t + \hat{t}$. Then S is called a finite-memory system whose memory is $\hat{t}$.

Proposition 3.3. Suppose a time system $S \subset A^T \times B^T$ is past-determined from $\hat{t}$ and stationary. Then S is a finite-memory system whose memory is $\hat{t}$.

PROOF. Let $\rho : S^{\hat{t}} \times X_{\hat{t}} \to Y_{\hat{t}}$ be the natural system response function of S. Let $h : X_{tt'} \times Y_{tt'} \to B$ $(t' = t + \hat{t})$ be such that

$$h(x_{tt'}, y_{tt'}) = \rho(F^{-t}(x_{tt'}, y_{tt'}), x_{\hat{t}})(\hat{t})$$

Since $\rho(F^{-t}(x_{tt'}, y_{tt'}), x_{\hat{t}})(\hat{t})$ is independent of $x_{\hat{t}}$ because of past-determinacy, $x_{\hat{t}}$ can be arbitrary in defining h. We shall show that S is a finite-memory system with respect to h. Let $(x, y) \in S$ be arbitrary. $(x_t, y_t) \in S_t$ and $S_t \subset F^t(S)$ imply $F^{-t}(x_t, y_t) \in S$. Consequently,

$$\rho(F^{-t}(x_t, y_t) \mid T^{\hat{t}}, F^{-t}(x_t) \mid T_{\hat{t}}) = F^{-t}(y_t) \mid T_{\hat{t}}$$

holds. Let $t' = t + \hat{t}$. Then

$$\begin{aligned}(F^{-t}(y_t) \mid T_{\hat{t}})(\hat{t}) &= y(t') \\ &= \rho(F^{-t}(x_t, y_t) \mid T^{\hat{t}}, F^{-t}(x_t) \mid T_{\hat{t}})(\hat{t}) \\ &= \rho(F^{-t}(x_{tt'}, y_{tt'}), F^{-t}(x_{t'}))(\hat{t}) \\ &= h(x_{tt'}, y_{tt'})\end{aligned}$$

Therefore, S is a finite-memory system whose memory is $\hat{t}$. Q.E.D.

The above proposition shows that a differential equation system described by Eqs. (5.1) and (5.2) is a finite-memory system.

The converse of the above proposition is given by Proposition 3.4 in the following.

Proposition 3.4. Let $S \subset A^T \times B^T$ be a time system, where T is the time set for stationary systems and, in addition, is well ordered. If S is a finite-memory system whose memory is $\hat{t}$, S is past-determined from $\hat{t}$.

PROOF. Let $(x, y) \in S$ and $(\hat{x}, \hat{y}) \in S$ be arbitrary. Since the system is of a finite memory, $y(t') = h(x_{tt'}, y_{tt'})$ and $\hat{y}(t') = h(\hat{x}_{tt'}, \hat{y}_{tt'})$ holds for any $t \in T$ and $t' = t + \hat{t}$. We shall show that $(x^{\hat{t}}, y^{\hat{t}}) = (\hat{x}^{\hat{t}}, \hat{y}^{\hat{t}})$ and $x^t = \hat{x}^t$ imply $\bar{y}^t = \hat{\bar{y}}^t$, where $t \geq \hat{t}$. Let $T_n = \{\tau : \bar{y}^t(\tau) \neq \hat{y}^t(\tau)\}$. If $T_n = \phi$, then $\bar{y}^t = \bar{\hat{y}}^t$. Suppose $T_n \neq \phi$. Since T is well ordered, $\min_{\tau \in T_n} \tau = \tau_0$ exists. Since $(x^{\hat{t}}, y^{\hat{t}}) = (\hat{x}^{\hat{t}}, \hat{y}^{\hat{t}})$,

$$y(\hat{t}) = h(x^{\hat{t}}, y^{\hat{t}}) = h(\hat{x}^{\hat{t}}, \hat{y}^{\hat{t}}) = \hat{y}(\hat{t})$$

should hold, which implies that $\hat{t} < \tau_0$. Since τ_0 is the minimum element of T_n, $(x^{\tau_0}, y^{\tau_0}) = (\hat{x}^{\tau_0}, \hat{y}^{\tau_0})$ holds. Suppose $\sigma = \tau_0 - \hat{t}$ is positive. Then we have

$$y(\tau_0) = h(x_{\sigma\tau_0}, y_{\sigma\tau_0}) = h(\hat{x}_{\sigma\tau_0}, \hat{y}_{\sigma\tau_0}) = \hat{y}(\tau_0)$$

which implies that $\tau_0 \notin T_n$. This contradicts the fact that τ_0 is the minimum of T_n. Therefore, T_n is empty. Q.E.D.

Chapter VI

STATIONARITY AND TIME INVARIANCE

The class of stationary systems is important because, in application, many systems can be considered as stationary with a sufficient degree of accuracy and because in theoretical studies the additional property of stationarity allows a deeper analysis of such a class of systems, yielding a better understanding of their behavior. A concept closely related with stationarity is that of time invariance; while the stationarity is defined in reference to the system itself and its restriction on appropriate time subsets, the time invariance is defined in reference to the auxiliary functions.

The class of systems considered in this chapter will be assumed to have state-space representation. This will simplify the developments, but will require a slight modification of earlier definitions. In practice, dynamical systems are usually defined in reference to a given state space.

Realization theory for stationary and time-invariant systems is developed in this chapter both for the general case and when the linearity property holds. Conditions for a strongly nonanticipatory realization are given, and the decomposition of a realizable linear input-response function in two parts, namely, mapping inputs to states and states to output values, is proven.

After considering the effect of past-determinacy, it is shown on an axiomatic basis how the developed theory can be linked directly with the classical linear systems theory.

1. STATE-SPACE STATIONARITY AND TIME INVARIANCE

For the class of systems considered in this chapter, it will be assumed that the time set is stationary and that the input and output objects are such that $X_t = F^t(X)$ and $Y_t = F^t(Y)$ for every $t \in T$. It will also be assumed at the outset that a state space is associated with the system under consideration. We have then the following definition.

Definition 1.1. A time system $S \subset X \times Y$ is stationary if and only if

$$(\forall t)(S_t \subset F^t(S))$$

The case when $S_t = F^t(S)$ will be referred to as full stationarity.

Definition 1.2. A time system S is time invariant if and only if there exists a response-function family $\bar{\rho}$ for S defined on a state space C,

$$\bar{\rho} = \{\rho_t : C \times X_t \to Y_t\}$$

such that

$$(\forall t)(\forall c)(\forall x_t)(\rho_t(c, x_t) = F^t(\rho_0(c, F^{-t}(x_t))))$$

Notice that because the state space is used directly in definition, the state object S_t is, in general, a proper subset of S_t^{ρ}:

$$S_t \subset S_t^{\rho} = \{(x_t, y_t) : (\exists c)(y_t = \rho_t(c, x_t))\}$$

Definition 1.3. A dynamical system $(\bar{\rho}, \bar{\phi})$ is time invariant if and only if

(i) the response family $\bar{\rho}$ is time invariant;
(ii) the state-transition family $\bar{\phi}$ is such that

$$(\forall t)(\forall t')(\forall c)(\forall x_{tt'})(\phi_{tt'}(c, x_{tt'}) = \phi_{0\tau}(c, F^{-t}(x_{tt'})))$$

where $\tau = t' - t$.

Linearity obviously does not affect the concepts of stationarity and time invariance. However, in reference to the linearity of a system, the time invariance can be defined in a special form.

Definition 1.4. A dynamical linear system is time invariant if and only if

(i) $(\forall t)(\forall c)[\rho_{1t}(c) = F^t(\rho_{10}(c))]$
(ii) $(\forall t)(\forall x_t)[\rho_{2t}(x_t) = F^t(\rho_{20}(F^{-t}(x_t)))]$
(iii) $(\forall t)(\forall t')(\forall c)[t' \geq t \to \phi_{1tt'}(c) = \phi_{10\tau}(c)]$ where $\tau = t' - t$
(iv) $(\forall t)(\forall t')(\forall x_{tt'})[t' \geq t \to \phi_{2tt'}(x_{tt'}) = \phi_{20\tau}(F^{-t}(x_{tt'}))]$

If $\bar{\rho}$ satisfies (i) and (ii), $\bar{\rho}$ is called a time-invariant linear-response family.

2. REALIZATION THEORY OF TIME-INVARIANT SYSTEMS

(a) General Time System

Theorem 2.1. Let $\bar{\rho} = \{\rho_t : \rho_t : C \times X_t \to Y_t \;\&\; t \in T\}$ be a family of time-invariant maps. Then $\bar{\rho}$ is realizable by a time-invariant dynamical system if and only if for all $t \in T$, $\bar{\rho}$ satisfies the condition

$$(\forall c)(\forall x^t)(\exists c')(\forall x_t)[\rho_t(c', x_t) = \rho_0(c, x^t \cdot x_t) \mid T_t]$$

PROOF. The proof is similar to Theorem 3.1, Chapter III, except that every state-transition function has to be shown time invariant. The *only if* part is clear. Let us consider the *if* part. Let $E \subset C \times C$ be such that

$$(c, c') \in E \leftrightarrow (\forall x)(\rho_0(c, x) = \rho_0(c', x'))$$

Since $\bar{\rho}$ is time invariant, we can define $\hat{\rho}_t : (C/E) \times X_t \to Y_t$ as $\hat{\rho}_t([c], x_t) = \rho_t(c, x_t)$. Let $f_{tt'} \subset (C/E \times X_{tt'}) \times (C/E)$ be such that

$$(([c], x_{tt'}), [c']) \in f_{tt'} \leftrightarrow (\forall x_{t'})(\hat{\rho}_{t'}([c'], x_{t'}) = \hat{\rho}_t([c], x_{tt'} \cdot x_{t'}) \mid T_{t'})$$

Then $f_{tt'}$ is a mapping such that $f_{tt'} : C/E \times X_{tt'} \to C/E$ and

$$\hat{\rho}_{t'}(f_{tt'}([c], x_{tt'}), x_{t'}) = \hat{\rho}_t([c], x_{tt'} \cdot x_{t'}) \mid T_{t'}$$

for every $x_{t'}$. We will show that $\bar{f} = \{f_{tt'}\}$ is time invariant. Since $\bar{\rho}$ is time invariant,

$$F^{-t}(\rho_{t'}(c', x_{t'})) = \rho_\tau(c', F^{-t}(x_{t'}))$$

and

$$F^{-t}(\rho_t(c, x_{tt'} \cdot x_{t'}) \mid T_{t'}) = \rho_0(c, F^{-t}(x_{tt'}) \cdot F^{-t}(x_{t'})) \mid T_\tau$$

hold where $\tau = t' - t$. Consequently, we have that

$$(\forall x_{t'})(\rho_{t'}(c', x_{t'}) = \rho_t(c, x_{tt'} \cdot x_{t'}) \mid T_{t'}) \leftrightarrow (\forall x_\tau)(\rho_\tau(c', x_\tau) = \rho_0(c, F^{-t}(x_{tt'}) \cdot x_\tau) \mid T_\tau)$$

that is,

$$f_{tt'}([c], x_{tt'}) = f_{0\tau}([c], F^{-t}(x_{tt'}))$$

$\bar{f}$ satisfies the semigroup property. Let $\mu : C/E \to C$ be such that $\mu([c]) \in [c]$. Let $\phi_{tt'} : C \times X_{tt'} \to C$ be such that

$$\phi_{tt'}(c, x_{tt'}) = \mu(f_{tt'}([c], x_{tt'}))$$

Then $\bar{\phi} = \{\phi_{tt'}\}$ satisfies the semigroup property and is consistent with $\bar{\rho}$.

Furthermore, since $\bar{f}$ is time invariant, we have that

$$\begin{aligned}\phi_{tt'}(c, x_{tt'}) &= \mu(f_{tt'}([c], x_{tt'})) \\ &= \mu(f_{o\tau}([c], F^{-t}(x_{tt'}))) \\ &= \phi_{o\tau}(c, F^{-t}(x_{tt'}))\end{aligned}$$

that is, $\bar{\phi}$ is time invariant. Q.E.D.

Theorem 2.2. A time system S is stationary if and only if it has a time-invariant dynamical system representation $(\bar{\rho}, \bar{\phi})$.

PROOF. Let us consider the *if* part first. Let $(x_t, y_t) \in S_t$. Since $S_t \subset S_t^{\rho}$ holds, there exists $c \in C$ such that $y_t = \rho_t(c, x_t)$. Since $\bar{\rho}$ is time invariant, $y_t = F^t(\rho_0(c, F^{-t}(x_t)))$ holds, that is, $F^{-t}(x_t, y_t) \in S$. Hence, S is stationary. Conversely, suppose S is stationary. If we can find for S a systems–response function $\rho_0: C \times X \to Y$ such that the following holds:

$$(\forall c)(\forall x^t)(\exists c')(\forall x)[\rho_0(c, x^t \cdot F^t(x)) \mid T_t = F^t(\rho_0(c', x))]$$

then from ρ_0 we can derive $\bar{\rho}$ which satisfies the condition of Theorem 2.1, where $\rho_t: C \times X_t \to Y_t$ is given by

$$\rho_t(c, x_t) = F^t(\rho_0(c, F^{-t}(x_t)))$$

Then we have the desired result from Theorem 2.1. We shall show that a stationary system has a systems-response function which satisfies the above condition. Let

$$C = \{c : c : X \to Y \,\&\, c \subseteq S\}$$

Let $\rho_0: C \times X \to Y$ be $\rho_0(c, x) = c(x)$. Suppose $c \in C$ and x^t are arbitrary. Let

$$f = F^{-t}[c(x^t \cdot F^t(-)) \mid T_t] : X \to Y$$

We shall show that $f \subseteq S$. Let $x \in X$ be arbitrary. Then

$$(x^t \cdot F^t(x), c(x^t \cdot F^t(x))) \in S$$

implies

$$(F^t(x), c(x^t \cdot F^t(x)) \mid T_t) \in S_t$$

Since $S_t \subset F^t(S)$ holds, we have

$$(x, F^{-t}(c(x^t \cdot F^t(x)) \mid T_t)) \in S$$

that is, $f \subseteq S$. Then it follows from the definition of C that there exists $c' \in C$ such that $c' = f$. Therefore, the above condition is satisfied. Q.E.D.

Notice that Theorem 2.2 does not take into consideration the nonanticipation conditions. This case will be dealt with later.

(b) Linear Time System

Proposition 2.1. Suppose a family of time-invariant linear maps

$$\bar{\rho} = \{\rho_t : \rho_t : C \times X_t \to Y_t \;\&\; t \in T\}$$

is reduced and realizable. Then a family of linear state-transition functions $\bar{\phi}$ associated with $\bar{\rho}$ is uniquely given by

$$\phi_{1tt'}(c) = \rho_{1t'}^{-1}(\rho_{1t}(c) \mid T_{t'})$$

$$\phi_{2tt'}(x_{tt'}) = \rho_{1t'}^{-1}(\rho_{2t}(x_{tt'} \cdot o) \mid T_{t'})$$

Furthermore, $(\bar{\rho}, \bar{\phi})$ is time invariant.

PROOF. The first part is given in Theorem 3.4, Chapter IV. We shall show that $(\bar{\rho}, \bar{\phi})$ is time invariant. Let $\phi_{1tt'}(c) = c'$. Then

$$\phi_{1tt'}(c) = c' \to \rho_{1t'}(c') = \rho_{1t}(c) \mid T_{t'} \to F^{t'}(\rho_{10}(c'))$$

$$= (F^t(\rho_{10}(c))) \mid T_{t'} = F^t(\rho_{10}(c) \mid T_\tau) \qquad \text{where} \quad \tau = t' - t$$

$$F^\tau(\rho_{10}(c')) = \rho_{10}(c) \mid T_\tau \to \rho_{1\tau}(c') = \rho_{10}(c) \mid T_\tau \to c' = \phi_{10\tau}(c)$$

Similarly, let $c' = \phi_{2tt'}(x_{tt'})$. Then

$$c' = \phi_{2tt'}(x_{tt'}) \to \rho_{1t'}(c') = \rho_{2t}(x_{tt'} \cdot o) \mid T_{t'} \to F^{t'}(\rho_{10}(c'))$$

$$= F^t(\rho_{20}(F^{-t}(x_{tt'} \cdot o))) \mid T_{t'} \to F^{t'}(\rho_{10}(c')) = F^t(\rho_{20}(F^{-t}(x_{tt'} \cdot o)) \mid T_\tau)$$

$$\text{where} \quad \tau = t' - t$$

$$\to \rho_{1\tau}(c') = \rho_{20}(F^{-t}(x_{tt'}) \cdot o) \mid T_\tau \to c' = \phi_{20\tau}(F^{-t}(x_{tt'}))$$

Hence, the dynamical system representation is time invariant. Q.E.D.

Theorem 2.3. Let

$$\bar{\rho} = \{\rho_t : \rho_t : C \times X_t \to Y_t \;\&\; t \in T\}$$

be a family of time-invariant linear maps. Then $\bar{\rho}$ is realizable by a linear time-invariant dynamical system if and only if $\bar{\rho}$ satisfies the input-response consistency and

$$(\forall (c, x^t))(\exists c')(\rho_{1t}(c') = \rho_o(c, x^t \cdot o) \mid T_t) \tag{6.1}$$

which is one of the state-response consistency conditions.

PROOF. Since the proof is similar to that of Theorem 3.2, Chapter IV, we shall sketch the procedure. For the case of time-invariant systems, $S_t = S_o^\rho \mid T_t$ may be a proper subset of S_t^ρ. Hence, the state-response consistency need not be satisfied in the complete form. Furthermore, due to the time invariance,

$$S_t \subset S_t^{\rho} \leftrightarrow \text{condition (6.1)} \leftrightarrow (\forall(c, x_{tt'}))(\exists c')\,(\rho_{1t'}(c') = \rho_t(c, x_{tt'} \cdot o) \mid T_{t'})$$

holds. The main procedure for the *if* part is the following. First, derive the reduced-response family and apply Proposition 2.1. Next, define a state-transition family as done in Theorem 3.2, Chapter IV, and show that it is time invariant. The details are left for the reader. Q.E.D.

Classical theory of realization is mainly concerned with the realization of an input-response function [5]. The counterpart in our general systems theory is given by the following theorem.

Theorem 2.4. Let

$$\bar{\rho}_2 = \{\rho_{2t} : \rho_{2t} : X_t \to Y_t \,\&\, t \in T\}$$

be a family of linear maps. $\bar{\rho}_2$ is realizable by a time-invariant strongly non-anticipatory linear dynamical system if and only if the following conditions are satisfied:

(i) $\bar{\rho}_2$ satisfies the input-response consistency;

(ii) $\bar{\rho}_2$ satisfies the strongly nonanticipatory condition, i.e.,

$$(\forall x_t)(\rho_{20}(o \cdot x_t) \mid \bar{T}^t = o)$$

(iii) $\bar{\rho}_2$ is time invariant, i.e.,

$$(\forall x_t)(\rho_{2t}(x_t) = F^t(\rho_{20}(F^{-t}(x_t))))$$

If $\bar{\rho}_2$ is realized, a state space C and a state-response function $\rho_{1t} : C \to Y_t$ are given by

$$C = \{F^{-t}(\rho_{20}(x^t \cdot o) \mid T_t) : x^t \in X^t \,\&\, t \in T\}$$

$$\rho_{1t}(c) = F^t(c)$$

PROOF. The *only if* part is clear. Let us consider the *if* part. We shall show that the state space C mentioned in Theorem 2.4 is a linear algebra. Let $c = F^{-t}(\rho_{20}(x^t \cdot o) \mid T_t)$ and $\alpha \in \mathscr{A}$ be arbitrary. Then we have

$$\alpha c = F^{-t}(\rho_{20}(\alpha x^t \cdot o) \mid T_t) \in C$$

Let $\hat{c} = F^{-\hat{t}}(\rho_{20}(\hat{x}^{\hat{t}} \cdot o) \mid T_{\hat{t}})$ be arbitrary. Suppose $t \geq \hat{t}$ such that $t = \hat{t} + \tau$ $(\tau \geq o)$. Conditions (i) and (iii) imply that

$$\rho_{20}(o \cdot F^{\tau}(\hat{x}^{\hat{t}}) \cdot o) \mid T_{\tau} = \rho_{2\tau}(F^{\tau}(\hat{x}^{\hat{t}}) \cdot o) = F^{\tau}(\rho_{20}(\hat{x}^{\hat{t}} \cdot o))$$

Consequently, we have that

$$\begin{aligned}
\rho_{20}(\hat{x}^{\hat{t}} \cdot o) \mid T_{\hat{t}} &= F^{-\tau}(\rho_{20}(o \cdot F^{\tau}(\hat{x}^{\hat{t}}) \cdot o) \mid T_{\tau}) \mid T_{\hat{t}} \\
&= F^{-\tau}((\rho_{20}(o \cdot F^{\tau}(\hat{x}^{\hat{t}}) \cdot o) \mid T_{\tau}) \mid T_t) \\
&= F^{-\tau}(\rho_{20}(o \cdot F^{\tau}(\hat{x}^{\hat{t}}) \cdot o) \mid T_t)
\end{aligned}$$

Therefore,

$$\begin{aligned} c + \hat{c} &= F^{-t}(\rho_{20}(x^t \cdot \mathrm{o}) \mid T_t) + F^{-t}(\rho_{20}(\mathrm{o} \cdot F^{\tau}(\hat{x}^{\hat{t}}) \cdot \mathrm{o}) \mid T_t) \\ &= F^{-t}(\rho_{20}((x^t + \mathrm{o} \cdot F^{\tau}(\hat{x}^{\hat{t}})) \cdot \mathrm{o}) \mid T_t) \in C \end{aligned}$$

Hence, C is a linear algebra. Let $\rho_{1t}: C \to Y_t$ be such that $\rho_{1t}(c) = F^t(c)$. We shall show that the conditions of Theorem 2.3 are satisfied. Let

$$c = F^{-\hat{t}}(\rho_{20}(x^{\hat{t}} \cdot \mathrm{o}) \mid T_{\hat{t}}) \in C$$

and x^t be arbitrary. Let $t + \hat{t} = \tau$. Then we have

$$\rho_{20}(x^t \cdot \mathrm{o}) = F^{-\hat{t}}(\rho_{20}(\mathrm{o} \cdot F^{\hat{t}}(x^t) \cdot \mathrm{o}) \mid T_{\hat{t}})$$

Consequently,

$$\begin{aligned} (\rho_{10}(c) + \rho_{20}(x^t \cdot \mathrm{o})) \mid T_t &= (c + \rho_{20}(x^t \cdot \mathrm{o})) \mid T_t \\ &= (F^{-\hat{t}}(\rho_{20}(x^{\hat{t}} \cdot \mathrm{o}) \mid T_{\hat{t}}) + F^{-\hat{t}}(\rho_{20}(\mathrm{o} \cdot F^{\hat{t}}(x^t) \cdot \mathrm{o}) \mid T_{\hat{t}})) \mid T_t \\ &= F^{-\hat{t}}(\rho_{20}(x^{\hat{t}} \cdot F^{\hat{t}}(x^t) \cdot \mathrm{o}) \mid T_{\hat{t}}) \mid T_t \\ &= F^{-\hat{t}}(\rho_{20}(x^{\hat{t}} \cdot F^{\hat{t}}(x^t) \cdot \mathrm{o}) \mid T_{\tau}) \end{aligned}$$

Therefore, for

$$c' = F^{-\tau}(\rho_{20}(x^{\hat{t}} \cdot F^{\hat{t}}(x^t) \cdot \mathrm{o}) \mid T_{\tau}) \in C$$

we have $\rho_{1t}(c') = \rho_o(c, x^t \cdot \mathrm{o}) \mid T_t$. Q.E.D.

Definition 2.1. A mapping $\hat{F}^t: X \to X$ (or $\hat{F}^t: Y \to Y$), which is defined by

$$\hat{F}^t(x) = \mathrm{o}^t \cdot F^t(x) \qquad \text{or} \qquad \hat{F}^t(y) = \mathrm{o}^t \cdot F^t(y)$$

where $\mathrm{o}^t = \mathrm{o} \mid T^t$, is called an extended shift operator.

Corollary 2.1. A linear map $\rho_{20}: X \to Y$ is realizable by a time-invariant strongly nonanticipatory linear dynamical system if and only if the following diagram is commutative:

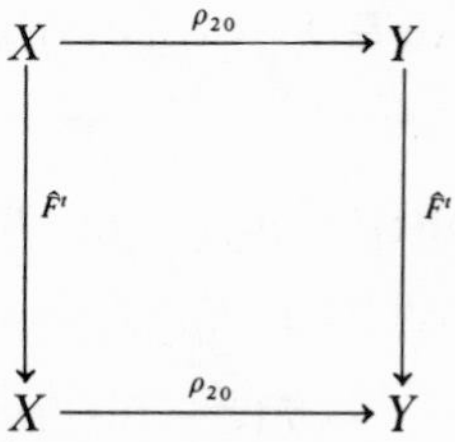

that is, ρ_{20} is a homomorphism with respect to the extended shift operator $\hat{F}^t$, i.e.,

$$(\forall t)(\forall x)(\hat{F}^t(\rho_{20}(x)) = \rho_{20}(\hat{F}^t(x)))$$

PROOF. From any given $\rho_{20}: X \to Y$ we can derive a family $\bar{\rho}_2$ of time-invariant maps by

$$\rho_{2t}(x_t) = F^t(\rho_{20}(F^{-t}(x_t)))$$

Then Theorem 2.4 says that $\bar{\rho}_2$ is realizable by a time-invariant strongly nonanticipatory linear dynamical system if and only if the following conditions are satisfied:

$$(\forall x_t)(F^t(\rho_{20}(F^{-t}(x_t))) = \rho_{20}(\mathrm{o} \cdot x_t) \mid T_t) \qquad \text{(input-response consistency)}$$

$$(\forall x_t)(\rho_{20}(\mathrm{o} \cdot x_t) \mid \bar{T}^t = \mathrm{o}) \qquad \text{(strong nonanticipation)}$$

Those two conditions imply and are implied by $(\forall x)(\hat{F}^t(\rho_{20}(x)) = \rho_{20}(\hat{F}^t(x)))$.

Q.E.D.

The following corollary is an expression of the classical realization theory [5] in the framework of general systems theory.

Corollary 2.2. A linear map $\rho_{20}: X \to Y$ is realizable by a time-invariant strongly nonanticipatory linear dynamical system if and only if the following conditions are satisfied:

(i) ρ_{20} is strongly nonanticipatory.

(ii) There exist a linear algebra C over the field $\mathscr{A}$ and two linear maps $F(t): C \to B$ and $G(t): X^t \to C$ such that

$$\rho_{20}(x^t \cdot \mathrm{o})(\tau) = F(\tau - t)[G(t)x^t]$$

holds for t and $\tau (\geq t)$, where $G(t)$ satisfies the following property:

$$G(t')(\mathrm{o} \cdot x_{tt'}) = G(t' - t)(F^{-t}(x_{tt'}))$$

PROOF. If ρ_{20} is realizable, let the state space and ρ_{1t} be as given in Theorem 2.4. Then let $F(t): C \to B$ and $G(t): X^t \to C$ be as follows:

$$F(t)(c) = \rho_{10}(c)(t) = c(t)$$

$$G(t)(x^t) = F^{-t}(\rho_{20}(x^t \cdot \mathrm{o}) \mid T_t)$$

Then for $\tau \geq t$, we have

$$\begin{aligned}\rho_{20}(x^t \cdot \mathrm{o})(\tau) &= (\rho_{20}(x^t \cdot \mathrm{o}) \mid T_t)(\tau) \\ &= F^{-t}(\rho_{20}(x^t \cdot \mathrm{o}) \mid T_t)(\tau - t) \\ &= F(\tau - t)[G(t)\,(x^t)]\end{aligned}$$

Furthermore, the input-response consistency implies that

$$\begin{aligned}
G(t')(0 \cdot x_{tt'}) &= F^{-t'}(\rho_{20}(0 \cdot x_{tt'} \cdot 0) \mid T_{t'}) \\
&= F^{-t'}(\rho_{2t}(x_{tt'} \cdot 0) \mid T_{t'}) \\
&= F^{-t'}(F^{t}(\rho_{20}(F^{-t}(x_{tt'}) \cdot 0)) \mid T_{t'}) \\
&= F^{-\tau}(\rho_{20}(F^{-t}(x_{tt'}) \cdot 0) \mid T_{\tau}) \\
&= G(\tau)(F^{-t}(x_{tt'}))
\end{aligned}$$

where $\tau = t' - t$.
The *if* part is clear. Q.E.D.

The following theorem is a basic result on a linear stationary system.

Theorem 2.5. Let S be a stationary linear system. There exists then a linear dynamical time-invariant representation of S if and only if there exists a linear map $\rho_{20}: X \to Y$ such that

(i) $(\forall x)[(x, \rho_{20}(x)) \in S]$
(ii) $(\forall t)(\forall x_t)[F^t(\rho_{20}(F^{-t}(x_t))) = \rho_{20}(0 \cdot x_t) \mid T_t]$

PROOF. Suppose a linear mapping $\rho_{20}: X \to Y$, which satisfies the above two conditions, can be found. We shall apply the natural realization to S, where the input-response function is ρ_{20}. Then in the process of natural realization, the response function ρ_{2t} is given by $\rho_{2t}: X_t \to Y_t$ such that $\rho_{2t}(x_t) = \rho_{20}(0 \cdot x_t) \mid T_t$. Hence, $\rho_{2t}(x_t) = F^t(\rho_{20}(F^{-t}(x_t)))$. Let $C = \{y : (0, y) \in S\}$. Let $\rho_{10}: C \to Y$ be such that $\rho_{10}(c) = c$. Let $\rho_{1t}: C \to Y$ be such that $\rho_{1t}(c) = F^t(\rho_{10}(c))$. Then $\bar{\rho} = \{\rho_t = \langle \rho_{1t}, \rho_{2t} \rangle\}$ clearly satisfies the input-response consistency. Let $c \in C$ and x^t be arbitrary. Let $y_t = \rho_0(c, x^t \cdot 0) \mid T_t$. Since $(0, y_t) \in S_t \subset F^t(S)$, we have $(0, F^{-t}(y_t)) \in S$; that is, there exists $c' \in C$ such that $F^{-t}(y_t) = \rho_{10}(c')$. Consequently, we have $\rho_{1t}(c') = \rho_0(c, x^t \cdot 0) \mid T_t$. Therefore, it follows from Theorem 2.3 that $\bar{\rho}$ is realizable by a linear time-invariant dynamical system. Apparently, $S_0{}^{\rho} = S$.

Conversely, suppose S has a time-invariant linear dynamical system representation $(\bar{\rho}, \bar{\phi})$. Then by definition of time invariance, $\rho_{2t}(x_t) = F^t(\rho_{20}(F^{-t}(x_t)))$. Since $\{\langle \rho_{1t}, \rho_{2t} \rangle\}$ is realizable, $\rho_{2t}(x_t) = \rho_{20}(0 \cdot x_t) \mid T_t$ for every x_t. Hence,

$$F^t(\rho_{20}(F^{-t}(x_t))) = \rho_{20}(0 \cdot x_t) \mid T_t$$

for every x_t. Q.E.D.

3. STATIONARY PAST-DETERMINED SYSTEMS

The representation of a stationary system by a time-invariant dynamical system depends on the availability of a family of response functions that possesses the required property. Although the existence of such a family of functions is already expected by the very fact of stationarity, there is no general procedure for constructing such a function. The objective of this section is to prove that for the class of past-determined systems, the natural systems-response function (as defined for that class of systems in Chapter V) satisfies the required conditions and can be used for a dynamic time-invariant representation of a stationary system.

Theorem 3.1. Suppose a stationary time system $S \subset X \times Y$ is past-determined from $\hat{t}$. Then $S_{\hat{t}}$ can be represented by a time-invariant strongly nonanticipatory dynamical system.

PROOF. Let $\rho_{\hat{t}}: S^{\hat{t}} \times X_{\hat{t}} \to Y_{\hat{t}}$ be the natural systems-response function. Let $S^{\hat{t}} = C$. For any $t \geq \hat{t}$ let $\rho_t: C \times X_t \to Y_t$ be such that

$$\rho_t(c, x_t) = F^{t-\hat{t}}(\rho_{\hat{t}}(c, F^{\hat{t}-t}(x_t)))$$

Then $\bar{\rho}_{\hat{t}} = \{\rho_t : t \geq \hat{t}\}$ is time invariant. If we show that

$$(\forall c)(\forall x_{\hat{t}t})(\exists c')(\forall x_t)[\rho_t(c', x_t) = \rho_{\hat{t}}(c, x_{\hat{t}t} \cdot x_t) \mid T_t]$$

the desired result follows from Theorem 2.1. Let x_t and $c = (x^{\hat{t}}, y^{\hat{t}}) \in C$ be arbitrary. Since the system is past-determined, there exists a unique $y_{\hat{t}t}$ such that $(x^{\hat{t}} \cdot x_{\hat{t}t}, y^{\hat{t}} \cdot y_{\hat{t}t}) \in S^t$. Let $t - \hat{t} = \sigma \geq$ o. Let $x^{\hat{t}} \cdot x_{\hat{t}t} = x^t$ and $y^{\hat{t}} \cdot y_{\hat{t}t} = y^t$. Since S is stationary, we have that

$$(x^t, y^t) \mid T_{\sigma t} \in S^t \mid T_{\sigma t} = S_\sigma \mid T_{\sigma t} \subset F^\sigma(S) \mid T_{\sigma t} = F^\sigma(S \mid T^{\hat{t}})$$

Consequently, $F^{-\sigma}((x^t, y^t) \mid T_{\sigma t}) \in S^{\hat{t}}$. Let $(x^t, y^t) \mid T_{\sigma t} = (x_{\sigma t}, y_{\sigma t})$. Let $c' = F^{-\sigma}(x_{\sigma t}, y_{\sigma t}) \in C$. Let x_t be arbitrary. Then $\rho_{\hat{t}}(c, x_{\hat{t}t} \cdot x_t) = y_{\hat{t}t} \cdot \hat{y}_t$ for some $\hat{y}_t$. The definition of $\rho_{\hat{t}}$ implies then that $(x^t \cdot x_t, y^t \cdot \hat{y}_t) \in S$. We have from the stationarity that $(x^t \cdot x_t, y^t \cdot \hat{y}_t) \mid T_\sigma \in S_\sigma \subset F^\sigma(S)$, that is,

$$F^{-\sigma}[(x^t \cdot x_t, y^t \cdot \hat{y}_t) \mid T_\sigma] = F^{-\sigma}(x_{\sigma t} \cdot x_t, y_{\sigma t} \cdot \hat{y}_t) \in S$$

On the other hand, let $y_{\hat{t}} = \rho_{\hat{t}}(c', F^{-\sigma}(x_t))$. Then we have that $(F^{-\sigma}(x_{\sigma t} \cdot x_t), F^{-\sigma}(y_{\sigma t}) \cdot y_{\hat{t}}) \in S$. Consequently, since S is past-determined from $\hat{t}$, we have $F^{-\sigma}(\hat{y}_t) = y_{\hat{t}}$, that is,

$$\rho_t(c', x_t) = F^\sigma(y_{\hat{t}}) = \hat{y}_t = \rho_{\hat{t}}(c, x_{\hat{t}t} \cdot x_t) \mid T_t \qquad \text{Q.E.D.}$$

Theorem 3.2. Let a stationary linear system $S \subset X \times Y$ be past-determined from $\hat{t}$. Then $S_{\hat{t}}$ can be represented by a time-invariant strongly nonanticipatory linear dynamical system.

PROOF. Let $\rho_{\hat{t}}:S^{\hat{t}} \times X_{\hat{t}} \to Y_{\hat{t}}$ be the natural systems-response function. Then $\rho_{2\hat{t}}:X_{\hat{t}} \to Y_{\hat{t}}$ is given by

$$\rho_{2\hat{t}}(x_{\hat{t}}) = y_{\hat{t}} \leftrightarrow (\text{o} \cdot x_{\hat{t}}, \text{o} \cdot y_{\hat{t}}) \in S$$

If we can show that

$$(\forall t)(\forall x_t)[F^{t-\hat{t}}(\rho_{2\hat{t}}(F^{\hat{t}-t}(x_t))) = \rho_{2\hat{t}}(\text{o} \cdot x_t) \mid T_t]$$

the desired result follows from Theorem 2.5. Let x_t be arbitrary. Let $y_{\hat{t}} = \rho_{2\hat{t}}(F^{-\sigma}(x_t))$ where $\sigma = t - \hat{t} \geq \text{o}$. Then $(\text{o} \cdot F^{-\sigma}(x_t), \text{o} \cdot y_{\hat{t}}) \in S$ holds. Notice that S is past-determined from $\hat{t}$ and that $(\text{o}, \text{o}) \in S$. Then $\rho_{2\hat{t}}(\text{o} \cdot x_t) = \text{o} \cdot \hat{y}_t$ for some $\hat{y}_t$, that is, $(\text{o} \cdot x_t, \text{o} \cdot \hat{y}_t) \in S$. Since S is stationary, we have that

$$(\text{o} \cdot x_t, \text{o} \cdot \hat{y}_t) \mid T_\sigma \in S_\sigma \subset F^\sigma(S)$$

that is

$$(\text{o} \cdot F^{-\sigma}(x_t), \text{o} \cdot F^{-\sigma}(\hat{y}_t)) \in S$$

Consequently, $y_{\hat{t}} = F^{-\sigma}(\hat{y}_t)$. Hence,

$$F^\sigma(\rho_{2\hat{t}}(F^{-\sigma}(x_t))) = F^\sigma(y_{\hat{t}}) = \hat{y}_t = \rho_{2\hat{t}}(\text{o} \cdot x_t) \mid T_t \qquad \text{Q.E.D.}$$

4. AXIOMATIC CONSTRUCTION OF A CLASS OF DYNAMICAL SYSTEMS

The general systems theory we are developing in this book obviously covers a full spectrum of specialized theories which are concerned with various classes of systems having deeper and more specific mathematical structures. However, it is of interest to establish specific bridges linking the present theory with the classical theories in a precise and rigorous manner. This task requires a careful treatment. It is important because such links show precisely which of the results in the general systems theory are valid abstractions of the specific, more limited results derived earlier, and in that way show which the general structural properties are that the system ought to possess in order to behave in a certain way.

We shall consider in this section the link between general systems and dynamical systems described by the constant coefficient linear ordinary differential equations. In addition to the interest in application, these systems actually play a central role in the classical theory of linear systems. The classical theory of linear systems can very well be considered as the theory of the class of linear differential equation systems.

The class of dynamical systems considered in this section will be described by a set of constant coefficient linear ordinary differential equations of the following form:

$$d\mathbf{c}/dt = F\mathbf{c} + G\mathbf{x} \tag{6.2}$$

$$\mathbf{y} = H\mathbf{c} \tag{6.3}$$

where F, G, and H are constant matrices, while $\mathbf{x}$, $\mathbf{y}$, and $\mathbf{c}$ are vector-valued functions.

We shall arrive at such a description on an axiomatic basis starting from the general linear time system $S \subset A^T \times B^T$.

Axiom 4.1. The input and output alphabets, A, B, the time set T, and the field $\mathscr{A}$ needed for the specification of a linear system are defined as: $A = E^m$; $B = E^r$; $T = R^t$ (the set of nonnegative real numbers); $\mathscr{A} = R$; $X = L_2(\text{o}, \infty)$.

Axiom 4.2. S is stationary and past-determined from $\hat{t}$.

Axiom 4.3. $S_{\hat{t}} = \{y_{\hat{t}} : (\text{o}, y_{\hat{t}}) \in S_{\hat{t}}\} = C$ is a finite-dimensional vector space with the dimension n.

Let $\rho_{2\hat{t}} : C \times X_{\hat{t}} \to Y_{\hat{t}}$ be the input-response function associated with the natural systems-response function. Let $\{c_1, \ldots, c_n\}$ be a basis of C and let $\phi : C \to E^n$ be such that

$$\phi(c) = (\alpha_1, \ldots, \alpha_n) \leftrightarrow c = \sum_{i=1}^{n} \alpha_i c_i$$

Notice that ϕ is linear. Let $\rho_{1\hat{t}}(c) = c$. Since the time-invariant systems-response family derived from $\rho_{2\hat{t}}$ and $\rho_{1\hat{t}}$ is realizable, the following condition (state-response consistency) holds:

$$(\forall x_{\hat{t}t})(\exists c)(\rho_{1t}(c) = F^{t-\hat{t}}(\rho_{1\hat{t}}(c)) = \rho_{2\hat{t}}(x_{\hat{t}t} \cdot \text{o}) \mid T_t)$$

that is,

$$\{F^{\hat{t}-t}(\rho_{2\hat{t}}(x_{\hat{t}t} \cdot \text{o}) \mid T_t) : x_{\hat{t}t} \in X_{\hat{t}t} \ \&\ t \in T\}$$

is a linear subspace of C. This fact will be used in the subsequent arguments.

The next axiom assures continuity in an appropriate sense.

Axiom 4.4.

(i) For each $x \in X$, $\rho_{2\hat{t}}(x_{\hat{t}}) : T_{\hat{t}} \to B(= E^r)$ is differentiable.
(ii) For each $t \in T_{\hat{t}}$, $\rho_{2\hat{t}}(-)(t) : X_{\hat{t}} \to B$ is continuous.
(iii) $\phi\rho_{1\hat{t}}(-) : C \to E^n$ is continuous.
(iv) $c \in C$ is an analytic function.

Notice that the system described by Eqs. (6.2) and (6.3) is past-determined from $\hat{t}$, where $\hat{t}$ is any positive number as proven in Chapter V.

It is clear that these four axioms are satisfied by the system of Eqs. (6.2) and (6.3). We shall show now how these equations can be derived from these axioms.

For notational convenience, $\hat{t}$ is replaced by o, which can be interpreted to mean that the time axis is shifted left by $\hat{t}$. In other words, we shall wrte ρ_{20} and ρ_{10} for $\rho_{2\hat{t}}$ and $\rho_{1\hat{t}}$. $T_{\hat{t}}$, $X_{\hat{t}}$, and other restrictions should be replaced accordingly. No generality is lost by doing so.

Since $\rho_{20}(-)(t): X \to E^r$ is continuous and $X = L_2(\mathrm{o}, \infty)$, the representation theorem of linear functionals on H space implies that there exists $w(t) \in L_2(\mathrm{o}, \infty)$ such that

$$\rho_{20}(x)(t) = \int_0^\infty w(t, \tau)x(\tau)\, d\tau$$

where $w(t, \tau) = w(t)(\tau): E^m \to E^r$ is a matrix. Axiom 4.2 implies that $\bar{\rho}_2$ is time invariant and satisfies the input-response consistency, i.e., $\rho_{20}(\mathrm{o} \cdot F^t(x)) \mid T_t = F^t(\rho_{20}(x))$ holds. Consequently, for each $\tau \geq t$, we have

$$\begin{aligned}
[\rho_{20}(\mathrm{o} \cdot F^t(x)) \mid T_t](\tau) &= \rho_{20}(\mathrm{o} \cdot F^t(x))(\tau) \\
&= \int_0^\infty w(\tau, \sigma)(\mathrm{o} \cdot F^t(x))(\sigma)\, d\sigma \\
&= \int_t^\infty w(\tau, \sigma)[F^t(x)](\sigma)\, d\sigma \\
&= \int_t^\infty w(\tau, \sigma)x(\sigma - t)\, d\sigma \\
&= \int_0^\infty w(\tau, \sigma + t)x(\sigma)\, d\sigma
\end{aligned}$$

On the other hand,

$$\begin{aligned}
F^t(\rho_{20}(x))(\tau) &= \rho_{20}(x)(\tau - t) \\
&= \int_\mathrm{o}^\infty w(\tau - t, \sigma)x(\sigma)\, d\sigma
\end{aligned}$$

Therefore, we have

$$\int_\mathrm{o}^\infty w(\tau, \sigma + t)x(\sigma)\, d\sigma = \int_\mathrm{o}^\infty w(\tau - t, \sigma)x(\sigma)\, d\sigma$$

Since the above equality is satisfied for every $x \in X$, we have $w(\tau, \sigma + t) = w(\tau - t, \sigma)(\tau - t \geq 0)$. Hence, for $\sigma = 0$ we have $w(\tau, t) = w(\tau - t, 0)$. Let $w(\tau - t, 0) \equiv w_0(\tau - t)$. Then

$$\rho_{20}(x)(t) = \int_0^\infty w_0(t - \sigma)x(\sigma)\, d\sigma \tag{6.4}$$

Furthermore, since ρ_{20} satisfies the strong nonanticipation, we have that for each $x_t \in X_t$

$$\rho_{20}(0 \cdot x_t)(t) = \int_t^\infty w_0(t - \tau)x_t(\tau)\, d\tau = 0$$

that is, $t < 0 \rightarrow w_0(t) = 0$. This is the nonanticipatory condition usually used for weighting functions. In summary, we have

$$\rho_{20}(x)(t) = \int_0^t w_0(t - \tau)x(\tau)\, d\tau$$

Following Corollary 2.2 of Theorem 2.4, let $F(t): E^n \rightarrow E^r$ be such that

$$F(t)(\alpha_1, \ldots, \alpha_n) = \sum_{i=1}^n \alpha_i c_i(t)$$

Since

$$\phi F^{-t}(\rho_{20}(- \cdot 0) \mid T_t): X^t \rightarrow E^n$$

is continuous, there exists a matrix $G(t, \sigma): E^m \rightarrow E^n$ such that

$$\phi F^{-t}(\rho_{20}(x^t \cdot 0) \mid T_t) = \int_0^t G(t, \sigma)x^t(\sigma)\, d\sigma$$

Since Corollary 2.2 of Theorem 2.4 says that G is time invariant and since the system is strongly nonanticipatory, we have from the same argument used for w that $G(t, \sigma) = G(t - \sigma, 0) \equiv G_0(t - \sigma)$ and $G_0(t) = 0$ for $t < 0$. Hence

$$\rho_{20}(x^t \cdot 0)(\tau) = F(\tau - t) \int_0^t G_0(t - \sigma)x^t(\sigma)\, d\sigma$$

On the other hand, we have from Eq. (6.4) that

$$\rho_{20}(x^t \cdot 0)(\tau) = \int_0^t w_0(\tau - \sigma)x^t(\sigma)\, d\sigma$$

Let $\tau - t = \tau'$ and $t - \sigma = -\sigma'$. Then we have $w(\tau', \sigma') = F(\tau') \cdot G_0(-\sigma')$. Since c_i is an analytic function, $F(t)$ is continuously differentiable. Therefore, the classical realization theory is applicable so that the system can be represented by Eqs. (6.2) and (6.3).

5. ABSTRACT TRANSFER FUNCTION†

When a dynamical system is expressed by a constant coefficient ordinary linear differential equation or by a constant coefficient linear difference equation, one of the most powerful techniques used in the analysis of such a system is based on a transfer function. It is, therefore, important to consider how a transfer function can be represented in the abstract framework as developed in this book.

In order to attack the problem, we have to introduce a deep structure for the input object X and the output object Y. Let V be a linear algebra such that $A = V^m$ and $B = V^r$, where m and r are positive integers. Let $U \subset V^T$ be a time object such that the following conditions are satisfied:

(i) U is a linear algebra over $\mathscr{A}$.

(ii) U is a commutative ring whose multiplication operation $*$ satisfies the condition: for u and $u' \in U$, $u * u' = \text{o} \leftrightarrow u = \text{o} \vee u' = \text{o}$.

(iii) The input object X and the output object Y are expressed by U as $X = U^m$ and $Y = U^r$.

U will be referred to as a basic time object.

As a concrete example, let $T = [\text{o}, \infty)$, $V = R$, and $U = C(\text{o}, \infty)$. For each $t \in T$, let

$$(u * \hat{u})(t) = \int_{\text{o}}^{t} u(t - \sigma) \cdot \hat{u}(\sigma)\, d\sigma$$

It is easy to show that the operation $*$ is commutative. Furthermore, it can be shown that $u * \hat{u} = \text{o} \leftrightarrow u = \text{o} \vee \hat{u} = \text{o}$.† Then $C(\text{o}, \infty)$ is a basic time object for such X and Y in the previous section.

Suppose a linear input-response function $\rho_{20}: X \to Y$ is expressed as

$$\rho_{20}(x) = y \leftrightarrow y_i = \sum_{j=1}^{m} w_{ij} * x_j \qquad \text{for each} \quad i\,(= 1, \ldots, r) \tag{6.5}$$

where $w_{ij} \in U$. The right-hand side of the above expression is an abstract form of the expression of a dynamical system by a weighting function. Let us introduce an equivalence relation $E \subset U^2 \times U^2$ such that

$$((u, u'), (\hat{u}, \hat{u}')) \in E \leftrightarrow (u * \hat{u}' = \hat{u} * u')$$

† See Mikusinski [6].

The following is a well-known result [7]:

(i) U^2/E is a field called the quotient field whose operations are defined by

$$\text{o element} = [\text{o}, u] \quad \text{where} \quad u \neq \text{o is arbitrary}$$

$$\text{identity} = [u, u] \quad \text{where} \quad u \neq \text{o is arbitrary}$$

$$[u, u'] + [\hat{u}, \hat{u}'] = [u * \hat{u}' + \hat{u} * u', u' * \hat{u}']$$

$$[u, u'] \times [\hat{u}, \hat{u}'] = [u * \hat{u}, u' * \hat{u}']$$

(ii) Let $u_o \in U$ be a nonzero fixed element of U. Let $h: U \to U^2/E$ be such that $h(u) = [u * u_o, u_o]$. Then h is a homomorphism in the sense that

$$h(u + u') = h(u) + h(u'), \qquad \text{and} \qquad h(u * u') = h(u) \times h(u')$$

An abstract transfer function is, then, given by applying h to Eq. (6.5), that is,

$$h(y_i) = h\left(\sum_{j=1}^{m} w_{ij} * x_j\right)$$

$$= \sum_{j=1}^{m} h(w_{ij}) \times h(x_j)$$

Therefore, the abstract transfer function $\mathrm{TF}(\rho_{20})$ of ρ_{20} is given by the following matrix form:

$$\mathrm{TF}(\rho_{20}) = \begin{bmatrix} h(w_{11}), \ldots, h(w_{1m}) \\ \vdots \\ h(w_{r1}), \ldots, h(w_{rm}) \end{bmatrix}$$

In particular, if $r = m = 1$, then

$$h(w_{11}) = h(y_1)/h(x_1)$$

In order to demonstrate that the abstract transfer function is a valid abstraction of the usual concept of a transfer function, we shall consider how a differential or integral operator can be represented in the quotient field U^2/E.

Let $D: U \to U$ be a linear operator defined by

$$D(u) = u' \leftrightarrow u = I * u' + \alpha(u) \tag{6.6}$$

where $I \in U$ is a fixed element and $\alpha: U \to U$ is a linear operator. In the usual calculus, D represents the differential operator, while relation (6.6) represents the following:

$$du/dt = u' \leftrightarrow u(t) = \int_0^t 1 \times u'(\tau)\, d\tau + u(0)$$

where 1 is a constant function such that $1(t) = 1$ for all t. When the homomorphism h is applied to relation (6.6), we have

$$h(D(u)) = h(u') \leftrightarrow h(u) = h(1) \times h(u') + h(\alpha(u))$$

that is

$$h(D(u)) = h(u') = (1/h(1)) \times (h(u) - h(\alpha(u)))$$

If $1/h(1)$ is replaced by s, the above relation is expressed by

$$h(D(u)) = s(h(u) - h(\alpha(u))) \tag{6.7}$$

If h is considered as a Laplace transformation, Eq. (6.7) is exactly the basic formula for the differential operator in the Laplace transformation theory.

Similarly, if the integral operator $I: U \rightarrow U$ is represented by the following relation:

$$I(u) = u' \leftrightarrow u' = 1 * u$$

the operator I is represented by the following:

$$h(I(u)) = h(u)/s \tag{6.8}$$

Equation (6.8) is also a basic formula in the Laplace transformation theory. Furthermore, it is easy to show that h is an injection. Hence, h has the "inverse transformation."

Chapter VII

CONTROLLABILITY

Some basic controllability-type concepts are defined in this chapter in a general setting. These concepts involve both a system and a performance function; classical controllability-type notions such as state controllability, functional controllability, or reproducibility are shown to be special cases of the more general notions.

Necessary conditions are given for the general controllability of a multivalued system; i.e., a system evaluated not by a single element but by an n-tuple, e.g., a system with more than one state variable for the case of state controllability. For the class of linear systems, both necessary and sufficient conditions for general controllability are derived. It is shown then that the classical conditions for the state controllability and functional controllability of differential equation systems can be obtained by the application of the general results.

The chapter concludes with the investigation of mutual relationships between various properties of the time systems as they relate to controllability and realizability conditions.

1. BASIC CONCEPTS

To formalize the controllability notions on a general level, two remarks are of interest:

(i) The ability of the system to perform in a certain way is evaluated usually by the output, but depends ultimately upon the input. The conditions for a system having given property will be expressed, therefore, in reference to the existence of certain inputs.

(ii) To specify a notion within this category, it is necessary, in general, to introduce a performance or evaluation function in terms of which the desired behavior is specified. This ought to be emphasized because in the classical theory (e.g., controllability of differential equation systems in the state space) such a function is not shown explicitly. We shall assume, therefore, that in the addition to the system $S \subset X \times Y$, there is given a map

$$G: X \times Y \to V$$

termed the evaluation function.

We shall define the system in this chapter in terms of a map on two objects M and U, i.e.,

$$S: M \times U \to Y$$

Several interpretations of S are possible; e.g., S is the initial-response function of a general system which is actually a relation; the system itself is a function; S is the union of a family of response functions, etc. When a specific interpretation for S is intended, it will be indicated as such.

The evaluation function is now $G: M \times U \times Y \to V$. The composition of S and G is a mapping $g: M \times U \to V$, such that for all $(m, u) \in M \times U$

$$g(m, u) = G(m, u, S(m, u))$$

There are three generic notions in this category.

Definition 1.1. A set $V' \subset V$ is reproducible (attainable, reachable) relative to g if and only if

$$(\forall v)[v \in V' \to (\exists(m, u))(g(m, u) = v)]$$

When the mapping g is clear from the context, the explicit reference to g will be omitted.

A point $v \in V$ is reproducible if and only if there exists a reproducible set V' such that $v \in V'$.

Definition 1.2. A set $V' \subset V$ is completely controllable relative to g (G and S) if and only if

$$(\forall v)(\forall u)[v \in V' \,\&\, u \in U \to (\exists m)(g(m, u) = v)]$$

When the mapping g is clear from the context, the explicit reference will be omitted.

A variation of Definition 1.2 is worth noticing, although it will not be explored in any detail in this book. Namely, the requirement in Definition 1.2

might appear too restrictive for some practical applications where it would be sufficient to have the performance within a subset V' for any $u \in U$ rather than requiring that the performance attain a given fixed value; the condition for complete controllability is then

$$(\forall u)[u \in U \rightarrow (\exists m)(g(m, u) \in V')]$$

The third notion is given for the case when the value object V has more than one component, i.e.,

$$V = V_1 \times \cdots \times V_k = \times\{V_j : j \in I_k\}$$

where $I_k = \{1, \ldots, k\}$. For notational convenience, we shall use the symbol V^k for V to indicate how many component sets there are in V. When V has more than one component, the system is referred to as *multivalued.*

For each $i \in I_k$, p_i will denote the projection map $p_i : V^k \rightarrow V_i$; i.e., $p_i(v)$ is the ith component of v.

We shall refer to V', a subset $V' \subseteq V^k$, as being a *Cartesian set* if and only if

$$V' = p_1(V') \times \cdots \times p_k(V')$$

i.e., if V' is the Cartesian product of its component sets. Given an arbitrary subset $V' \subset V^k$, the Cartesian set $\hat{V}'$ generated by V' is the set $\hat{V}' = p_1(V') \times \cdots \times p_k(V')$. Obviously, a set is Cartesian if and only if $\hat{V}' = V'$.

We can now introduce the following concept.

Definition 1.3. A set $V' \subset V^k$ is cohesive (for a given S and G) if and only if the Cartesian set $\hat{V}'$ generated by V' is not reproducible; i.e., V' is cohesive if and only if the following statement is true:

$$(\exists v)[v \in \hat{V}' \;\&\; (\forall(m, u))(g(m, u) \neq v)]$$

Alternately, V' is a noncohesive set if and only if

$$(\forall v)[v \in \hat{V}' \rightarrow (\exists(m, u))(g(m, u) = v)]$$

Interpretation of the reproducibility is obvious: Any element of a reproducible set can be generated when desired. Controllability is a stronger condition requiring that the given value, $v \in V'$, be achieved over a range of different conditions, i.e., for all $u \in U$. Cohesiveness is a more special concept since it is defined only for the systems evaluated in terms of several components. A system is cohesive relative to a Cartesian set $V' = V_1' \times \cdots \times V_k'$ if not all of the possible combinations of component values can be achieved simultaneously; i.e., a given value component, say $\hat{v}_i \in p_i(V')$, can be achieved only in conjunction with some of the values from the remaining component sets. It appears as if there were an internal relationship between the components of the value set, hence the term "cohesiveness." If the set V' is cohesive, there exists a nontrivial relation

$$\Psi \subset p_1(V') \times \cdots \times p_k(V')$$

such that

$$v = (v_1, \ldots, v_k) \in \Psi \leftrightarrow (\exists(m, u))[g(m, u) = v]$$

Ψ will be referred to as the *cohesiveness relation*. When Ψ is a function, i.e.,

$$\Psi : V_1 \times \cdots \times V_i \to V_{i+1} \times \cdots \times V_k$$

it is referred to as the *cohesiveness function*, and the set will be termed *functionally cohesive*.

Several additional conventions will be adopted in this chapter:

(i) When the selection of m and/or u is restricted, e.g., to a set $M' \times U' \subset M \times U$, we shall talk about reproducibility, controllability relative to $M' \times U'$, and the set V' will be denoted by $V'[M' \times U']$.

(ii) If $V' = \mathscr{R}(g)$, the system itself will be referred to as cohesive or as controllable depending on which property is considered. The system is then noncohesive if and only if the range $\mathscr{R}(g)$ is a Cartesian set.

(iii) If $V' \subseteq g(M \times \{\hat{u}\})$ where $\hat{u}$ is a given element of U, we shall say that the system is controllable from $\hat{u}$ or simply *$\hat{u}$-controllable* in V'. Analogously, if V' is a unit set $V' = \{\hat{v}\}$, we shall say that the system is *controllable to $\hat{v}$* or simply *$\hat{v}$-controllable.*

(a) Relationship between Basic Controllability-Type Concepts

Relationships between various concepts introduced in the previous pages can be easily derived from the definitions themselves.

Proposition 1.1. If V' is a completely controllable set, it is also a reproducible set.

The reverse of Proposition 1.1 is not true, however.

Proposition 1.2. If V' is a noncohesive set, it is also reproducible.

The relationship between the complete controllability and cohesiveness is slightly more subtle. Noncohesiveness does not imply complete controllability nor *vice versa*, as one might intuitively expect. To show the validity of this assertion, let V' be noncohesive and furthermore there exist $\hat{u} \in U$

and $\hat{v} \in V$ such that $g(m, \hat{u}) = \hat{v}$ for all $m \in M$. This does not violate noncohesiveness, but surely precludes complete controllability, since an arbitrary $v^*, v^* \neq \hat{v}$, cannot be reproduced regardless of the selected $m \in M$ as long as $u = \hat{u}$. Alternatively, let V' be completely controllable. There is no reason why $\hat{V}'$ should be reproducible. Of course, if $V' = \hat{V}'$, then complete controllability does imply noncohesiveness of V' by definition. From this conclusion, some sufficient conditions for the absence of controllability can be expressed in terms of cohesiveness.

Proposition 1.3. Let $V' \subset V^k$ be a Cartesian set, i.e., $V' = \hat{V}'$. If V' is cohesive, it cannot be completely controllable.

The preceding proposition is important in various applications: *If the given set of interest is Cartesian, in order to show that it cannot be completely controllable, it is sufficient to show that it is cohesive.*

Proposition 1.4. Let $V' \subset V^k$ be a Cartesian set; V' is completely controllable over U' if and only if $V'[M \times \{u\}]$ is a noncohesive set for every $u \in U'$.

PROOF. Let V' be noncohesive within $M \times \{u\}$ for every $u \in U'$; then $V' \subseteq g(M \times \{u\})$ for every $u \in U'$; i.e., V' is completely controllable by definition. Alternatively, let V' be completely controllable over U'; then $V' \subset g(M \times \{u\})$ for every $u \in U'$. V' is then reproducible by definition, and since V' is a Cartesian set, it is noncohesive. Q.E.D.

2. SOME GENERAL CONDITIONS FOR CONTROLLABILITY

Controllability depends upon the character of the mapping g and the constraints in the domain $M \times U$, while the conditions for controllability depend upon the more specific definition of these sets and mappings. However, for the case of a multivalued system, i.e., when the value object has more than one component, it is possible to derive rather general conditions for controllability which reflect the interdependence between the value components. Obviously, this is the question intimately related with the concept of cohesiveness. Indeed, the generic problem we shall consider here is the following:

Given a multivalued system g. What are the conditions for the noncohesiveness of g? That is, is the set $\hat{g}(M \times U)$ reproducible? From this, the controllability conditions follow immediately. Controllability in this case is dependent strictly upon the interdependence between the value components, and we shall refer to it as *algebraic (or structural) controllability*. The importance of this case can be seen from the fact that almost all controllability-type conditions developed so far for various specific systems are of this type.

We can now give sufficient conditions for cohesiveness.

Theorem 2.1. Let $g: M \times U \to V^k$ be a multivalued system. If there exists a positive integer $j < k$ and a function $F: V^j \to V^k$ such that diagram

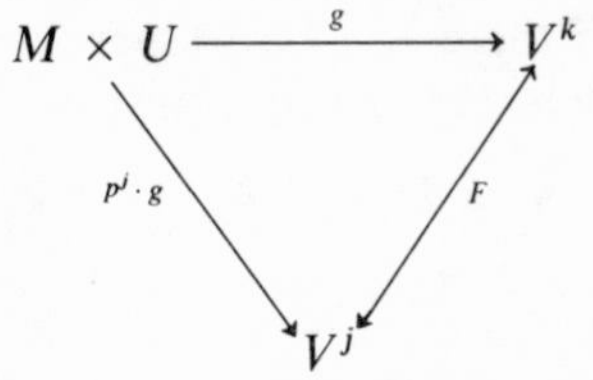

is commutative and $p^{k-j}g(M \times U)$ is nonempty and not a unit set, the system is cohesive [i.e., $g(M \times U)$ is not a Cartesian set], where $V^j = V_1 \times \cdots \times V_j$, p^j and p^{k-j} are projections from V^k into V^j and V^k into $V_{j+1} \times \cdots \times V_k$, respectively.

PROOF. Let $q = k - j$. Since $p^q \cdot g(M \times U)$ has more than two points, let v^q and $\hat{v}^q$ be two distinct points in $p^q \cdot g(M \times U)$ such that $v^q = p^q \cdot g(m, u)$ and $\hat{v}^q = p^q \cdot g(\hat{m}, \hat{u})$. Suppose the system is noncohesive. Then since $(p^j \cdot g(m, u), p^q \cdot g(\hat{m}, \hat{u})) \in \hat{g}(M \times U)$, there exists $(m', u') \in M \times U$ such that $g(m', u') = (p^j g(m, u), p^q \cdot g(\hat{m}, \hat{u}))$. Since the diagram is commutative, it follows that

$$F \cdot p^j \cdot g(m, u) = g(m', u') \neq g(m, u) = F \cdot p^j \cdot g(m, u)$$

This is a contradiction. Q.E.D.

To derive necessary and sufficient conditions, more structure has to be added to the specification of the system. In particular, this can be achieved by introducing linear structure in the value object V. This, however, has to be done so that V can be recognized as a multicomponent set, which is necessary for the cohesiveness and algebraic controllability. There are two ways that this can be done.

(i) Let V be linear and $\bar{p} = (p_1, \ldots, p_k)$ a family of projection operators, $p_i: V \to V_i$ such that V_i is a linear subspace and V is a direct sum, $V = V_1 \oplus \cdots \oplus V_k$. Component sets of the direct sum V are now considered as the components of the value set V. Notice that starting from a given linear value set V, the component sets and the number of those sets depend upon the family of projection operators $\bar{p}$ which, in general, is not unique for a given space, since there are many different ways to decompose a linear space into subspaces. *Cohesiveness in V has to be defined, therefore, in reference to a given family of projections.*

(ii) Let $V = V_1 \times \cdots \times V_k = V^k$ and each component set V_i is a linear space. The set V is then a linear space too.

We can now give necessary and sufficient conditions for cohesiveness.

Theorem 2.2. Let V be a linear space on a field $\mathscr{A}$ and $g(M \times U)$ a nonzero linear subspace of V. There exists a projection family for which the system is cohesive if and only if there exists a proper linear subspace $V_1 \subset V$ and a function $F: V_1 \to V$ such that the diagram

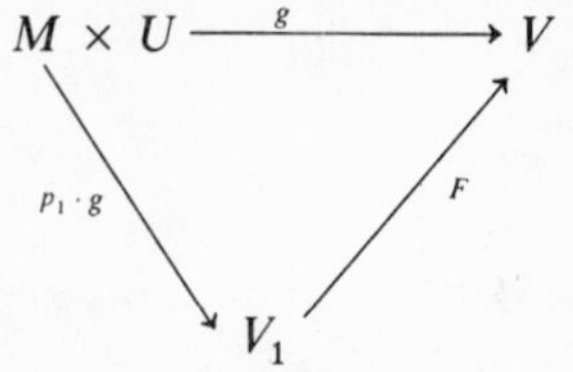

is commutative, where p_1 is the projection from V into V_1.

PROOF. Let us consider the *if* part first. If $(I - p_1) \cdot g(M \times U)$ has a nonzero element, it has more than two elements. Hence, the cohesiveness of the system follows from Theorem 2.1. Suppose $(I - p_1) \cdot g(M \times U) = \{o\}$, that is, $g(M \times U) \subset V_1$. Let $\hat{V}_1 = g(M \times U)$. Let V_2 be a linear subspace such that $V = \hat{V}_1 \oplus V_2$. Then another linear subspace $\hat{V}_1'$ can be constructed by the same method used in Theorem 1.2, Chapter II, such that $V = \hat{V}_1' \oplus V_2$ and $\hat{V}_1 \not\subset \hat{V}_1'$. Consequently, there exists $\hat{v} \neq o$ such that $\hat{v} \in \hat{V}_1$ and $\hat{v} \notin \hat{V}_1'$. Let the projection family of $\{\hat{V}_1', V_2\}$ be $\{p_1, p_2\}$. Then $p_2\hat{v} \neq o$. Suppose the system is noncohesive with respect to $\{p_1, p_2\}$. Then, since for any $\alpha \in \mathscr{A}$ and $\alpha' \in \mathscr{A}$, $p_1 \alpha\hat{v} \in \hat{V}_1'$, and $p_2 \alpha'\hat{v} \in V_2$, we have

$$\alpha p_1 \hat{v} + \alpha' p_2 \hat{v} = \alpha\hat{v} + (\alpha' - \alpha) p_2 \hat{v} \in \hat{V}_1$$

Since $(\alpha' - \alpha) p_2 \hat{v} \in V_2$, we have a contradiction with respect to $V = \hat{V}_1 \oplus V_2$.

Let us consider the *only if* part. The relation $V \supset \hat{g}(M \times U) \supset g(M \times U)$ always holds. If the system is cohesive, there exists a nonzero element $\hat{v} \in V$ such that $\hat{v} \notin g(M \times U)$. Let $V_2 = \{\alpha\hat{v} : \alpha \in \mathscr{A}\}$. Then, since $g(M \times U)$ is a linear subspace and hence since $V_2 \cap g(M \times U) = \{o\}$, V_2 is a proper subspace of V. Then there exists another subspace V_1 such that $V = V_1 \oplus V_2$ and $g(M \times U) \subset V_1$. Let $F: V_1 \to V$ such that $F(v_1) = v_1$. If $v = g(m, u)$, then $v \in V_1$. Hence, $p_1 \cdot g(m, u) = g(m, u)$; that is, $F \cdot p_1 \cdot g(m, u) = g(m, u)$.

Q.E.D.

Theorem 2.3. Let V^k be a finite-dimensional linear space on a field $\mathscr{A}$, where each V_i is generated by $\hat{v}_i \neq o$; that is, $V_i = \{\alpha\hat{v}_i : \alpha \in \mathscr{A}\}$. Let $g(M \times U)$ be a linear subspace of V^k. The system then is cohesive if and only if there exists a proper linear subspace $V^j \subset V^k$ and a function $F: V^j \to V^k$ such that the diagram

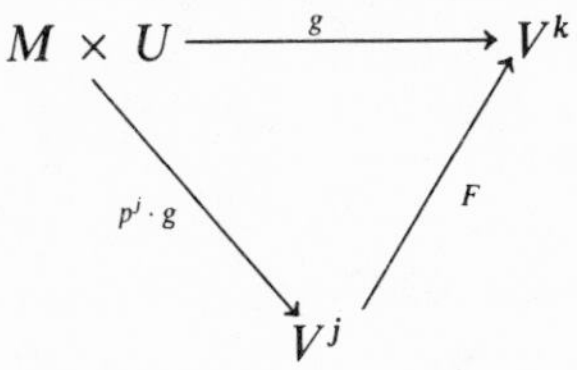

is commutative and $p^{k-j} \cdot g(M \times U)$ has a nonzero element.

PROOF. The proof of the *if* part comes from Theorem 2.1. Let us prove the *only if* part. Let $V' = g(M \times U)$ and $V_i' = p_i g(M \times U)$. Since V' is cohesive, V' is a proper subset of $V_1' \oplus \cdots \oplus V_k'$. Then there exists one component set V_i' such that $V_i' \not\subset V'$, or otherwise $V_1' \oplus \cdots \oplus V_k' \subset V'$. Hence, suppose $V_k' \not\subset V'$. Then there exists one element v_k' such that $0 \neq v_k' \in V_k'$ and $v_k' \notin V'$. Since V_k is generated by $\hat{v}_k$, we can assume $v_k' = \hat{v}_k$ without losing generality. Let $\lambda \subset V^{k-1} \times V_k$ such that

$$(v^{k-1}, v_k) \in \lambda \leftrightarrow (\exists (m, u))(v^{k-1} = p^{k-1} \cdot g(m, u) \quad \text{and} \quad v_k = p_k g(m, u)).$$

We shall show that λ is a function, $\lambda: \mathscr{D}(\lambda) \to V_k$. Since V' is a proper linear subspace, there exists a linear function $f: V \to \mathscr{A}$ such that $f(\hat{v}_k) = 1$ and $f(v') = 0$ if $v' \in V'$. Suppose $(v^{k-1}, v_k) \in \lambda$ and $(v^{k-1}, v_k') \in \lambda$, where $v_k = \alpha \hat{v}_k$ and $v_k' = \alpha' \cdot \hat{v}_k$. Then $f(v^{k-1} + v_k) = 0 = f(v^{k-1} + v_k')$. Consequently, $\alpha f(\hat{v}_k) = \alpha' f(\hat{v}_k)$, or $\alpha = \alpha'$. Hence $v_k = v_k'$; that is, λ is a function. Let $F: V^{k-1} \to V^k$ such that $F(v^{k-1}) = v^{k-1} + \lambda(v^{k-1})$. Then

$$\begin{aligned} F \cdot p^{k-1} \cdot g(m, u) &= p^{k-1} \cdot g(m, u) + \lambda(p^{k-1} \cdot g(m, u)) \\ &= p^{k-1} \cdot g(m, u) + p_k g(m, u) = g(m, u) \end{aligned}$$ Q.E.D.

Notice that no direct requirement on the linearity of g is imposed. Of course, the easiest way to satisfy the conditions for both Theorems 2.2 and 2.3 is when M and U are linear spaces and g is a linear function.

Since by Proposition 1.3 cohesiveness precludes controllability, we have the following immediate corollaries.

Corollary 2.1. The set $\hat{g}(M \times U)$ cannot be completely controllable if the conditions from Theorem 2.1 are satisfied.

Corollary 2.2. Suppose $g(M \times U)$ is a linear subspace of a linear space V. There exists a family of projections such that $\hat{g}(M \times U)$ cannot be completely controllable if the conditions from Theorem 2.2 are satisfied.

Corollary 2.3. Suppose $g(M \times U)$ is a linear subspace of a finite-dimensional linear space V^k. $\hat{g}(M \times U)$ cannot be completely controllable if the conditions from Theorem 2.3 are satisfied.

Corollary 2.4. Suppose $g(M \times \{o\})$ is a linear subspace of a linear space V where $U = \{o\}$. The set V cannot be completely controllable if and only if the conditions from Theorem 2.2 are satisfied.

PROOF. The proof of the *if* part comes from Corollary 2.2. Let us consider the *only if* part. Since V is not controllable, $g(M \times \{o\})$ is a proper subspace of V. Let $V_1 = g(M \times \{o\})$. Then there exists another subspace V_2 such that $V = V_1 \oplus V_2$. Let $F: V_1 \to V$ such that $F(v_1) = v_1$. Then we have the desired result. Q.E.D.

To appreciate the conceptual importance of the corollaries, it should be noticed that the function $F: V^j \to V^k$ satisfies the relation:

$$p^{k-j}F((v_1, \ldots, v_j)) = (v_{j+1}, \ldots, v_k) \,\&\, (\exists(m, u))(p^j g(m, u) = (V_1, \ldots, V_j))$$
$$\leftrightarrow (\exists(m, u))[g(m, u) = (v_1, \ldots, v_j, v_{j+1}, \ldots, v_k)] \qquad (7.1)$$

i.e., $p^{k-j} \cdot F$ is the cohesiveness function. g is a function with a multicomponent range, and the existence of F means that there exists a relationship between these components; namely, if j components are fixed, the remaining $(k - j)$ are determined by F. Statement (7.1) suggests then the following test for controllability. One starts from the set of equations

$$\begin{aligned} v_1 &= g_1(m, u) \\ &\vdots \\ v_k &= g_k(m, u) \end{aligned} \qquad (7.2)$$

where $g_i(m, u) = p_i \cdot g(m, u)$. If it is possible to derive from (7.2) a set of equations of the form

$$\begin{aligned} v_1 &= F_1(v^j) \\ &\vdots \\ v_k &= F_k(v^j) \end{aligned}$$

where $v^j \in V^j$ and $j < k$, the set $\hat{g}(M \times U)$ cannot be controllable. In other words, a sufficient condition for noncontrollability is that it is possible to eliminate the inputs m and u from the systems equations (expressed in terms of g) and to derive a relationship solely between the components of the value sets.

In connection with this procedure, the following remarks should be noted:

(i) Only the proof of the existence of the functions $F_1, \ldots, F_k$ is needed in order to prove noncontrollability, the actual form of these functions being irrelevant. *In this sense, the mapping F plays a similar role in the controllability theory as the Lyapunov functions in the stability theory.*

(ii) Starting from a given system, S, the character of the function F depends upon G; e.g., when S is a differential equation system, F can be the relationship between the values of the solutions at a (or any) given time, e.g., $F(y^r(\hat{t})) = y^k(\hat{t})$, or it might be the relationship between the entire functions, e.g., $y^k = F(y^r)$ where y^r and y^k correspond to v^r and v^k.

Theorem 2.2 clearly indicates the role of linearity in the controllability theory. In essence, it makes sufficient conditions both necessary and sufficient. For linear systems, several other results can be derived, however. For example, under certain conditions it is sufficient to test controllability only with respect to a single element of U.

Theorem 2.4. Let M be monoid, V^k and U Abelian groups, and g is such that $g(m, u) = g^1(m) + g^2(u)$, where g^2 is a homomorphism; i.e., $g^2(u + u') = g^2(u) + g^2(u')$.

Let $V' = \bigcap\{g(M, \{u\}) : u \in U\}$. If there exists $u' \in U$ such that $\hat{V}'[M \times \{u'\}]$ is completely controllable, then $\hat{V}'[M \times U]$ is completely controllable.

PROOF. Let $\hat{v}$ be an arbitrary element of $\hat{V}'$. Let $\hat{v}' = \hat{v} + g^2(u')$. We shall show $\hat{v}' \in \hat{V}'$. It follows from the definition of $\hat{V}'$ that $\hat{v}_i' - g_i^2(u') = \hat{v}_i \in \bigcap_u g_i(M, u)$, where $\hat{v}_i' = p_i\hat{v}'$, $g_i^2(u') = p_i \cdot g^2(u')$, $\hat{v}_i = p_i \cdot \hat{v}$, and $g_i(m, u) = p_i \cdot g(m, u)$. Hence,

$$\hat{v}_i' \in \bigcap_u g_i(M, u) + g_i^2(u') = \bigcap_u g_i(M, u + u') = \bigcap_u g_i(M, u)$$

which implies that $\hat{v}' \in \hat{V}'$. Hence, if $\hat{V}'[M \times \{u'\}]$ is completely controllable, $\hat{v}' = \hat{v} + g^2(u')$ can be expressed as $\hat{v}' = g(\hat{m}', u')$ for some $(\hat{m}', u') \in M \times \{u'\}$; that is, $\hat{v} = g^1(\hat{m}') + g^2(o)$. Consequently, $\hat{V}'[M \times \{o\}]$ is completely controllable. Similarly, we can show that if $\hat{V}'[M \times \{o\}]$ is completely controllable, $\hat{v}'[M \times \{u\}]$ is completely controllable for any $u \in U$. Therefore, if $\hat{V}'[M \times \{u'\}]$ is completely controllable for some $u' \in U$, $\hat{V}'[M \times U]$ is completely controllable. Q.E.D.

Although Theorem 2.4 is proven for Abelian groups rather than for linear algebras, it certainly holds for a linear algebra also since such an algebra is an Abelian group.

3. CONTROLLABILITY OF TIME SYSTEMS

(a) State-Space Controllability

In this section, we shall be concerned with a special kind of controllability suitable for dynamical systems. Specifically, the following assumptions will be made:

(i) S is a dynamical system with a canonical representation $(\bar{\phi}, \bar{\lambda})$ and the associated state space C.

(ii) $U = C$ and $V = C$.

(iii) The evaluation map g is defined in terms of the state-transition family in the manner indicated in definitions of specific notions given below.

We shall be primarily concerned in this section with the following concepts.

Definition 3.1. A dynamical system is (state-space) completely controllable if and only if

$$(\forall c)(\forall \hat{c})(\exists x^t)[\hat{c} = \phi_{ot}(c, x^t)]$$

Definition 3.2. A dynamical system is (state) controllable from the state c_0 if and only if

$$(\forall \hat{c})(\exists x^t)[\hat{c} = \phi_{ot}(c_0, x^t)]$$

For simplicity, when c_0 is the zero element of the linear space, the system will be referred to as (state) controllable.

Definition 3.3. A dynamical system is (state) controllable to the state c_0 or zero-state controllable if and only if

$$(\forall c)(\exists x^t)[c_0 = \phi_{ot}(c, x^t)]$$

The importance of the above state-space controllability notions for dynamical systems is due to the canonical representation; namely, according to such a representation, dynamic behavior of a system can be fully explored in terms of its state-transition family. The output of the system is determined by the output function, which is static and has no effect on dynamic evolution of the system. It should also be noted that state-space controllability is the controllability-type notion, which was historically introduced first and as a matter of fact the only one explored in detail in classical systems theory and for the class of linear systems [8].

The concepts given in Definitions 3.1–3.3 are obviously rather close, but they are not equivalent even when the system is linear. To make these concepts equivalent, additional conditions have to be assumed as shown in the following.

Proposition 3.1. Let S be a linear time-invariant dynamical system such that there exists $\hat{t}$ such that

$$(\forall c)(\forall x^t)(\exists \hat{x}^{\hat{t}})(c = \phi_{ot}(0, x^t) \to c = \phi_{o\hat{t}}(o, \hat{x}^{\hat{t}})) \tag{7.3}$$

$$(\forall c)(\forall x^t)(\exists \hat{x}^{\hat{t}})(o = \phi_{ot}(c, x^t) \to o = \phi_{o\hat{t}}(c, \hat{x}^{\hat{t}})) \tag{7.4}$$

$$(\forall t)(\forall c')(\exists c)(c' = \phi_{ot}(c, 0)) \tag{7.5}$$

Then the three definitions of state space controllability are equivalent where c_0 is o of the linear space C.

PROOF

(i) Definition 3.2 → Definition 3.3: Let c be arbitrary. Let $c' = \phi_{10\hat{t}}(c)$. Since the system is state controllable, $c' = \phi_{20t'}(\hat{x}^{t'})$ for some $\hat{x}^{t'}$. However, condition (7.3) implies that $c' = \phi_{20\hat{t}}(x^{\hat{t}})$ for some $x^{\hat{t}}$. Hence,

$$o = \phi_{10\hat{t}}(c) + \phi_{20\hat{t}}(-(x^{\hat{t}})) = \phi_{o\hat{t}}(c, -x^{\hat{t}})$$

Hence, Definition 3.3 is satisfied.

(ii) Definition 3.3 → Definition 3.2: Let c' be arbitrary. It follows from condition (7.5) that $c' = \phi_{10\hat{t}}(c)$ for some $c \in C$. Since the system is controllable to the state o, $0 = \phi_{ot'}(c, \hat{x}^{t'})$ for some $\hat{x}^{t'}$. Then $o = \phi_{o\hat{t}}(c, x^{\hat{t}})$ for some $x^{\hat{t}}$, which comes from condition (7.4). Since

$$o = \phi_{10\hat{t}}(c) + \phi_{20\hat{t}}(x^{\hat{t}}) = c' + \phi_{20\hat{t}}(x^{\hat{t}})$$

$c' = \phi_{o\hat{t}}(o, -(x^{\hat{t}}))$.

(iii) Definitions 3.2 and 3.3 → Definition 3.1: Let c and c' be arbitrary. Since the system is state controllable and simultaneously controllable to the state o, $0 = \phi_{ot}(c, x^t)$ and $c' = \phi_{ot'}(0, \hat{x}^{t'})$ hold for some x^t and $\hat{x}^{t'}$. Since the system is time invariant, $c' = \phi_{tt''}(0, F^t(\hat{x}^{t'}))$ where $t'' = t' + t$. Notice that it follows from the properties of a dynamical system that $0 = \phi_{ot}(c, x^t) = \phi_{ot''}(c, x^t \cdot 0)$ and $c' = \phi_{tt''}(0, F^t(\hat{x}^{t'})) = \phi_{ot''}(0, 0 \cdot F^t(\hat{x}^{t'}))$. (Refer to Chapter IV.) Hence, $c' = \phi_{ot''}(c, x^t \cdot F^t(\hat{x}^t))$.

(iv) Definition 3.1 → Definition 3.2: This is obvious. Q.E.D.

The desired result comes from the combination of (i), (ii), (iii), and (iv).

As it will be shown later, conditions (7.3), (7.4), and (7.5) follow from some basic properties of linear systems.

Apparently, conditions (7.3), (7.4), and (7.5) are not unique for making the three definitions equivalent. However, it can be proven that those conditions are satisfied by usual linear time-invariant differential equation systems, while condition (7.5) may not be satisfied by a linear time-invariant difference equation. This fact shows one point of difference with respect to the system's behavior between a differential equation system and a difference equation system.

The state space of a dynamical system most often has several components, and it is meaningful to talk about algebraic-type state-space controllability, i.e., about the interdependence between the state components implied by the system.

Let $M = \bigcup_{t \in T} X^t$ and $\bar{\phi}: C \times M \to C$ such that

$$\bar{\phi}(c, x^t) = \phi_{0t}(c, x^t)$$

Then Definitions 3.1–3.3 become

(i) complete controllability: $(\forall c)(\forall \hat{c})(\exists m)(\hat{c} = \bar{\phi}(c, m))$
(ii) controllability from c_0: $(\forall \hat{c})(\exists m)(\hat{c} = \bar{\phi}(c_0, m))$
(iii) controllability to c_0: $(\forall c)(\exists m)(c_0 = \bar{\phi}(c, m))$

Apparently, $\bar{\phi}$ takes the role of g in Section 1.

The basic conditions for cohesiveness and algebraic controllability, as given in Theorems 2.1 and 2.2, then apply directly here; specifically, Corollaries 2.1 and 2.2 take on the following form.

Corollary 3.1. Let S be a dynamical system $(\bar{\rho}, \bar{\phi})$ with a multicomponent state space $C^k = C_1 \times \cdots \times C_k$. If there exist a positive integer $j < k$ and a function $F: C^j \to C^k$ such that for every $t \in T$ the diagram

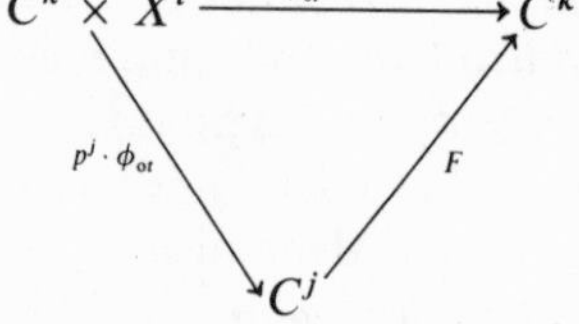

is commutative, the system cannot be completely controllable in C^k where $C^j = C_1 \times \cdots \times C_j$ and $p^{k-j}\bar{\phi}(C \times M)$ has more than two elements.

Corollary 3.2. Let S be a linear time-invariant dynamical system. Then the system is not state controllable in C if and only if there exists a proper linear subspace C_1 in the state space C and a function $F: C_1 \to C$ such that for

every $t \in T$ the diagram

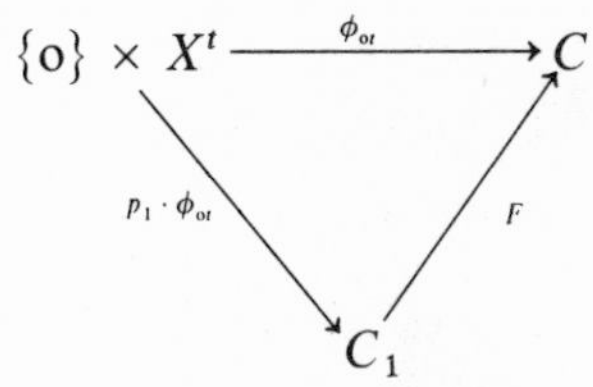

is commutative.

PROOF. We shall show that $\bar{\phi}(\{o\} \times M)$ is a linear subspace of C. Let $c = \bar{\phi}(o, x^t)$ and $\hat{c} = \bar{\phi}(o, \hat{x}^{t'})$ be arbitrary. Then, since ϕ_{ot} is linear, $\alpha c = \bar{\phi}(o, \alpha x^t)$. Suppose $t' \geq t$. Let $t' = t + \tau$. Then, since ϕ_{ot} is linear and time invariant, $c = \phi_{ot}(o, x^t) = \phi_{ot'}(o, o \cdot F^\tau(x^t))$, and hence

$$c + c' = \phi_{ot'}(o, o \cdot F^\tau(x^t) + \hat{x}^{t'}) = \bar{\phi}(o, o \cdot F^\tau(x^t) + \hat{x}^{t'})$$

Consequently, $\bar{\phi}(\{o\} \times M)$ is a linear subspace. Then Corollary 2.4 is directly applicable to the present case except that in Corollary 3.2, $(I - p_1)\bar{\phi}(\{o\} \times M)$ may not have nonzero element. However, if $(I - p_1)\bar{\phi}(\{o\} \times M)$ does not have nonzero element, $\bar{\phi}(\{o\} \times M)$ is apparently included in a proper subspace C_1, and hence the system is not controllable. Q.E.D.

Corollary 3.2 can be reformulated using the linear space properties of C. When the state space C is a linear algebra on a field $\mathscr{A}$, a linear functional f^c on C is a linear function such that $f^c : C \to \mathscr{A}$. Let F^c be the class of all linear functionals on C. An addition and scalar multiplication by the elements from $\mathscr{A}$ can be defined on F^c as

$$(f^c + \hat{f}^c)(c) = f^c(c) + \hat{f}^c(c), \qquad \text{and} \quad (\alpha f^c)(c) = \alpha \cdot f^c(c)$$

If $f^c(c) = o$ for every $c \in C$, f^c will be denoted by o. Furthermore, since

$$f^c(c + c') = f^c(c) + f^c(c'), \qquad \text{and} \qquad f^c(\alpha c) = \alpha f^c(c)$$

hold, F^c itself can be considered a linear algebra on the field $\mathscr{A}$, and hence F^c will be referred to as the algebraic conjugate space of C.

The following lemma is basic for the later arguments [9].

Lemma 3.1. Let C_o be a proper subspace of the linear space C, and suppose $c_1 \in C \setminus C_o$. Then there exists an element $f^c \in F^c$ such that $f^c(c) = o$ if $c \in C_o$ and $f^c(c_1) = 1$, where 1 is the unit element of $\mathscr{A}$.

We can now give another necessary and sufficient condition for the state-space controllability of linear dynamical systems.

Theorem 3.1. Let S be a linear time-invariant dynamical system and F^c the algebraic conjugate space of C. The system is then state controllable if and only if for any $f^c \in F^c$

$$(\forall x^t)[f^c(\phi_{20t}(x^t)) = \mathrm{o}] \rightarrow [f^c = \mathrm{o}] \tag{7.6}$$

PROOF. As shown in Corollary 3.2,

$$C_o = \{c : c = \phi_{20t}(x^t) \,\&\, x^t \in X^t \,\&\, t \in T\}$$

is a linear subspace of the state space C. Then it follows from Lemma 3.1 that the system is not state controllable iff C_o is a proper subspace of C, i.e., iff

$$(\exists f^c)(f^c \neq \mathrm{o} \,\&\, (\forall x^t)(f^c(\phi_{20t}(x^t)) = \mathrm{o}))$$

Notice that it follows from the definition that $f^c \neq \mathrm{o}$ implies $(\exists c_1 \in C)$ $(f^c(c_1) \neq \mathrm{o})$. Q.E.D.

Application of Theorem 3.1 yields directly the well-known conditions for the controllability of linear differential equation systems. Namely, let the state-transition family be defined by a linear vector–matrix differential equation

$$d\mathbf{z}/dt = F\mathbf{z} + G\mathbf{x}, \qquad \mathbf{y} = H\mathbf{z} \tag{7.7}$$

where F, G, and H are time invariant. Condition (7.6) from the preceding theorem takes on, then, the following form.

The system is not state controllable if and only if

$$(\exists f^c \neq \mathrm{o})(\forall x^t)(f^c(\phi_{20t}(x^t)) = \mathrm{o})$$

Using explicit expression for the state transition obtained from Eq. (7.7), the condition becomes

$$(\exists \xi \in C)\left(\xi^T \cdot \int_0^t e^{F(t-\sigma)} G x(\sigma)\, d\sigma = \mathrm{o}\right)$$

for any $t \in T$ and x and $\xi \neq \mathrm{o}$; that is,

$$(\forall t \in T)(\xi^T e^{Ft} G = \mathrm{o})$$

This, finally, holds if and only if

$$\operatorname{rank}[G, FG, \ldots, F^{n-1}G] < n$$

where n is the dimension of F, which is a well-known condition for the controllability of differential equation systems [10].

Regarding zero-state controllability, Theorem 3.1 can be generalized to the systems that are time variant.

Theorem 3.2. Let S be a linear dynamical system. S is zero-state controllable if and only if for any $f^c \in F^c$

$$(\forall c_0)[c_0 \in C_0 \to f^c(c_o) = o] \to [f^c = o]$$

where

$$C_0 = \{c : (\exists x^t)(\phi_{ot}(c, x^t) = o)\}$$

PROOF. We shall show that C_0 is a linear subspace. Suppose $\phi_{ot}(c, x^t) = o$. Then $\phi_{ot}(\alpha c, \alpha x^t) = o$. Suppose $\phi_{ot}(c, x^t) = o$ and $\phi_{ot'}(c', \hat{x}^{t'}) = o$. Suppose $t' > t$. Then, since

$$\phi_{ot'}(c, x^t \cdot o) = \phi_{tt'}(\phi_{ot}(c, x^t), o) = \phi_{tt'}(o, o) = o$$

$$\phi_{ot'}(c, x^t \cdot o) + \phi_{ot'}(c', \hat{x}^{t'}) = \phi_{ot'}(c + c', x^t \cdot o + \hat{x}^{t'}) = o$$

Hence, C_0 is a linear subspace of C. The desired result follows by the same argument as used in Theorem 3.1. Q.E.D.

(b) Output-Functional Controllability

In this section, we shall be concerned again with a dynamical system but with the case when the system is evaluated by its entire output function rather than state at certain time. Specifically, the following assumptions will be made to particularize the concept of reproducibility:

(i) S is a dynamical system.
(ii) U is the initial state object, $U = C_0$, while V is the output object, $V = Y$, and $M = X$.
(iii) g is the initial-response function, $\rho_0 : C_0 \times X \to Y$.

We shall illustrate the application of the general results from Section 2 only for the concept of cohesiveness.

Definition 3.4. Let S be a dynamical, multivariable system, $Y = Y_1 \times \cdots \times Y_k = Y^k$. S is functionally output cohesive if and only if $\rho_0(C_0 \times X)$ is not a Cartesian set.

Conditions for cohesiveness follow now directly from the general conditions given in Section 2.

Corollary 3.3. Let S be a multivariable dynamical system, $Y = Y_1 \times \cdots \times Y_k = Y^k$. If there exists a positive integer $j < k$ and a function $F : Y^j \to Y^k$

such that the diagram

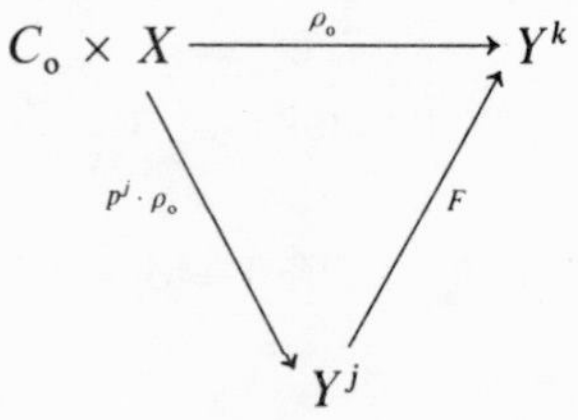

is commutative and $p^{k-j}\rho_0(C_0 \times X)$ has more than two elements, the system is functionally output cohesive.

Corollary 3.4. Let S be a linear time-invariant dynamical system. Then the system is functionally output cohesive if and only if there exists a proper linear subspace $Y' \subset Y$ and a function $F: Y' \to Y$ such that the diagram

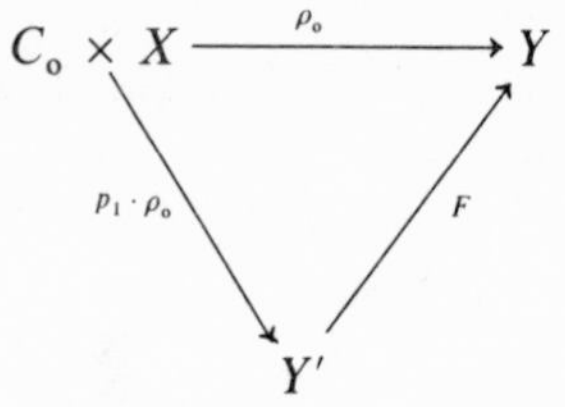

is commutative and $(I - p_1)\rho_0(C \times X)$ has a nonzero element, where p_1 is the projection operator $p_1 : Y \to Y'$.

Further application of the general results lead to the well-known results on functional reproducibility for linear differential equation systems [11].

4. OVERVIEW OF SOME BASIC LINEAR TIME SYSTEMS PROPERTIES RELATED TO CONTROLLABILITY

The existence of various auxiliary functions and various systems properties depends upon the conditions that the systems-response family has to satisfy. For example, the realizability depends upon (P1)–(P4) from Chapter III, Section 1, while the state-space controllability for linear systems depends upon the conditions given in Theorem 3.1. It is of interest to provide an overview of all the conditions considered so far that involve the response-function family and, in particular, for the class of linear time-invariant systems where they can be expressed in terms of the state- and input-response families separately. Table 7.1 is then a list of such conditions either considered so far or of potential interest when studying some additional properties. It is also indicated in the table what system property depends upon given conditions.

TABLE 7.1

BASIC PROPERTIES OF THE SYSTEMS-RESPONSE FUNCTION

Number	Form	Name
P1	$(\forall c)(\forall t)(\exists c')(\rho_{1t}(c') = \rho_{10}(c) \mid T_t)$	existence of ϕ_{10t}
P2	$(\forall t)(\forall x^t)(\exists c')(\rho_{1t}(c') = \rho_{20}(x^t \cdot \mathrm{o}) \mid T_t)$	existence of ϕ_{20t}
P3	$(\forall c')(\exists t)(\exists x^t)(\rho_{1t}(c') = \rho_{20}(x^t \cdot \mathrm{o}) \mid T_t)$	controllability
P4	$(\exists \hat{t})(\forall c')(\exists t)(\exists x^t)(t \leq \hat{t} \,\&\, \rho_{1t}(c') = \rho_{20}(x^t \cdot \mathrm{o}) \mid T_t)$	strong controllability
P5	$(\forall c)(\exists t)(\exists x^t)(\rho_{10}(c) \mid T_t = \rho_{20}(x^t \cdot \mathrm{o}) \mid T_t)$	zero-state controllability
P6	$(\exists \hat{t})(\forall c)(\exists t)(\exists x^t)(t \leq \hat{t} \,\&\, \rho_{10}(c) \mid T_t = \rho_{20}(x^t \cdot \mathrm{o}) \mid T_t)$	strong zero-state controllability
P7	$(\forall t)(\forall c')(\exists c)(\rho_{1t}(c') = \rho_{10}(c) \mid T_t)$	free-response completeness
P8′	$(\exists \hat{t})(\forall c)(\forall x^t)(\exists \hat{x}^{\hat{t}})(\rho_{1t}(c) = \rho_{20}(x^t \cdot \mathrm{o}) \mid T_t \rightarrow \rho_{1\hat{t}}(c) = \rho_{20}(\hat{x}^{\hat{t}} \cdot \mathrm{o}) \mid T_{\hat{t}})$	finite connectedness from zero
P8″	$(\exists \hat{t})(\forall c)(\forall x^t)(\exists \hat{x}^{\hat{t}})(\rho_{\mathrm{o}}(c, x^t \cdot \mathrm{o}) \mid T_t = \mathrm{o} \rightarrow \rho_{\mathrm{o}}(c, \hat{x}^{\hat{t}} \cdot \mathrm{o}) \mid T_{\hat{t}} = \mathrm{o})$	finite connectedness to zero
P9	$(\forall c')(\forall t)(\exists c)(\exists x^t)(\rho_{1t}(c') = (\rho_{10}(c) + \rho_{20}(x^t \cdot \mathrm{o})) \mid T_t)$	state-response consistency
P10	$(\exists \hat{t})(\forall c')(\forall c)(\exists t)(\exists x^t)(t \leq \hat{t} \,\&\, \rho_{1t}(c') = (\rho_{10}(c) + \rho_{20}(x^t \cdot \mathrm{o})) \mid T_t)$	strong complete controllability
P11	$(\exists \hat{t})(\forall c)(\forall c')(\rho_{10}(c) \mid T^{\hat{t}} = \rho_{10}(c') \mid T^{\hat{t}} \rightarrow c = c')$	observability
P12	$(\exists \hat{t})(\forall c)(\forall t \geq \hat{t})(\forall x_t)(\rho_{\mathrm{o}}(c, \mathrm{o} \cdot x_t) \mid T^{\hat{t}} = \mathrm{o} \rightarrow \rho_{\mathrm{o}}(c, \mathrm{o} \cdot x_t) \mid \overline{T}^t = \mathrm{o})$	past-determinacy
P13	$(\forall c')(\forall c)(\exists t)(\exists x^t)(\rho_{1t}(c') = (\rho_{10}(c) + \rho_{20}(x^t \cdot \mathrm{o})) \mid T_t)$	complete controllability
P14	$(\forall c)(\rho_{10}(c) = \mathrm{o} \rightarrow c = \mathrm{o})$	reducibility
P15′	$(\exists \hat{t})(\forall c)(\rho_{10}(c) \mid T^{\hat{t}} = \mathrm{o} \rightarrow \rho_{10}(c) = \mathrm{o})$	left analyticity from $\hat{t}$
P15″	$(\forall c)(\forall t)(\rho_{10}(c) \mid T_t = \mathrm{o} \rightarrow \rho_{10}(c) = \mathrm{o})$	right analyticity
P16	$\{y : (\exists c)(c \in C \,\&\, y = \rho_{10}(c))\}$ is finite dimensional	finite-dimensional state space
P17′	$(\forall x)(\forall \hat{x})(\forall t)(x \mid \overline{T}^t = \hat{x} \mid \overline{T}^t \rightarrow \rho_{20}(x) \mid \overline{T}^t = \rho_{20}(\hat{x}) \mid \overline{T}^t)$	nonanticipation
P17″	$(\forall x)(\forall \hat{x})(\forall t)(x \mid T^t = \hat{x} \mid T^t \rightarrow \rho_{20}(x) \mid \overline{T}^t = \rho_{20}(\hat{x}) \mid \overline{T}^t)$	strong nonanticipation

Most of the above properties are intimately related to the state transition. Indeed, the meaning of any properties listed in Table 7.1 can be grasped more easily when they are expressed in terms of state transitions than in terms of a response family. In particular, if ρ_{1t} has the inverse, or if ρ_{1t} is reduced, the above properties can be described on the state space as given in Table 7.2.

TABLE 7.2

BASIC PROPERTIES OF THE STATE-TRANSITION FAMILY

Number	Form
P1	$(\forall c)(\forall t)(\exists c')(c' = \phi_{10t}(c))$
P2	$(\forall t)(\forall x^t)(\exists c')(c' = \phi_{20t}(x^t))$
P3	$(\forall c')(\exists t)(\exists x^t)(c' = \phi_{20t}(x^t))$
P4	$(\exists \hat{t})(\forall c')(\exists t)(\exists x^t)(t \le \hat{t}\ \&\ c' = \phi_{20t}(x^t))$
P5	$(\forall c)(\exists t)(\exists x^t)(\phi_{10t}(c) = \phi_{20t}(x^t))$
P6	$(\exists \hat{t})(\forall c)(\exists t)(\exists x^t)(t \le \hat{t}\ \&\ \phi_{10t}(c) = \phi_{20t}(x^t))$
P7	$(\forall t)(\forall c')(\exists c)(c' = \phi_{10t}(c))$
P8′	$(\exists \hat{t})(\forall c)(\forall x^t)(\exists \hat{x}^{\hat{t}})(c = \phi_{20t}(x^t) \rightarrow c = \phi_{20\hat{t}}(\hat{x}^{\hat{t}}))$
P8″	$(\exists \hat{t})(\forall c)(\forall x^t)(\exists \hat{x}^{\hat{t}})(\mathrm{o} = \phi_{\mathrm{o}t}(c, x^t) \rightarrow \mathrm{o} = \phi_{\mathrm{o}\hat{t}}(c, \hat{x}^{\hat{t}}))$
P9	$(\forall c')(\forall t)(\exists c)(\exists x^t)(c' = \phi_{10t}(c) + \phi_{20t}(x^t))$
P10	$(\exists \hat{t})(\forall c')(\forall c)(\exists t)(\exists x^t)(t \le \hat{t}\ \&\ (c' = \phi_{10t}(c) + \phi_{20t}(x^t)))$
P11	No corresponding relation
P12	No corresponding relation
P13	$(\forall c')(\forall c)(\exists t)(\exists x^t)(c' = \phi_{10t}(c) + \phi_{20t}(x^t))$
P14	No corresponding relation
P15′	No corresponding relation
P15″	No corresponding relation
P16	Finite-dimensional state space
P17′	Nonanticipation
P17″	Strong nonanticipation

Obviously, the properties listed in Table 7.2 are interdependent, and while some are basic, some others are induced by the primary ones.

In order to clarify the interrelationships among them, we shall first show that P14–P16 induce past-determinacy, finite connectedness, and free-response completeness. The relations among past-determinacy, strong nonanticipation, and left analyticity are given in Chapter V.

Proposition 4.1. Suppose a time-invariant linear dynamical system $(\bar{\rho}, \bar{\phi})$ is reduced. Then if its state space C is finite dimensional (P16), it satisfies finite connectedness from zero (P8′).

PROOF. Let

$$C_{\mathrm{o}} = \{c : \rho_{1t}(c) = \rho_{20}(x^t \cdot \mathrm{o}) \mid T_t\ \&\ c \in C\ \&\ x^t \in X^t\ \&\ t \in T\}$$

C_{o} is the set of states reachable from the origin. We shall show that C_{o} is a linear subspace of C. Let $c \in C_{\mathrm{o}}$ and $\alpha \in \mathscr{A}$ be arbitrary, where $\rho_{1t}(c) = \rho_{20}(x^t \cdot \mathrm{o}) \mid T_t$. Since $\bar{\rho}$ is linear, $\rho_{1t}(\alpha c) = \rho_{20}(\alpha(x^t) \cdot \mathrm{o}) \mid T_t$ holds; that is, $c \in C_{\mathrm{o}} \rightarrow \alpha c \in C_{\mathrm{o}}$. Let $\hat{c} \in C_{\mathrm{o}}$ be arbitrary, where $\rho_{1\hat{t}}(\hat{c}) = \rho_{20}(\hat{x}^{\hat{t}} \cdot \mathrm{o}) \mid T_{\hat{t}}$. Let

$t' = t + \hat{t}$. Then we have that

$$\begin{aligned} \rho_{1t'}(c) &= F^{\hat{t}}(\rho_{20}(x^t \cdot o) \mid T_t) \\ &= F^{\hat{t}}(\rho_{20}(x^t \cdot o)) \mid T_{t'} \\ &= \rho_{2\hat{t}}(F^{\hat{t}}(x^t) \cdot o) \mid T_{t'} \\ &= \rho_{20}(o \cdot F^{\hat{t}}(x^t) \cdot o) \mid T_{t'} \qquad \text{(input-response consistency)} \end{aligned}$$

Similarly, we have that

$$\rho_{1t'}(\hat{c}) = \rho_{20}(o \cdot F^t(\hat{x}^{\hat{t}}) \cdot o) \mid T_{t'}$$

Consequently, we have that

$$\rho_{1t'}(c + \hat{c}) = \rho_{20}((o \cdot F^{\hat{t}}(x^t) + o \cdot F^t(\hat{x}^{\hat{t}})) \cdot o) \mid T_{t'}$$

that is, $c + \hat{c} \in C_o$. Since C is finite dimensional, $C_o \subset C$ is also finite dimensional. Let $\{c_1, \ldots, c_n\}$ be a basis of C_o which is assumed n-dimensional. where for each i, $\rho_{1t_i}(c_i) = \rho_{20}(x_i^{t_i} \cdot o) \mid T_{t_i}$. Let $\hat{t} = \max_i \{t_1, \ldots, t_n\}$. Then for each i,

$$\rho_{1\hat{t}}(c_i) = \rho_{20}(o \cdot F^{\hat{t} - t_i}(x_i^{t_i}) \cdot o) \mid T_{\hat{t}}$$

holds. Let $c \in C_o$ be arbitrary. Then $c = \sum_{i=1}^n \lambda_i c_i$ for some $\lambda_i \in \mathscr{A}$ $(i = 1, \ldots, n)$. Consequently,

$$\rho_{1\hat{t}}(c) = \rho_{20}\left(\sum_{i=1}^{n} \lambda_i(o \cdot F^{\hat{t} - t_i}(x_i^{t_i}) \cdot o)\right) \mid T_{\hat{t}}$$

holds, that is,

$$c = \phi_{o\hat{t}}\left(o, \sum_{i=1}^{n} \lambda_i(o \cdot F^{\hat{t} - t_i}(x_i^{t_i}) \cdot o)\right) \qquad \text{Q.E.D.}$$

Proposition 4.2. Suppose a time-invariant linear dynamical system $(\bar{\rho}, \bar{\phi})$ is reduced. Then, if its state space C is finite dimensional (P16), it satisfies finite connectedness to zero (P8″).

PROOF. Let

$$C_o = \{c : \rho_o(c, x^t \cdot o) \mid T_t = o \,\&\, c \in C \,\&\, x^t \in X^t \,\&\, t \in T\}$$

We shall show that C_o is a linear subspace of C. Let $\alpha \in \mathscr{A}$ and $c \in C_o$ be arbitrary, where $\rho_o(c, x^t \cdot o) \mid T_t = o$. Then, $\rho_o(\alpha c, \alpha(x^t) \cdot o) \mid T_t = o$; that is, $c \in C_o \rightarrow \alpha c \in C_o$. Let $\hat{c} \in C_o$ be arbitrary where $\rho_o(\hat{c}, \hat{x}^{\hat{t}} \cdot o) \mid T_{\hat{t}} = o$. Let $t' = t + \hat{t}$. Then, since

$$[\rho_o(c, x^t \cdot o) \mid T_t] \mid T_{t'} = \rho_o(c, x^t \cdot o) \mid T_{t'} = o$$

and $\rho_0(\hat{c}, \hat{x}^{\hat{t}} \cdot o) \mid T_{t'} = o$ holds, we have

$$\rho_0(c + \hat{c}, x^t \cdot o + \hat{x}^{\hat{t}} \cdot o) \mid T_{t'} = o$$

that is, $c + \hat{c} \in C_0$. Let $\{c_1, \ldots, c_n\}$ be a basis of C_0 as before, where $\rho_0(c_i, x_i^{t_i} \cdot o) \mid T_{t_i} = o$ $(i = 1, \ldots, n)$. Let $\hat{t} = \max_i \{t_1, \ldots, t_n\}$. Then $\rho_0(c_i, x_i^{t_i} \cdot o) \mid T_{\hat{t}} = o$. Consequently, for $c = \sum_{i=1}^n \lambda_i c_i \in C_0$,

$$\rho_0\left(c, \sum_{i=1}^n \lambda_i(x_i^{t_i} \cdot o)\right) \mid T_{\hat{t}} = o$$

holds, that is,

$$\phi_{0\hat{t}}\left(c, \sum_{i=1}^n \lambda_i(x_i^{t_i} \cdot o)\right) = o \qquad \text{Q.E.D.}$$

Finite connectedness (P8) does not necessarily hold only for linear systems. A finite automaton is a typical example of nonlinear systems that possess this property. However, we can show an example that a linear system with an infinite-dimensional space does not satisfy this property in general. In this sense, finite connectedness can be considered as a basic property of "finite" state systems.

Proposition 4.3. Suppose a time-invariant linear dynamical system $(\bar{\rho}, \bar{\phi})$ is reduced. If its state space is finite dimensional and right analytic, then it satisfies free response completeness.

PROOF. Let $\phi_{10t}(c) = o$. Then

$$\begin{aligned} \phi_{10t}(c) = o &\rightarrow \rho_{1t}(\phi_{10t}(c)) = o \\ &\rightarrow \rho_{10}(c) \mid T_t = o \\ &\rightarrow \rho_{10}(c) = o \\ &\rightarrow c = o \end{aligned}$$

That is, $\phi_{10t}: C \rightarrow C$ is a one-to-one mapping. Furthermore, since C is finite dimensional, ϕ_{10t} should be a one-to-one correspondence, that is, ϕ_{10t}^{-1} exists.
Q.E.D.

The above result is well known for a constant coefficient linear ordinary differential equation system. However, traditional proof is based only on a mathematical argument, that is, convergence of an operator sequence rather than system theoretic arguments.

Propositions 4.1–4.3 result in the following corollaries with the help of Proposition 3.1.

Corollary 4.1. Suppose a time-invariant finite-dimensional linear dynamical system $(\bar{\rho}, \bar{\phi})$ is reduced and satisfies right analyticity. Then the three kinds of state-space controllability introduced in Section 3 are mutually equivalent.

Corollary 4.2. The three kinds of state-space controllability are mutually equivalent for a constant-coefficient linear ordinary differential equation system, i.e.,

$$d\mathbf{c}/dt = F\mathbf{c} + G\mathbf{x}$$

where F and G are matrices, while $\mathbf{x}$ and $\mathbf{c}$ are vectors.

Interrelationships among the properties in Tables 7.1 and 7.2 are summarized in Fig. 4.1. The interpretation of the relationships pictorially represented

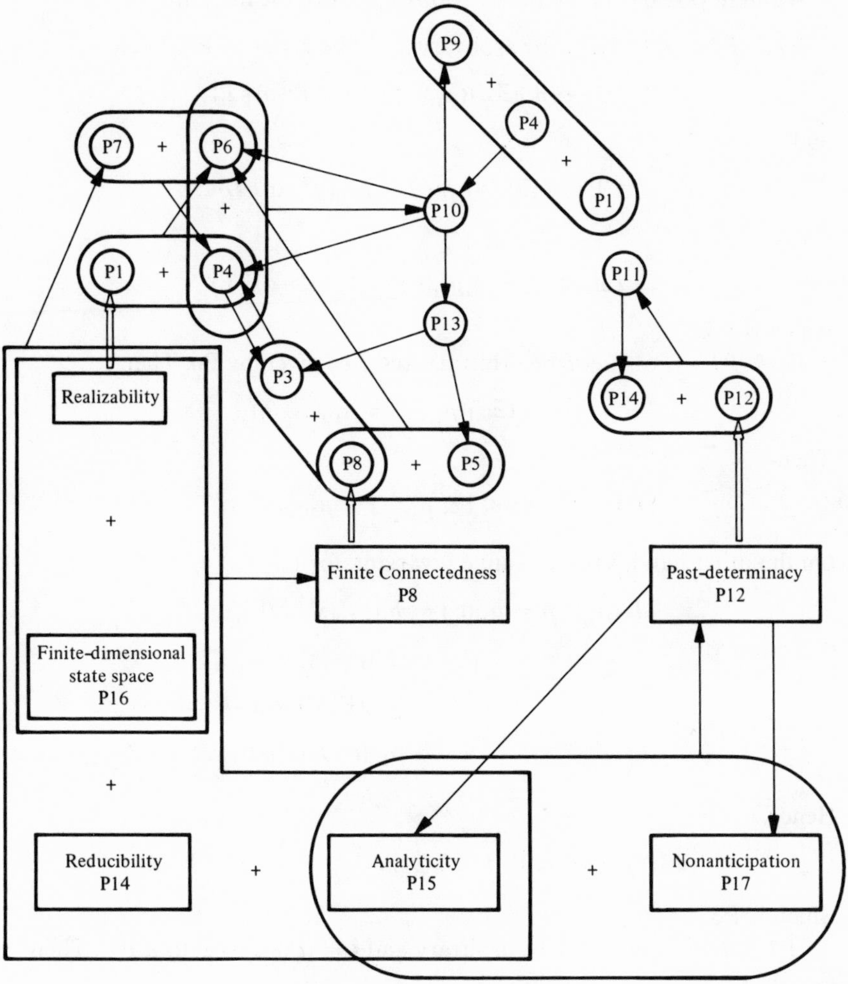

FIG. 4.1

in the diagram is straightforward: An arrow denotes which conditions can be derived by starting from given assumptions. For instance, the diagram shows that if a linear response family $\{\rho_t\}$ satisfies P3 and P8, then P4 is satisfied also; or when P9, P4, and P1 are given, P10 can be derived, etc.

The essential independent assumptions on which the diagram is based are the following: (1) realizability, (2) analyticity, (3) finite-dimensional state space, (4) past-determinacy. They can be, therefore, considered the most basic properties for linear dynamical systems. Any other property listed can be derived from some of those basic assumptions.

We shall prove some of the relationships from the diagram.

P3 + P8 → P4: Let c' be arbitrary and $\hat{t}$ be given by P8. Then

$$\text{P3} \rightarrow (\exists t)(\exists x^t)(\rho_{1t}(c') = \rho_{20}(x^t \cdot \text{o}) \mid T_t)$$

and

$$\text{P8} \rightarrow (\exists x'^{\hat{t}})(\rho_{1\hat{t}}(c') = \rho_{20}(x'^{\hat{t}} \cdot \text{o}) \mid T_{\hat{t}})$$

Therefore,

$$(\exists \hat{t})(\forall c')(\exists x'^{\hat{t}})(\rho_{1\hat{t}}(c') = \rho_{20}(x'^{\hat{t}} \cdot \text{o}) \mid T_{\hat{t}})$$

which is P4.

P4 + P1 → P6: Let c be arbitrary and $\hat{t}$ be given by P4. Then

$$\text{P1} \rightarrow (\exists c')(\rho_{1\hat{t}}(c') = \rho_{10}(c) \mid T_{\hat{t}})$$

Then

$$\text{P4} \rightarrow (\exists x^t)(t \leq \hat{t} \;\&\; \rho_{1t}(c') = \rho_{20}(x^t \cdot \text{o}) \mid T_t)$$

On the other hand, since $\bar{\rho}$ is time invariant, for $\tau = \hat{t} - t$,

$$\begin{aligned} F^\tau(\rho_{1t}(c')) = \rho_{1\hat{t}}(c') &= F^\tau(\rho_{20}(x^t \cdot \text{o}) \mid T_t) \\ &= (F^\tau(\rho_{20}(x^t \cdot \text{o}))) \mid T_{\hat{t}} \\ &= (\rho_{2\tau}(F^\tau(x^t) \cdot \text{o})) \mid T_{\hat{t}} \\ &= \rho_{20}(\text{o} \cdot F^\tau(x^t) \cdot \text{o}) \mid T_{\hat{t}} \end{aligned}$$

Hence,

$$(\exists \hat{t})(\forall c)(\exists t)(\exists x^t)(t \leq \hat{t} \;\&\; \rho_{10}(c) \mid T_t = \rho_{20}(x^t \cdot \text{o}) \mid T_t)$$

which is P6.

P7 + P6 → P4: Let c' be arbitrary and $\hat{t} = (\hat{t}$ of P6$) + (\hat{t}$ of P7$)$. Then

$$\text{P7} \rightarrow (\exists c)(\rho_{1\hat{t}}(c') = \rho_{10}(c) \mid T_{\hat{t}})$$

Then, since $\bar{\rho}$ is time invariant,

$$\text{P6} \to (\exists x^{\hat{t}})(\rho_{10}(c) \mid T_{\hat{t}} = \rho_{20}(x^{\hat{t}} \cdot \text{o}) \mid T_{\hat{t}})$$

Hence,

$$(\exists \hat{t})(\forall c')(\exists t)(\exists x^t)(t \leq \hat{t} \;\&\; \rho_{1t}(c') = \rho_{20}(x^t \cdot \text{o}) \mid T_t)$$

which is P4.

P6 + P4 → P10: Let $\hat{t}$ = ($\hat{t}$ of P4) + ($\hat{t}$ of P6). Let c and c' be arbitrary. Let t' = ($\hat{t}$ of P6). Then

$$\text{P6} \to (\exists x^{t'})(\text{o} = \rho_{10}(c) \mid T_{t'} + \rho_{20}(x^{t'} \cdot \text{o}) \mid T_{t'})$$

Let t'' = ($\hat{t}$ of P4). Since $\bar{\rho}$ is time invariant,

$$\text{P4} \to (\exists x^{t''})(\rho_{1t''}(c') = \rho_{20}(x^{t''} \cdot \text{o}) \mid T_{t''})$$

Let $x^{\hat{t}} = x^{t'} \cdot F^{t'}(x^{t''})$. Then

$$\begin{aligned}(\rho_{10}(c) + \rho_{20}(x^{\hat{t}} \cdot \text{o})) \mid T_{\hat{t}} &= ((\rho_{10}(c) + \rho_{20}(x^{t'} \cdot \text{o})) \mid T_{t'}) \mid T_{\hat{t}} \\ &\quad + \rho_{20}(\text{o} \cdot F^{t'}(x^{t''}) \cdot \text{o}) \mid T_{\hat{t}} \\ &= \rho_{2t'}(F^{t'}(x^{t''}) \cdot \text{o}) \mid T_{\hat{t}} \\ &= \rho_{20}(x^{t''} \cdot \text{o}) \mid T_{t''} = \rho_{1t''}(c')\end{aligned}$$

Therefore,

$$(\exists \hat{t})(\forall c')(\forall c)(\exists t)(\exists x^t)(t \leq \hat{t} \;\&\; \rho_{1t}(c') = (\rho_{10}(c) + \rho_{20}(x^t \cdot \text{o})) \mid T_t)$$

which is P10.

P9 + P1 + P4 → P10: Let $\hat{t}$ be the $\hat{t}$ given by P4. Let c' and c be arbitrary. Then

$$\begin{aligned}\text{P9} &\to (\exists \hat{c})(\exists \hat{x}^{\hat{t}})(\rho_{1\hat{t}}(c') = (\rho_{10}(\hat{c}) + \rho_{20}(\hat{x}^{\hat{t}} \cdot \text{o})) \mid T_{\hat{t}}) \\ \text{P1} &\to (\exists \tilde{c})(\rho_{1\hat{t}}(\tilde{c}) = \rho_{10}(\hat{c} - c) \mid T_{\hat{t}}) \\ \text{P4} &\to (\exists \tilde{x}^{\hat{t}})(\rho_{1\hat{t}}(\tilde{c}) = \rho_{20}(\tilde{x}^{\hat{t}} \cdot \text{o}) \mid T_{\hat{t}})\end{aligned}$$

because $\bar{\rho}$ is time invariant. Hence,

$$\begin{aligned}\rho_{1\hat{t}}(c') &= (\rho_{10}(\hat{c}) + \rho_{20}(\hat{x}^{\hat{t}} \cdot \text{o})) \mid T_{\hat{t}} \\ &= (\rho_{10}(c) + \rho_{10}(\hat{c} - c) + \rho_{20}(\hat{x}^{\hat{t}} \cdot \text{o})) \mid T_{\hat{t}} \\ &= \rho_{10}(c) \mid T_{\hat{t}} + \rho_{1\hat{t}}(\tilde{c}) + \rho_{20}(\hat{x}^{\hat{t}} \cdot \text{o}) \mid T_{\hat{t}} \\ &= \rho_{10}(c) \mid T_{\hat{t}} + \rho_{20}(\tilde{x}^{\hat{t}} \cdot \text{o}) \mid T_{\hat{t}} + \rho_{20}(\hat{x}^{\hat{t}} \cdot \text{o}) \mid T_{\hat{t}} \\ &= \rho_{10}(c) \mid T_{\hat{t}} + \rho_{20}((\tilde{x}^{\hat{t}} + \hat{x}^{\hat{t}}) \cdot \text{o}) \mid T_{\hat{t}}\end{aligned}$$

Therefore,

$$(\exists\hat{t})(\forall c')(\forall c)(\exists t)(\exists x^t)(t \leq \hat{t} \;\&\; \rho_{1t}(c') = (\rho_{10}(c) + \rho_{20}(x^t \cdot \mathrm{o})) \mid T_t)$$

which is P10.

P5 + P8 → P6: Let $\hat{t}$ be equal to $\hat{t}$ of P8. Let c be arbitrary. Then

$$\mathrm{P5} \rightarrow (\exists t)(\exists x^t)(\rho_{10}(c) \mid T_t = \rho_{20}(x^t \cdot \mathrm{o}) \mid T_t)$$

That is, $\rho_{1t}(\mathrm{o}) = \rho_{\mathrm{o}}(c, -(x^t \cdot \mathrm{o})) \mid T_t$. Hence

$$\mathrm{P8} \rightarrow (\exists \hat{x}^{\hat{t}})(\rho_{1\hat{t}}(\mathrm{o}) = \rho_{\mathrm{o}}(c, \hat{x}^{\hat{t}} \cdot \mathrm{o}) \mid T_{\hat{t}})$$

That is,

$$\rho_{10}(c) \mid T_{\hat{t}} = \rho_{20}(-(\hat{x}^{\hat{t}} \cdot \mathrm{o})) \mid T_{\hat{t}}$$

Hence

$$(\exists\hat{t})(\forall c)(\exists t)(\exists x^t)(t \leq \hat{t} \;\&\; \rho_{10}(c) \mid T_t = \rho_{20}(x^t \cdot \mathrm{o}) \mid T_t)$$

which is P6.

Chapter VIII

MINIMAL REALIZATION

Starting from the primary concept of a system defined on the input and output objects, we can introduce many different dynamical realizations and in reference to a variety of state spaces. It is of practical interest to find a state space that is the "smallest" in an appropriate sense and to have a procedure to construct such a space starting from some initial information about the system. This is the problem to be considered in this chapter.

Several different concepts of minimal realizations are used in specialized systems theories, and we have first introduced the related set of concepts in our general framework in a precise manner. The relationships between these concepts is then investigated and the conditions given for the existence of such minimal realizations. The uniqueness of some minimal realizations is also established. It is also shown how minimal realizations can be characterized by systems properties, in particular controllability and reducibility.

1. CONCEPTS OF MINIMAL REALIZATIONS

A given time system S can have many different dynamic realizations, and two pairs $(\bar{\rho}, \bar{\phi})$ and $(\bar{\bar{\rho}}, \bar{\bar{\phi}})$ can be dynamical realizations of the same system. Among the class of dynamic realizations, it is of interest to find out which are equivalent in an appropriate sense, which have some special property, and in particular which are the "simplest" or the "smallest" in a given sense. This leads to the concern for a minimal realization.

Since the dynamic behavior of a system is described in terms of changes of the states, the minimality of a realization refers to the state space itself.

Actually, there are two distinct concepts of state-space minimality used in different branches of systems theory, namely:

(i) In automata theory, a realization is minimal if the state space is of the smallest cardinality [12].

(ii) In dynamical systems theory or control theory, a minimal realization has the smallest number of state variables [5].

Both of these concepts refer to the space and use similar terminology, although they are apparently quite distinct. In order to develop a general theory of minimal realizations (which will include the specialized cases considered earlier), we shall have to consider quite carefully first the concept of minimality itself.

A precise procedure to introduce a concept of minimal realization consists of three steps. First, a class $\mathscr{S}_D$ of systems of interest is identified, e.g., stationary, linear, finite in a given sense, etc. Since a system is described as a rule by means of some auxiliary functions, the systems of interest are defined in terms of a class of auxiliary functions of a certain kind. Second, an equivalence relation is established within the class of interest. Third, an ordering is defined in the equivalence classes with respect to which the minimality of a realization is identified.

We shall consider three types of equivalences.

Definition 1.1. Let $\mathscr{S}_D$ be a class of dynamical realizations. Two realizations $(\bar{\rho}, \bar{\phi}), (\bar{\hat{\rho}}, \bar{\hat{\phi}}) \in \mathscr{S}_D$ are input–output equivalent if and only if

$$S_o^{\rho} = S_o^{\hat{\rho}}$$

that is,

$$(\forall c)(\forall x)(\exists \hat{c})[\rho_o(c, x) = \hat{\rho}_o(\hat{c}, x)] \;\&\; (\forall \hat{c})(\forall x)(\exists c)[\rho_o(c, x) = \hat{\rho}_o(\hat{c}, x)]$$

Definition 1.2. Let $\mathscr{S}_D$ be a class of dynamical realizations. Two realizations $(\bar{\rho}, \bar{\phi}), (\bar{\hat{\rho}}, \bar{\hat{\phi}}) \in \mathscr{S}_D$ are response equivalent if and only if

$$(\forall c)(\exists \hat{c})(\forall x)[\rho_o(c, x) = \hat{\rho}_o(\hat{c}, x)]$$

and

$$(\forall \hat{c})(\exists c)(\forall x)[\rho_o(c, x) = \hat{\rho}_o(\hat{c}, x)]$$

The third type of equivalence refers to the input part of the response function. For a linear system, the response function can be decomposed into state-response part (for zero input) and the input-response part (for zero initial state). Generalizing this idea for any c_o, the function $\rho_o(c_o, -): X \to Y$ will be referred to as the c_o input response. To identify c_o in question, we shall refer to it as the *reference point* since in interpretation it has, as a rule, a special role. We have then the following definition.

Definition 1.3. Let $\mathcal{S}_D$ be a class of dynamical realizations, $(\bar{\rho}, \bar{\phi})$, $(\bar{\hat{\rho}}, \bar{\hat{\phi}}) \in \mathcal{S}_D$, and c_o, $\hat{c}_o$ the reference points of $(\bar{\rho}, \bar{\phi})$ and $(\bar{\hat{\rho}}, \bar{\hat{\phi}})$, respectively. $(\bar{\rho}, \bar{\phi})$ and $(\bar{\hat{\rho}}, \bar{\hat{\phi}})$ are input-response equivalent if and only if

$$\rho_o(c_o, -) = \hat{\rho}_o(\hat{c}_o, -)$$

that is,

$$(\forall x)[\rho_o(c_o, x) = \hat{\rho}_o(\hat{c}_o, x)]$$

The first type of equivalence is apparently most general, but two other types are of special interest in practice. For example, for the class of linear systems, if one is primarily interested in the steady state, i.e., when the transient due to the initial condition is practically negligible, the system can be described solely in terms of input response. Actually, the analysis that uses Laplace and other transform techniques is based on this premise.

We shall consider two types of orderings.

Definition 1.4. Let $\mathcal{S}_D{}^E$ be an equivalence class of dynamical systems, $(\bar{\rho}, \bar{\phi})$, $(\bar{\hat{\rho}}, \bar{\hat{\phi}}) \in \mathcal{S}_D{}^E$, and C, $\hat{C}$ the corresponding state spaces with the cardinality $K(C)$ and $K(\hat{C})$, respectively. A relation $\leq$ in $\mathcal{S}_D{}^E$ is then defined by

$$(\bar{\rho}, \bar{\phi}) \leq (\bar{\hat{\rho}}, \bar{\hat{\phi}}) \leftrightarrow K(C) \leq K(\hat{C})$$

Definition 1.5. Let $\mathcal{S}_D{}^E$ be an equivalent class of dynamical systems, $(\bar{\rho}, \bar{\phi})$, $(\bar{\hat{\rho}}, \bar{\hat{\phi}}) \in \mathcal{S}_D{}^E$, C, and $\hat{C}$ the corresponding state spaces. A relation $\precsim$ in $\mathcal{S}_D{}^E$ is then defined by

$$(\bar{\rho}; \bar{\phi}) \precsim (\bar{\hat{\rho}}, \bar{\hat{\phi}}) \leftrightarrow \text{there exists an epimorphism } h: \hat{C} \to C$$

It can be shown readily that $\precsim$ is a partial ordering where

$$(\bar{\rho}, \bar{\phi}) = (\bar{\hat{\rho}}, \bar{\hat{\phi}}) \leftrightarrow (\bar{\rho}, \bar{\phi}) \precsim (\bar{\hat{\rho}}, \bar{\hat{\phi}}) \,\&\, (\bar{\hat{\rho}}, \bar{\hat{\phi}}) \precsim (\bar{\rho}, \bar{\phi})$$

$$(\bar{\rho}, \bar{\phi}) \prec (\bar{\hat{\rho}}, \bar{\hat{\phi}}) \leftrightarrow (\bar{\rho}, \bar{\phi}) \precsim (\bar{\hat{\rho}}, \bar{\hat{\phi}}) \,\&\, \neg [(\bar{\hat{\rho}}, \bar{\hat{\phi}}) \precsim (\bar{\rho}, \bar{\phi})]$$

If a state space has no algebraic structure, the ordering relation $\leq$ can be considered as a special case of the ordering relation $\precsim$ since in such a case any surjection can be considered as an epimorphism. The ordering $\leq$ is of the type used in the automata theory, while the ordering $\precsim$ is used in the control theory where the minimal realization problem is considered with respect to Euclidean spaces E^n. The ordering in the latter case is determined according to dimensions of the spaces involved, and since for any two positive integers m and n such that $m < n$, there is a linear surjection from E^n to E^m but not conversely, the ordering relation is of the type given in Definition 1.5.

Using different types of equivalences with different types of ordering, one has the following six minimality concepts.

Definition 1.6. $(\bar{\rho}, \bar{\phi})$ is a minimal state-space realization if and only if for any dynamical system $(\bar{\hat{\rho}}, \bar{\hat{\phi}})$ in the same class $\mathscr{S}_D$, the following holds:

$$S_o^{\rho} = S_o^{\hat{\rho}} \rightarrow K(C) \leq K(\hat{C})$$

Definition 1.7. $(\bar{\rho}, \bar{\phi})$ is a minimal dimension realization if and only if for any dynamical system $(\bar{\hat{\rho}}, \bar{\hat{\phi}})$ in the same class $\mathscr{S}_D$, the following holds:

$$S_o^{\rho} = S_o^{\hat{\rho}} \rightarrow [(\bar{\rho}, \bar{\phi}) \succeq (\bar{\hat{\rho}}, \bar{\hat{\phi}}) \rightarrow (\bar{\rho}, \bar{\phi}) \preceq (\bar{\hat{\rho}}, \bar{\hat{\phi}})]$$

Definition 1.8. $(\bar{\rho}, \bar{\phi})$ is a minimal state-space response realization if and only if for any dynamical system $(\bar{\hat{\rho}}, \bar{\hat{\phi}})$ in the same class $\mathscr{S}_D$, the following holds:

$$(\forall c)(\exists \hat{c})(\forall x)(\rho_o(c, x) = \hat{\rho}_o(\hat{c}, x)) \,\&\, (\forall \hat{c})(\exists c)(\forall x)(\rho_o(c, x) = \hat{\rho}_o(\hat{c}, x)) \rightarrow K(C) \leq K(\hat{C})$$

Definition 1.9. $(\bar{\rho}, \bar{\phi})$ is a minimal dimension response realization if and only if it satisfies the following: For any dynamical system $(\bar{\hat{\rho}}, \bar{\hat{\phi}})$ in the same class $\mathscr{S}_D$, if there exists an epimorphism $h: C \rightarrow \hat{C}$ such that h satisfies the following commutative diagram:

$$\begin{array}{ccc} C \times X_t & \xrightarrow{\rho_t} & Y_t \\ {\scriptstyle h}\downarrow \quad \downarrow{\scriptstyle I} & & \downarrow{\scriptstyle I} \\ \hat{C} \times X_t & \xrightarrow{\hat{\rho}_t} & Y_t \end{array}$$

then there exists an epimorphism $\hat{h}: \hat{C} \rightarrow C$ such that the following commutative diagram holds:

$$\begin{array}{ccc} \hat{C} \times X_t & \xrightarrow{\hat{\rho}_t} & Y_t \\ {\scriptstyle \hat{h}}\downarrow \quad \downarrow{\scriptstyle I} & & \downarrow{\scriptstyle I} \\ C \times X_t & \xrightarrow{\rho_t} & Y_t \end{array}$$

Definition 1.10. $(\bar{\rho}, \bar{\phi})$ is a minimal state-space input-response realization if and only if for any dynamical system $(\bar{\hat{\rho}}, \bar{\hat{\phi}})$ in the same class $\mathscr{S}_D$, the following holds:

$$(\forall x)(\rho_o(c_o, x) = \hat{\rho}_o(\hat{c}_o, x)) \rightarrow K(C) \leq K(\hat{C})$$

where c_o and $\hat{c}_o$ are the reference points of C and $\hat{C}$, respectively.

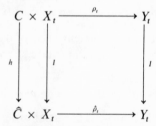

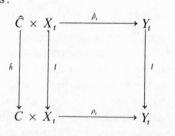

Definition 1.11. $(\bar{\rho}, \bar{\phi})$ is a minimal dimension input-response realization if and only if for any dynamical system $(\bar{\hat{\rho}}, \bar{\hat{\phi}})$ in the same class $\mathscr{S}_D$, the following holds:

$$(\forall x)(\rho_o(c_o, x) = \hat{\rho}_o(\hat{c}_o, x)) \rightarrow ((\bar{\rho}, \bar{\phi}) \geq (\bar{\hat{\rho}}, \bar{\hat{\phi}}) \rightarrow (\bar{\rho}, \bar{\phi}) \leq (\bar{\hat{\rho}}, \bar{\hat{\phi}}))$$

Definition 1.8 is used in the automata theory, while in the linear control theory the concepts in Definitions 1.7 and 1.11 are used. The interrelationships between various minimality concepts will be clarified in the next section.

2. CHARACTERIZATION OF THE MINIMAL REALIZATION OF STATIONARY SYSTEMS

In this section, the characterization of various minimal realizations will be considered, using the concepts of controllability and reducibility.

Controllability and reducibility are defined in Definition 3.1, Chapter VII, and Definition 2.8, Chapter II, respectively, as follows.

Definition 2.1. Let $(\bar{\rho}, \bar{\phi})$ be a time-invariant dynamical system, where $c_o \in C$ is the reference point. Then $(\bar{\rho}, \bar{\phi})$ is controllable if and only if

$$(\forall c)(\exists x^t)(c \in C \rightarrow c = \phi_{ot}(c_o, x^t))$$

As mentioned in the preceding section, if the state space C is linear, the origin is taken as the reference point.

Definition 2.2. Let $(\bar{\rho}, \bar{\phi})$ be a time-invariant dynamical system. Then $(\bar{\rho}, \bar{\phi})$ is reduced if and only if for any c and c'

$$(\forall x)(\rho_o(c, x) = \rho_o(c', x)) \rightarrow c = c'$$

Various forms of "finiteness" of a state space play an important role in the minimal realization theory. We shall, therefore, introduce two types of finite systems.

Definition 2.3. Let $(\bar{\rho}, \bar{\phi})$ be a dynamical system. $(\bar{\rho}, \bar{\phi})$ is a finite dynamical system if and only if the state space of $(\bar{\rho}, \bar{\phi})$ is a finite set.

Definition 2.4. Let $(\bar{\rho}, \bar{\phi})$ be a dynamical system with the linear state space C on the field $\mathscr{A}$. $(\bar{\rho}, \bar{\phi})$ is finite dimensional if and only if C has a finite base, i.e., there exists a set of k elements $\{\hat{c}_1, \ldots, \hat{c}_k\}$ in C such that for any $c \in C$ there exists the unique set of elements of the field $\mathscr{A}$, $\alpha_1, \ldots, \alpha_k \in \mathscr{A}$,

such that

$$c = \alpha_1 \hat{c}_1 + \cdots + \alpha_k \hat{c}_k$$

On the basis of the above definitions, we can characterize now minimal realizations of a system.

Proposition 2.1. Let $(\bar{\rho}, \bar{\phi})$ be a time-invariant finite linear dynamical system. The system $(\bar{\rho}, \bar{\phi})$ is a minimal realization in the sense of Definition 1.6 if and only if $(\bar{\rho}, \bar{\phi})$ is reduced.

PROOF. We shall consider the *only if* part first. Suppose $(\bar{\rho}, \bar{\phi})$ is not reduced. Then the equivalence (congruence) relation $E \subset C \times C$ where

$$(c, c') \in E \leftrightarrow (\forall x)(\rho_o(c, x) = \rho_o(c', x))$$

is not trivial. Let a new function $\hat{\rho}_t : (C/E) \times X_t \to Y_t$ be such that

$$\hat{\rho}_t([c], x_t) = \rho_t(c, x_t)$$

Then $\bar{\hat{\rho}} = \{\hat{\rho}_t : t \in T\}$ is a time-invariant linear realizable input–output response. (Since C is a linear algebra, C/E is a linear algebra in the usual sense.) Consequently, we have a time-invariant finite linear dynamical system $(\bar{\hat{\rho}}, \bar{\hat{\phi}})$ such that $S_o{}^{\rho} = S_o{}^{\hat{\rho}}$. Since E is not trivial, we have $K(C/E) < K(C)$. (Notice that C is finite.) Consequently, $(\bar{\rho}, \bar{\phi})$ is not a minimal realization in the sense of Definition 1.6.

Next, we shall consider the *if* part. Suppose $(\bar{\hat{\rho}}, \bar{\hat{\phi}})$ is of the same type as $(\bar{\rho}, \bar{\phi})$. Suppose $(\bar{\rho}, \bar{\phi})$ is reduced. It follows from $S_o{}^{\rho} = S_o{}^{\hat{\rho}}$ that

$$(\forall c)(\exists \hat{c})(\rho_{10}(c) = \hat{\rho}_{10}(\hat{c})), \qquad \text{and} \qquad (\forall \hat{c})(\exists c)(\rho_{10}(c) = \hat{\rho}_{10}(\hat{c}))$$

because ρ_o and $\hat{\rho}_o$ are linear. Let a relation $\phi \subset \hat{C} \times C$ be such that

$$(\hat{c}, c) \in \phi \leftrightarrow \hat{\rho}_{10}(\hat{c}) = \rho_{10}(c)$$

If $(\hat{c}, c) \in \phi$ and $(\hat{c}, c') \in \phi$ hold, then we have that $\rho_{10}(c) = \hat{\rho}_{10}(\hat{c}) = \rho_{10}(c')$. Consequently, $c = c'$ because $(\bar{\rho}, \bar{\phi})$ is reduced. Furthermore, since

$$(\forall \hat{c})(\exists c)(\rho_{10}(c) = \hat{\rho}_{10}(\hat{c}))$$

holds, we have $\mathscr{D}(\phi) = \hat{C}$. Therefore, ϕ is a mapping, i.e., $\phi : \hat{C} \to C$. Moreover, since $(\forall c)(\exists \hat{c})(\rho_{10}(c) = \hat{\rho}_{10}(\hat{c}))$ holds, ϕ is a surjection. Consequently, we have $K(C) \le K(\hat{C})$. Q.E.D.

In general, if a state space is finite, the system is a minimal realization in the sense of Definition 1.6 only if it is reduced, which can be proven in the same way as Proposition 2.1. However, when a system is nonlinear, the converse is not true, which is shown by the following example.

Example 2.1. Let us consider two finite automata, S_1 and S_2. S_1 is given by

the time set: $T = \{0, 1, 2, \ldots\}$, the input alphabet: $A = \{a_1, a_2\}$

the state space: $C_1 = \{1, 2, 3\}$, the output alphabet: $B = \{0, 1\}$

The state transition $\phi_1 : C_1 \times A \to C_1$ and the output function $\lambda_1 : C_1 \times A \to B$ for S_1 are given by the following table:

	ϕ_1		λ_1	
C_1 \ A	a_1	a_2	a_1	a_2
1	2	1	1	0
2	1	2	0	1
3	2	2	1	1

S_2 is given by

the time set: $T = \{0, 1, 2, \ldots\}$, the input alphabet: $A = \{a_1, a_2\}$

the state space: $C_2 = \{1, 2\}$, the output alphabet: $B = \{0, 1\}$

The state transition $\phi_2 : C_2 \times A \to C_2$ and the output function $\lambda_2 : C_2 \times A \to B$ for S_2 are given by

$$\phi_2 = \phi_1 \mid C_2 \times A, \qquad \lambda_2 = \lambda_1 \mid C_2 \times A$$

Let us denote the systems-response functions of S_1 and S_2 by $\rho_t^{\,1} : C_1 \times X_t \to Y_t$ and $\rho_t^{\,2} : C_2 \times X_t \to Y_t$, where X_t and Y_t are A^{T_t} and B^{T_t}, respectively.

It is clear from the definition of S_1 and S_2 that

$$S_1 = S_2 \cup \{(x, \rho_o^{\,1}(3, x)) : x \in X\}$$

where $X = X_o$.

Notice that x takes one of the following two forms:†

$$x = a_1 \cdot x_1 \quad \text{or} \quad x = a_2 \cdot x_1 \qquad \text{where} \quad x_1 \in A^{T_1}$$

If $x = a_1 \cdot x_1$, then

$$\begin{aligned} \rho_o^{\,1}(3, x) &= \rho_o^{\,1}(3, a_1 \cdot x_1) \\ &= \lambda_1(3, a_1) \cdot \rho_1^{\,1}(\phi_1(3, a_1), x_1) \\ &= \lambda_1(1, a_1) \cdot \rho_1^{\,1}(\phi_1(1, a_1), x_1) \\ &= \rho_o^{\,1}(1, a_1 \cdot x_1) \end{aligned}$$

† In order to simplify the notation, we write $a_1 \cdot x_1$ instead of $(o, a_1) \cdot x_1$.

If $x = a_2 \cdot x_1$, then

$$\begin{aligned}\rho_o{}^1(3, x) &= \rho_o{}^1(3, a_2 \cdot x_1)\\ &= \lambda_1(3, a_2) \cdot \rho_1{}^1(\phi_1(3, a_2), x_1)\\ &= \lambda_1(2, a_2) \cdot \rho_1{}^1(\phi_1(2, a_2), x_1)\\ &= \rho_o{}^1(2, a_2 \cdot x_1)\end{aligned}$$

Therefore, we have $S_o^{\rho 1} = S_o^{\rho 2}$. It is easy to show that $\bar{\rho}^1$ is reduced. Consequently, this example shows that the reducibility does not imply the minimal realization in the sense of Definition 1.6 even if a system is time invariant and finite.

A characterization of the minimal realization in the sense of Definition 1.7 is given by the following proposition.

Proposition 2.2. Let $(\bar{\rho}, \bar{\phi})$ be a time-invariant finite-dimensional linear dynamical system. Then $(\bar{\rho}, \bar{\phi})$ is a minimal realization in the sense of Definition 1.7 if and only if $(\bar{\rho}, \bar{\phi})$ is reduced.

PROOF. We shall consider the *only if* part first. Suppose $(\bar{\rho}, \bar{\phi})$ is not reduced. Then the congruence relation $E \subset C \times C$, which is defined by

$$(c, c') \in E \leftrightarrow \rho_{10}(c) = \rho_{10}(c')$$

is not trivial. Let $\hat{\rho}_{1t}: C/E \to Y_t$ be such that

$$\hat{\rho}_{1t}([c]) = \rho_{1t}(c)$$

This definition is proper because $\rho_{1t}(c) = F^t(\rho_{10}(c))$. Then

$$\bar{\hat{\rho}} = \{(\hat{\rho}_{1t}, \rho_{2t}) : t \in T\}$$

is a realizable systems response of the same type as $\bar{\rho}$. Furthermore, $S_o{}^{\rho} = S_o{}^{\hat{\rho}}$ holds. The natural mapping $h: C \to C/E$, where $h(c) = [c]$, is an epimorphism but not isomorphism because E is not trivial. If $(\bar{\rho}, \bar{\phi})$ is a minimum realization, there exists an epimorphism $\hat{h}: C/E \to C$. Then we have the following contradiction:

$$\dim(C) = \dim(\mathscr{R}(\hat{h})) \leq \dim(C/E) < \dim(C)$$

where $\dim(C)$ is the dimension of C. Consequently, $(\bar{\rho}, \bar{\phi})$ is not a minimal realization in the sense of Definition 1.7. We shall consider the *if* part next. Suppose $(\bar{\rho}, \bar{\phi})$ is reduced. Let $(\bar{\hat{\rho}}, \bar{\hat{\phi}})$ be a dynamical system of the same type as $(\bar{\rho}, \bar{\phi})$ such that $S_o{}^{\rho} = S_o{}^{\hat{\rho}}$ holds. Let $h: C \to \hat{C}$ be an epimorphism. Let $E \subset C \times C$ be a congruence relation where

$$(c, c') \in E \leftrightarrow h(c) = h(c')$$

For each $t \in T$ let a linear function $\hat{\rho}'_{1t}: C/E \to Y_t$ be such that

$$\hat{\rho}'_{1t}([c]) = \hat{\rho}_{1t}(h(c))$$

Since h is an epimorphism, $\bar{\hat{\rho}}' = \{(\hat{\rho}'_{1t}, \hat{\rho}_{2t}): t \in T\}$ is a realizable linear-response family, and $S_o^{\hat{\rho}} = S_o^{\hat{\rho}'}$ holds.

Since $S_o^{\rho} = S_o^{\hat{\rho}} = S_o^{\hat{\rho}'}$ holds, we have the following relations:

$$(\forall c)(\exists [c'])(\rho_{10}(c) = \hat{\rho}'_{10}([c'])) \tag{8.1}$$

$$(\forall [c'])(\exists c)(\rho_{10}(c) = \hat{\rho}'_{10}([c'])) \tag{8.2}$$

Since $\bar{\rho}$ is reduced and since Eq. (8.2) implies that $\rho_{10}(C) \supset \hat{\rho}'_{10}(C/E)$, we have a linear function $\rho_{10}^{-1} \cdot \hat{\rho}'_{10}: C/E \to C$, where $\rho_{10}^{-1}: \rho_{10}(C) \to C$ such that $\rho_{10}^{-1}(\rho_{10}(c)) = c$ for every $c \in C$. Furthermore, Eq. (8.1) implies that $\rho_{10}^{-1} \cdot \hat{\rho}'_{10}$ is an epimorphism. Since C/E is isomorphic to $\hat{C}$, therefore there exists an epimorphism $\hat{h}: \hat{C} \to C$. Q.E.D.

Proposition 2.2 holds when dynamical systems are finite dimensional. This requirement seems too restrictive. However, as the following example shows, Proposition 2.2 does not hold, in general, for infinite-dimensional systems even if they are linear and time invariant.

Example 2.2. Let $C = E^{\infty}$ whose typical element is represented by $c = (c_1, c_2, \ldots)$. Let

$$C_1 = \{(o, c_2, c_3, \ldots) : (o, c_2, c_3, \ldots) \in C\}$$

Let a projection $P: C \to C_1$ be such that $P((c_1, c_2, \ldots)) = (o, c_2, c_3, \ldots)$. Let $\theta: C_1 \to C$ be such that $\theta((o, c_2, c_3, \ldots)) = (c_2, c_3, \ldots)$. Notice that P and θ are linear and that θ is an isomorphism. Let a time-invariant linear dynamical system $(\bar{\rho}, \bar{\phi})$ be defined by

the time set:	$T = \{0, 1, 2, \ldots\}$
the state space:	$C = E^{\infty} = \{(c_1, c_2, \ldots)\}$
the input alphabet:	A = an arbitrary linear algebra
the output alphabet:	$B = E^{\infty}$

The state transition $\phi: C \times A \to C$ and the output function $\lambda: C \times A \to B$ are given by

$$\phi(c, a) = P(c), \qquad \lambda(c, a) = P(c)$$

The initial systems response $\rho_o: C \times X \to Y$ is then given by: For every $t \in T$

$$\rho_o(c, x)(t) = P(c)$$

$(\bar{\rho}, \bar{\phi})$ is not reduced. Let $(\bar{\rho}', \bar{\phi}')$ be the reduced system of $(\bar{\rho}, \bar{\phi})$; that is, $C' = C/E$ and $\rho_t' : C' \times X_t \to Y_t$ are given by

$$(c, c') \in E \leftrightarrow \rho_{10}(c) = \rho_{10}(c')$$

and $\rho_t'([c], x_t) = \rho_t(c, x_t)$. Let $(\bar{\hat{\rho}}, \bar{\hat{\phi}})$ be an arbitrary time-invariant linear dynamical system which satisfies the relation $S_o{}^{\rho} = S_o{}^{\hat{\rho}}$. Then, since $S_o{}^{\rho'} = S_o{}^{\rho} = S_o{}^{\hat{\rho}}$ holds, we have an epimorphism $h' : \hat{C} \to C'$. (Refer to the last part of the proof of Proposition 2.2.) Furthermore, since C' is isomorphic to C_1, where the isomorphism is expressed by $h'' : C' \to C_1$, there is an epimorphism $\theta \cdot h'' \cdot h' : \hat{C} \to C$. Consequently, although $(\bar{\rho}, \bar{\phi})$ is not reduced, $(\bar{\rho}, \bar{\phi})$ is minimal in the sense of Definition 1.7.

Next, we shall consider Definition 1.8. Since the ordering used in the definition is defined by the cardinality, finite state spaces are of practical interest in this case again.

Proposition 2.3. Let $(\bar{\rho}, \bar{\phi})$ be a time-invariant finite dynamical system. Then $(\bar{\rho}, \bar{\phi})$ is a minimal realization in the sense of Definition 1.8 if and only if $\bar{\rho}$ is reduced.

PROOF. The *only if* part is clear. We shall consider the *if* part. Let $(\bar{\hat{\rho}}, \bar{\hat{\phi}})$ be a finite time-invariant dynamical system. Suppose the condition

$$(\forall c)(\exists \hat{c})(\forall x)(\rho_o(c, x) = \hat{\rho}_o(\hat{c}, x)) \,\&\, (\forall \hat{c})(\exists c)(\forall x)(\rho_o(c, x) = \hat{\rho}_o(\hat{c}, x))$$

holds. Let a relation $\phi \subset \hat{C} \times C$ be such that

$$(\hat{c}, c) \in \phi \leftrightarrow (\forall x)(\rho_o(c, x) = \hat{\rho}_o(\hat{c}, x))$$

If $(\hat{c}, c) \in \phi$ and $(\hat{c}, c') \in \phi$ hold, we have

$$(\forall x)(\rho_o(c, x) = \hat{\rho}_o(\hat{c}, x) = \hat{\rho}_o(c', x))$$

Since $\bar{\rho}$ is reduced, $c = c'$ holds; that is, ϕ is functional. Furthermore, the condition that $(\bar{\rho}, \bar{\phi})$ and $(\bar{\hat{\rho}}, \bar{\hat{\phi}})$ are equivalent in the sense of Definition 1.2 yields that $\phi : \hat{C} \to C$ is an onto mapping. Therefore, $K(C) \leq K(\hat{C})$. Q.E.D.

As Example 2.1 shows, a time-invariant finite dynamical system is not in general a minimal realization in the sense of Definition 1.6 even if the system is reduced, while a reduced system is minimal in the sense of Definition 1.8. Definition 1.6 requires stronger properties of a system than Definition 1.8, and in reality, linearity condition is necessary for Proposition 2.1.

The minimal realization in the sense of Definition 1.9 is characterized by the following proposition.

Proposition 2.4. Let $(\bar{\rho}, \bar{\phi})$ be a time-invariant dynamical system. Suppose $(\bar{\rho}, \bar{\phi})$ is linear whose state space is of finite dimension, or the state space C

is a finite set without any algebraic structure. Then $(\bar{\rho}, \bar{\phi})$ is a minimal realization in the sense of Definition 1.9 if and only if it is reduced.

PROOF. We shall consider the *only if* part first. Suppose $(\bar{\rho}, \bar{\phi})$ is not reduced. Let a congruence relation $E \subset C \times C$ be

$$(c, c') \in E \leftrightarrow (\forall x)(\rho_0(c, x) = \rho_0(c', x))$$

Since E is a congruence relation, C/E has the same algebraic structure as C in the usual sense. Let $\hat{\rho}_t : (C/E) \times X_t \to Y_t$ be

$$\hat{\rho}_t([c], x_t) = \rho_t(c, x_t)$$

Since $(\bar{\rho}, \bar{\phi})$ is time invariant, $\hat{\rho}_t$ is well defined as above, and $\bar{\hat{\rho}} = \{\hat{\rho}_t : t \in T\}$ is realizable. Let $h: C \to C/E$ be the natural mapping; that is, $h(c) = [c]$. Then h is an epimorphism and satisfies the commutative diagram of Definition 1.9. However, since E is not trivial, h is not an isomorphism. The desired result follows from the similar arguments used in Proposition 2.2.

We shall consider the *if* part. Suppose $(\bar{\rho}, \bar{\phi})$ is reduced. Let $h: C \to \hat{C}$ be an epimorphism which satisfies the commutative diagram of Definition 1.9. We shall show that h is an injection. Suppose $h(c) = h(c')$ for some elements c, c' in C. Then it follows from the diagram of Definition 1.9 that for any $x \in X$

$$\rho_0(c, x) = \hat{\rho}_0(h(c), x) = \hat{\rho}_0(h(c'), x) = \rho_0(c', x)$$

Consequently, $(\forall x)(\rho_0(c, x) = \rho_0(c', x))$ holds, which implies that $c = c'$ because $(\bar{\rho}, \bar{\phi})$ is reduced. Q.E.D.

Proposition 2.4 is proven for the systems that are linear or whose state space does not have any algebraic structure. However, as the proof of that proposition shows, the proposition holds for any system if the relation $E \subset C \times C$ is a congruence relation with respect to the algebraic structure of the state space, where E is defined by

$$(c, c') \in E \leftrightarrow (\forall x)(\rho_0(c, x) = \rho_0(c', x))$$

When the equivalence relation used to define a minimal realization is given with respect to the input response only, the minimal realization is termed a minimal input-response realization of the respective kind. The importance of this type of minimal realization is due to the fact that most of the problems in the design of linear control systems area (or in the filters design) are specified by giving some weighting functions as an input-response function. We shall consider this type of minimal realization by using Definitions 1.10 and 1.11.

Proposition 2.5. Let $(\bar{\rho}, \bar{\phi})$ be a time-invariant finite dynamical system. $(\bar{\rho}, \bar{\phi})$ is a minimal realization in the sense of Definition 1.10 if and only if $(\bar{\rho}, \bar{\phi})$ is controllable and reduced.

PROOF. We shall consider the *only if* part first. If $(\bar{\rho}, \bar{\phi})$ is not reduced, then the equivalence $E \subset C \times C$, where

$$(c, c') \in E \leftrightarrow (\forall x)(\rho_o(c, x) = \rho_o(c', x))$$

is not trivial. Let $\hat{\rho}_t : (C/E) \times X_t \to Y_t$ be such that

$$\hat{\rho}_t([c], x_t) = \rho_t(c, x_t)$$

Since $\bar{\rho}$ is time invariant, the above $\bar{\rho}_t$ is well defined. Then $\bar{\bar{\rho}} = \{\hat{\rho}_t : t \in T\}$ is a realizable systems response. Suppose $c_0 \in C$ is the reference point of C. Let $[c_0]$ be the reference point of C/E. Then

$$\{(x, \rho_o(c_o, x)) : x \in X\} = \{(x, \hat{\rho}_o([c_o], x)) : x \in X\}$$

holds. Since E is not trivial and since C is finite, we have $K(C/E) < K(C)$. Consequently, $(\bar{\rho}, \bar{\phi})$ is not a minimal realization in the sense of Definition 1.10. Suppose $(\bar{\rho}, \bar{\phi})$ is not controllable. Let $C_o \subset C$ and $\hat{\rho}_t : C_o \times X_t \to Y_t$ be such that

$$C_o = \{c : (\exists x^t)(c = \phi_{ot}(c_o, x^t))\}, \qquad \hat{\rho}_t = \rho_t \mid C_o \times X_t$$

where $c_o \in C$ is the reference point. We shall show first that $\bar{\bar{\rho}} = \{\hat{\rho}_t : t \in T\}$ can be realizable by a time-invariant finite dynamical system. Let $c \in C_o$ and $x^t \in X^t$ be arbitrary. Then $c = \phi_{o\tau}(c_o, x^\tau)$ for some x^τ. Consequently,

$$\phi_{ot}(c, x^t) = \phi_{\tau t'}(c, F^\tau(x^t)) = \phi_{ot'}(c_o, x^\tau \cdot F^\tau(x^t)) \in C_o$$

where $t' = t + \tau$. Therefore, we have that $\phi_{ot}(C_o \times X^t) \subset C_o$; that is, $\hat{\phi}_{tt'} : C_o \times X_{tt'} \to C_o$ can be defined by $\hat{\phi}_{tt'} = \phi_{tt'} | C_o \times X_{tt'}$. It is clear that $(\bar{\bar{\rho}}, \bar{\hat{\phi}})$ is a time-invariant finite dynamical system, where $\bar{\hat{\phi}} = \{\hat{\phi}_{tt'} : t, t' \in T\}$. Since $c_o = \phi_{oo}(c_o, x^o)$, $c_o \in C_o$ and

$$\{(x, \rho_o(c_o, x)) : x \in X\} = \{(x, \hat{\rho}_o(c_o, x)) : x \in X\}$$

hold. Let c_o be the reference point of C_o. Then, since $K(C_o) < K(C)$ holds, $(\bar{\rho}, \bar{\phi})$ is not a minimal realization in the sense of Definition 1.10.

We shall consider the *if* part. Suppose $(\bar{\rho}, \bar{\phi})$ is reduced and controllable. Suppose $(\bar{\rho}, \bar{\phi})$ is not a minimal realization in the sense of Definition 1.10. Then there exists a finite set $\hat{C}$ and a time-invariant finite dynamical system $(\bar{\hat{\rho}}, \bar{\hat{\phi}})$, whose state space is $\hat{C}$ such that the following holds:

$$1.\ K(\hat{C}) < K(C), \qquad 2.\ (\forall x)(\rho_o(c_o, x) = \hat{\rho}_o(\hat{c}_o, x)) \tag{8.3}$$

where $c_o \in C$ and $\hat{c}_o \in \hat{C}$ are the reference points. Since $\bar{\rho}$ and $\bar{\hat{\rho}}$ are realizable, we have

$$\begin{aligned} &(\forall \hat{x}^t)(\exists c' \in C)(\forall x_t)(\rho_t(c', x_t) = \rho_o(c_o, \hat{x}^t \cdot x_t) \mid T_t) \\ &(\forall \hat{x}^t)(\exists c' \in \hat{C})(\forall x_t)(\hat{\rho}_t(c', x_t) = \hat{\rho}_o(\hat{c}_o, \hat{x}^t \cdot x_t) \mid T_t) \end{aligned} \tag{8.4}$$

Since $(\bar{\rho}, \bar{\phi})$ is controllable, we have

$$(\forall c \in C)(\exists \hat{x}^t)(c = \phi_{ot}(c_o, \hat{x}^t))$$

that is,

$$(\forall c \in C)(\exists \hat{x}^t)(\forall x_t)(\rho_t(c, x_t) = \rho_o(c_o, \hat{x}^t \cdot x_t) \mid T_t) \tag{8.5}$$

It follows from (8.3)–(8.5) that

$$(\forall c \in C)(\exists c' \in \hat{C})(\forall x_t)(\rho_t(c, x_t) = \hat{\rho}_t(c', x_t)) \tag{8.6}$$

Let $\hat{C}_o \subset \hat{C}$ be such that

$$\hat{C}_o = \{c' : (\exists c)(\forall x_t)(\rho_t(c, x_t) = \hat{\rho}_t(c', x_t))\} \tag{8.7}$$

Relation (8.6) implies that $\hat{C}_o \neq \phi$. Let $\phi \subset \hat{C}_o \times C$ be such that

$$(c', c) \in \phi \leftrightarrow (\forall x_t)(\rho_t(c, x_t) = \hat{\rho}_t(c', x_t))$$

Relation (8.7) implies that $\mathscr{D}(\phi) = \hat{C}_o$. If $(c', c) \in \phi$ and $(c', \hat{c}) \in \phi$ hold, then we have

$$(\forall x_t)(\rho_t(c, x_t) = \hat{\rho}_t(c', x_t) = \rho_t(\hat{c}, x_t))$$

Since $(\bar{\rho}, \bar{\phi})$ is reduced, $c = \hat{c}$ holds. Therefore, ϕ is a mapping; i.e., $\phi : \hat{C}_o \to C$. Furthermore, ϕ is a surjection due to (8.6). Consequently, we have that $K(C) \leq K(\hat{C}_o) \leq K(\hat{C})$, which contradicts the condition $K(\hat{C}) < K(C)$.

Q.E.D.

Before characterizing minimal realization in the sense of Definition 1.11, we shall introduce the following lemma.

Lemma 2.1. Suppose that $(\bar{\rho}, \bar{\phi})$ is a reduced time-invariant linear dynamical system. Let

$$C_o = \{c : (\exists x^t)(\phi_{20t}(x^t) = c) \,\&\, t \in T\}$$

Then

$$C_o = \{c : (\exists x^t)(\rho_{1t}(c) = \rho_{20}(x^t \cdot o) \mid T_t) \,\&\, t \in T\}$$

and C_o is a linear subspace of C.

PROOF. In general,

$$C_o \subset \{c : (\exists x^t)(\rho_{1t}(c) = \rho_{20}(x^t \cdot o) \mid T_t) \,\&\, t \in T\} \equiv C_o'$$

Suppose $\rho_{1t}(c) = \rho_{20}(x^t \cdot o) \mid T_t$. Since $\{\rho_t\}$ is reduced,

$$c = \rho_{1t}^{-1}(\rho_{20}(x^t \cdot o) \mid T_t) = \phi_{20t}(x^t)$$

(Refer to Chapter IV, Section 3.) Hence $C_0 = C_0'$. The latter part follows from the proof of Corollary 3.2, Chapter VII. Q.E.D.

We have now the following proposition.

Proposition 2.6. Let $(\bar{\rho}, \bar{\phi})$ be a time-invariant finite-dimensional linear dynamical system. Then $(\bar{\rho}, \bar{\phi})$ is a minimal realization in the sense of Definition 1.11 if and only if it is controllable and reduced.

PROOF We shall consider the *only if* part first. If $(\bar{\rho}, \bar{\phi})$ is not reduced, there exists a nontrivial congruence relation $E \subset C \times C$ where

$$(c, c') \in E \leftrightarrow \rho_{10}(c) = \rho_{10}(c')$$

Let $\hat{\rho}_{1t}: C/E \to Y_t$ be such that

$$\hat{\rho}_{1t}([c]) = \rho_{1t}(c)$$

Then $\bar{\hat{\rho}} = \{(\hat{\rho}_{1t}, \rho_{2t}) : t \in T\}$ can be realizable by a time-invariant finite-dimensional linear dynamical system. Let $h: C \to C/E$ be the natural mapping; i.e., $h(c) = [c]$. Then h is an epimorphism. Since E is not trivial, h is not an isomorphism. Therefore, the argument used in Proposition 2.2 implies that $(\bar{\rho}, \bar{\phi})$ is not a minimal realization in the sense of Definition 1.11. Suppose $(\bar{\rho}, \bar{\phi})$ is reduced but not controllable. Let $C_0 \subset C$ be such that

$$C_0 = \{c : (\exists x^t)(c = \phi_{20t}(x^t))\}$$

Then Lemma 2.1 says that C_0 is a subspace of C, and C_0 is a proper subspace due to uncontrollability. Let $\hat{\rho}_{1t}: C_0 \to Y_t$ such that $\hat{\rho}_{1t} = \rho_{1t} \mid C_0$. Then we shall show that $\{(\hat{\rho}_{1t}, \rho_{2t})\}$ is realizable. The input-response consistency is obviously satisfied because $\{(\rho_{1t}, \rho_{2t})\}$ is realizable. We want to show that

$$(\forall t)(\forall t')(\forall c \in C_0)(\forall x_{tt'})(\exists c' \in C_0)(\hat{\rho}_{1t'}(c') = (\hat{\rho}_{1t}(c) + \rho_{2t}(x_{tt'} \cdot o)) \mid T_{t'})$$

Since $\{\langle \rho_{1t}, \rho_{2t} \rangle\}$ is realizable, the following holds:

$$(\forall t)(\forall x^t)(\forall c \in C_0)(\exists c' \in C)(\rho_{1t}(c') = \rho_{10}(c) \mid T_t + \rho_{20}(x^t \cdot o) \mid T_t)$$

We shall show that $c' \in C_0$. Since $c \in C_0$, $\rho_{1t'}(c) = \rho_{20}(\hat{x}^{t'} \cdot o) \mid T_{t'}$ for some $t' \in T$ and $\hat{x}^{t'}$. Let $t + t' = \tau$. Then, since $\{\rho_t\}$ is time invariant, we have

$$\begin{aligned} \rho_{1\tau}(c') &= \rho_{1t'}(c) \mid T_\tau + \rho_{2t'}(F^{t'}(x^t) \cdot o) \mid T_\tau \\ &= \rho_{20}(\hat{x}^{t'} \cdot o) \mid T_\tau + \rho_{2t'}(F^{t'}(x^t) \cdot o) \mid T_\tau \\ &= \rho_{20}(\hat{x}^{t'} \cdot F^{t'}(x^t) \cdot o) \mid T_\tau \end{aligned}$$

Hence, $c' \in C_0$. Consequently,

$$(\forall t)(\forall x^t)(\forall c \in C_0)(\exists c' \in C_0)(\hat{\rho}_{1t}(c') = \hat{\rho}_{10}(c) \mid T_t + \rho_{20}(x^t \cdot o) \mid T_t)$$

Since $\{\rho_t\}$ is time invariant, we have the result. Therefore, $\{(\hat{\rho}_{1t}, \rho_{2t})\}$ is realizable by a time-invariant finite-dimensional linear dynamical system. Let $h: C \to C_o$ be the projection from C onto C_o. Then h is an epimorphism but not an isomorphism because C_o is a proper subspace of C. Consequently, the argument used in Proposition 2.2 implies that $(\bar{\rho}, \bar{\phi})$ is not a minimal realization in the sense of Definition 1.11. Next, we shall consider the *if* part. Suppose $(\bar{\rho}, \bar{\phi})$ is reduced and controllable. Suppose $(\bar{\hat{\rho}}, \bar{\hat{\phi}})$ is a time-invariant finite-dimensional linear dynamical system such that $(\forall x)(\rho_o(\text{o}, x) = \hat{\rho}_o(\text{o}, x))$ holds, and, moreover, there exists an epimorphism $h: C \to \hat{C}$. Let a congruence relation $E \subset C \times C$ be such that $(c, c') \in E \leftrightarrow h(c) = h(c')$. Let a linear function $\hat{\rho}'_{1t}: C/E \to Y_t$ be such that

$$\hat{\rho}'_{1t}([c]) = \hat{\rho}_{1t}(h(c))$$

Then $\bar{\rho}' = \{(\hat{\rho}'_{1t}, \hat{\rho}_{2t}): t \in T\}$ is realizable by a time-invariant finite-dimensional linear dynamical system. (Refer to the realization theory.) Since $(\forall x)(\rho_o(\text{o}, x) = \hat{\rho}_o(\text{o}, x))$ implies that $\rho_{20}(x) = \hat{\rho}_{20}(x)$ for every x, we have from the realizability condition that

$$(\forall t)(\forall x^t)(\exists [c])(\hat{\rho}'_{1t}([c]) = \rho_{20}(x^t \cdot \text{o}) \mid T_t) \tag{8.8}$$

Since $(\bar{\rho}, \bar{\phi})$ is controllable,

$$C = \{c : (\exists x^t)(c = \phi_{20t}(x^t))\}$$

that is,

$$(\forall c)(\exists t)(\exists x^t)(c \in C \to \rho_{1t}(c) = \rho_{20}(x^t \cdot \text{o}) \mid T_t) \tag{8.9}$$

holds. Relations (8.8) and (8.9) imply that

$$(\forall c)(\exists t)(\exists [c'])(c \in C \to \rho_{1t}(c) = \hat{\rho}'_{1t}([c']))$$

Since $\bar{\rho}' = \{(\hat{\rho}'_{1t}, \hat{\rho}_{2t}): t \in T\}$ is time invariant, the above relation yields

$$(\forall c)(\exists [c'])(c \in C \to \rho_{10}(c) = \hat{\rho}'_{10}([c']))$$

Let $\hat{C}' \subset C/E$ be such that

$$\hat{C}' = \{[c'] : (\exists c \in C)(\rho_{10}(c) = \hat{\rho}'_{10}([c']))\}$$

$\hat{C}'$ is a linear subspace of C/E. Since $(\bar{\rho}, \bar{\phi})$ is reduced, there exists a linear surjection $\rho_{10}^{-1} \cdot \hat{\rho}'_{10}: \hat{C}' \to C$. Q.E.D.

3. UNIQUENESS OF MINIMAL INPUT-RESPONSE REALIZATION

A minimal realization is, in general, not unique (as can easily be seen from the proofs of propositions in the previous section). In this section, we shall show, however, that a minimal input-response realization is unique within an isomorphism. Some other minimal realizations will be considered in Chapter XII.

We shall first consider the uniqueness problem for a general case.

Proposition 3.1. Let $(\bar{\rho}, \bar{\phi})$ and $(\bar{\hat{\rho}}, \bar{\hat{\phi}})$ be time-invariant dynamical systems. Suppose $(\bar{\rho}, \bar{\phi})$ and $(\bar{\hat{\rho}}, \bar{\hat{\phi}})$ are reduced and controllable. They are then realizations of the same input-response family; i.e., for every x, $\rho_o(c_o, x) = \hat{\rho}_o(\hat{c}_o, x)$ if and only if there exists a one-to-one correspondence $h: C \to \hat{C}$ such that $h(c_o) = \hat{c}_o$ and the following diagram is commutative for every $x_{tt'}$ and $x_{t'}$ and for $o \leq t \leq t'$:

$$
\begin{array}{ccccccc}
X^t & \xrightarrow{\ \phi_{ot}(c_o,\,-)\ } & C & \xrightarrow{\ \phi_{tt'}(-,\,x_{tt'})\ } & C & \xrightarrow{\ \rho_{t'}(-,\,x_{t'})\ } & Y_t \\
\downarrow I & & \downarrow h & & \downarrow h & & \downarrow I \\
X^t & \xrightarrow{\ \hat{\phi}_{ot}(\hat{c}_o,\,-)\ } & \hat{C} & \xrightarrow{\ \hat{\phi}_{tt'}(-,\,x_{tt'})\ } & \hat{C} & \xrightarrow{\ \hat{\rho}_{t'}(-,\,x_{t'})\ } & Y_t
\end{array}
$$

PROOF. We shall consider the *if* part first. Suppose there exists a one-to-one correspondence $h: C \to \hat{C}$ such that the above diagram is commutative. Then, for any $x \in X$ and $t' \geq o$,

$$
\begin{aligned}
\rho_o(c_o, x) \mid T_{t'} &= \rho_{t'}(\phi_{ot'}(c_o, x^{t'}), x_{t'}) \qquad \text{where} \quad x = x^{t'} \cdot x_{t'} \\
&= \hat{\rho}_{t'}(h \cdot \phi_{ot'}(c_o, x^{t'}), x_{t'}) \\
&= \hat{\rho}_{t'}(\hat{\phi}_{ot'}(\hat{c}_o, x^{t'}), x_{t'}) \\
&= \hat{\rho}_o(\hat{c}_o, x^{t'} \cdot x_{t'}) \mid T_{t'}
\end{aligned}
$$

Therefore, $\rho_o(c_o, x) = \hat{\rho}_o(\hat{c}_o, x)$ for every $x \in X$.

Next we shall consider the *only if* part. Let $h \subset C \times \hat{C}$ be a relation such that

$$(c, \hat{c}) \in h \leftrightarrow (\exists x^t)(c = \phi_{ot}(c_o, x^t) \,\&\, \hat{c} = \hat{\phi}_{ot}(\hat{c}_o, x^t))$$

Suppose $(c, \hat{c}) \in h$ and $(c, \hat{c}') \in h$ hold such that

$$c = \phi_{ot}(c_o, x^t) \,\&\, \hat{c} = \hat{\phi}_{ot}(\hat{c}_o, x^t)$$

and

$$c = \phi_{ot'}(c_o, \hat{x}^{t'}) \,\&\, \hat{c}' = \hat{\phi}_{ot'}(\hat{c}_o, \hat{x}^{t'})$$

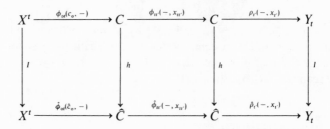

Then we have that for any x_t

$$\begin{aligned}\rho_t(c, x_t) &= \rho_t(\phi_{ot}(c_o, x^t), x_t)\\ &= \rho_o(c_o, x^t \cdot x_t) \mid T_t\\ &= \hat{\rho}_o(\hat{c}_o, x^t \cdot x_t) \mid T_t\\ &= \hat{\rho}_t(\hat{\phi}_{ot}(\hat{c}_o, x^t), x_t)\\ &= \hat{\rho}_t(\hat{c}, x_t)\end{aligned} \tag{8.10}$$

Similarly, we have that for any $x_{t'}$

$$\rho_{t'}(c, x_{t'}) = \hat{\rho}_{t'}(\hat{c}', x_{t'}) \tag{8.11}$$

Since $\bar{\rho}$ and $\bar{\hat{\rho}}$ are time invariant, Eqs. (8.10) and (8.11) imply that

$$(\forall x)(\rho_o(c, x) = \hat{\rho}_o(\hat{c}, x))$$

and

$$(\forall x)(\rho_o(c, x) = \hat{\rho}_o(\hat{c}', x))$$

Consequently, we have that

$$(\forall x)(\hat{\rho}_o(\hat{c}, x) = \hat{\rho}_o(\hat{c}', x))$$

Since $(\bar{\hat{\rho}}, \bar{\hat{\phi}})$ is reduced, $\hat{c} = \hat{c}'$ holds. Let $c \in C$ be arbitrary. Since $(\bar{\rho}, \bar{\phi})$ is controllable, there exists $\hat{x}^t$ such that $c = \phi_{ot}(c_o, \hat{x}^t)$ holds. Let $\hat{c} = \hat{\phi}_{ot}(\hat{c}_o, \hat{x}^t)$. Then we have that $(c, \hat{c}) \in h$. Consequently, $\mathscr{D}(h) = C$. Therefore, h is a surjection from C onto $\hat{C}$. Moreover, since C and $\hat{C}$ are related with h in a systematic way, the inverse h^{-1} of h exists. Hence, h is a one-to-one correspondence. The relation $h(c_o) = \hat{c}_o$ follows from the fact that $c_o = \phi_{oo}(c_o, x^o)$ and $\hat{c}_o = \hat{\phi}_{oo}(\hat{c}_o, x^o)$. We shall show that the diagram is commutative with respect to h.

Since

$$(\phi_{ot}(c_o, x^t), \hat{\phi}_{ot}(\hat{c}_o, x^t)) \in h$$

we have that

$$\hat{\phi}_{ot}(\hat{c}_o, x^t) = h \cdot \phi_{ot}(c_o, x^t), \qquad \text{for any } x^t$$

The relation that for any $c \in C$ and $x_{tt'}$

$$\hat{\phi}_{tt'}(h(c), x_{tt'}) = h \cdot \phi_{tt'}(c, x_{tt'})$$

is proven as follows. Since $(\bar{\rho}, \bar{\phi})$ is controllable, there exists x^τ such that

$c = \phi_{o\tau}(c_o, x^\tau)$. Let $\mu = (t' - t) + \tau$. Then

$$(\phi_{o\mu}(c_o, x^\tau \cdot F^{\tau - t}(x_{tt'})), \hat{\phi}_{o\mu}(\hat{c}_o, x^\tau \cdot F^{\tau - t}(x_{tt'}))) \in h$$
$$\rightarrow (\phi_{\tau\mu}(\phi_{o\tau}(c_o, x^\tau), F^{\tau - t}(x_{tt'})), \hat{\phi}_{\tau\mu}(\hat{\phi}_{o\tau}(\hat{c}_o, x^\tau), F^{\tau - t}(x_{tt'}))) \in h$$
$$\rightarrow (\phi_{\tau\mu}(c, F^{\tau - t}(x_{tt'})), \hat{\phi}_{\tau\mu}(h(c), F^{\tau - t}(x_{tt'}))) \in h$$
$$\rightarrow (\phi_{tt'}(c, x_{tt'}), \hat{\phi}_{tt'}(h(c), x_{tt'})) \in h$$
$$\rightarrow \hat{\phi}_{tt'}(h(c), x_{tt'}) = h \cdot \phi_{tt'}(c, x_{tt'})$$

Similarly, we can prove that for any $c \in C$ and $x_{t'}$

$$\hat{\rho}_{t'}(h(c), x_{t'}) = h \cdot \rho_{t'}(c, x_{t'})$$

Therefore, the diagram is commutative with respect to h. Q.E.D.

Proposition 3.1 states that if two dynamical systems $(\bar{\rho}, \bar{\phi})$ and $(\bar{\hat{\rho}}, \bar{\hat{\phi}})$, which are reduced and controllable, are realizations of the same input-response family, they are isomorphic in the sense that there exists a one-to-one correspondence $h: C \rightarrow \hat{C}$ such that $(\bar{\hat{\rho}}, \bar{\hat{\phi}})$ is determined by $(\bar{\rho}, \bar{\phi})$ as follows:

$$\hat{\rho}_t(\hat{c}, x_t) = \rho_t(h^{-1}(\hat{c}), x_t) \tag{8.12}$$

$$\hat{\phi}_{tt'}(\hat{c}, x_{tt'}) = h\phi_{tt'}(h^{-1}(\hat{c}), x_{tt'}) \tag{8.13}$$

h satisfies the relation $h(c_o) = \hat{c}_o$, and hence if C and $\hat{C}$ are considered algebra with null-ary operators c_o and $\hat{c}_o$, then h is considered an isomorphism.

Combining Proposition 3.1 with Proposition 2.5, we have the following corollary.

Corollary 3.1. Given two time-invariant finite dynamical systems $(\bar{\rho}, \bar{\phi})$ and $(\bar{\hat{\rho}}, \bar{\hat{\phi}})$, which are minimal realizations in the sense of Definition 1.10. They are then realizations of the same input-response family if and only if the diagram of Proposition 3.1 is commutative where h is a one-to-one correspondence and $h(c_o) = \hat{c}_o$ holds.

According to the above corollary, a minimal realization in the sense of Definition 1.10 of a time-invariant finite dynamical system is uniquely determined within an isomorphism. [Refer to Eqs. (8.12) and (8.13).]

Regarding a linear minimal realization, an equivalent proposition to Proposition 3.1 can be proven. However, in order to emphasize the intimate relationship between the present result and the classical theory, we shall present that fact in the form of a uniqueness proposition as follows.

Proposition 3.2. Given two strongly nonanticipatory time-invariant linear dynamical systems $S = (\bar{\rho}, \bar{\phi})$ and $\hat{S} = (\bar{\hat{\rho}}, \bar{\hat{\phi}})$, which are reduced and controllable. They are then realizations of the same input-response family; i.e., for every x_t, $\rho_{2t}(x_t) = \hat{\rho}_{2t}(x_t)$ if and only if there exists a linear one-to-one correspondence $h: C \to \hat{C}$ such that the following diagram is commutative:

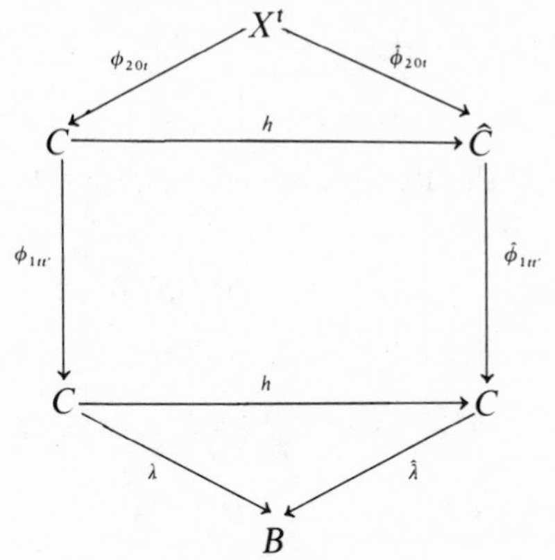

where λ and $\hat{\lambda}$ are the output functions of $(\bar{\rho}, \bar{\phi})$ and $(\bar{\hat{\rho}}, \bar{\hat{\phi}})$.

PROOF. Let us consider the *only if* part first. Notice that the output function $\lambda: C \to B$ of a strongly nonanticipatory time-invariant linear dynamical system $(\bar{\rho}, \bar{\phi})$ is given by: For any t, x_t, and c

$$\lambda(c) = \rho_t(c, x_t)(t) = \rho_{1t}(c)(t) = \rho_{10}(c)(\text{o})$$

For any t and x^t, we have

$$\begin{aligned} \rho_{1t}(\phi_{20t}(x^t)) &= \rho_{\text{o}}(\text{o}, x^t \cdot \text{o}) \mid T_t \\ &= \rho_{2t}(x^t \cdot \text{o}) \mid T_t \\ &= \hat{\rho}_{2t}(x^t \cdot \text{o}) \mid T_t \\ &= \hat{\rho}_{1t}(\hat{\phi}_{20t}(x^t)) \end{aligned}$$

Consequently, for every $t' \geq t$, we have

$$\begin{aligned} \lambda(\phi_{1tt'}(\phi_{20t}(x^t))) &= \rho_{1t'}(\phi_{1tt'}(\phi_{20t}(x^t)))(t') \\ &= (\rho_{1t}(\phi_{20t}(x^t)) \mid T_{t'})(t') \\ &= \rho_{1t}(\phi_{20t}(x^t))(t') \\ &= \hat{\rho}_{1t}(\hat{\phi}_{20t}(x^t))(t') \\ &= \hat{\lambda}(\hat{\phi}_{1tt'}(\hat{\phi}_{20t}(x^t))) \end{aligned}$$

Since S and $\hat{S}$ are reduced, ρ_{1t} and $\hat{\rho}_{1t}$ are one to one, that is, have the inverse on $\rho_{1t}(C)$ and $\hat{\rho}_{1t}(\hat{C})$. Furthermore, since S and $\hat{S}$ are controllable,

$$C = \bigcup_t \phi_{ot}(X^t) \qquad \text{and} \qquad \hat{C} = \bigcup_t \hat{\phi}_{ot}(X^t)$$

hold. Let $h_c : C \to \hat{C}$ be such that

$$h_c(c) = \hat{\rho}_{1t}^{-1} \cdot \rho_{1t}(c)$$

where t is arbitrary if the right-hand side is properly defined. Notice that since the systems are time invariant, for any t and t',

$$\hat{\rho}_{1t}^{-1}(\rho_{1t}(c)) = \hat{\rho}_{1t'}^{-1}(\rho_{1t'}(c))$$

holds if both sides are properly defined and that

$$C = \bigcup_t \phi_{ot}(X^t)$$

and

$$\rho_{1t}(\phi_{20t}(x^t)) = \hat{\rho}_{1t}(\hat{\phi}_{20t}(x^t))$$

for every x^t imply $\mathscr{D}(h_c) = C$. Consequently, h_c is a one-to-one correspondence (i.e., $h_c^{-1} = \rho_{1t}^{-1}\hat{\rho}_{1t}$ for some t), and for every t and $t' \geq t$

$$h_c(\phi_{1tt'}(\phi_{20t}(x^t))) = \hat{\phi}_{1tt'}(\hat{\phi}_{20t}(x^t))$$

holds, and in particular, if $t' = t$,

$$h_c(\phi_{20t}(x^t)) = \hat{\phi}_{20t}(x^t)$$

holds. The *if* part is apparent. Q.E.D.

By combining Proposition 3.2 with Proposition 2.6, we have the following result.

Corollary 3.2. Given two strongly nonanticipatory time-invariant finite-dimensional linear dynamical systems $(\bar{\rho}, \bar{\phi})$ and $(\bar{\hat{\rho}}, \bar{\hat{\phi}})$, which are minimal realizations in the sense of Definition 1.11 and whose state spaces satisfy the conditions from Proposition 2.6, they are then realizations of the same linear input-response family if and only if the diagram of Proposition 3.2 is commutative, where h is a linear one-to-one correspondence.

Chapter IX

STABILITY

Stability is a vast and important subject in systems theory, and in this chapter we shall only open that subject to the considerations on the general systems level. It will be shown how several concepts of stability can be formalized within the general systems theory framework developed in this book, but only one of them, namely, the stability of a subset of a state space, will be investigated in detail. In order to show how some conventional notions and problems deeply rooted in the analysis framework can be successfully dealt with even on the general systems level, the concept of a Lyapunov-type function is introduced for a general dynamical system, and the necessary and sufficient conditions for stability are given in terms of that function.

Conceptually, the stability depends on the system and the way its behavior is evaluated to determine the "inertness" (i.e., stability) of a given mode of operation. In addition to the definition of a system, one, therefore, also needs a way to determine whether the system's behavior has changed appreciably after a perturbation. The assessment is done in terms of a properly defined concept of a neighborhood, and the system is then stable in reference to the given notion of neighborhood if the changes in the system's behavior are small enough for small variations in operating conditions.

The generality of the stability considerations and results derived in this chapter should be noticed. All the systems objects used are defined solely as abstract sets, and only one auxiliary set, which is ordered for the sake of a proper definition of the Lyapunov-type function, is introduced. The information of the system is condensed in the form of a preordering. Using only that much structure, we have been able to prove the basic Lyapunov-type theorem, giving necessary and sufficient conditions for stability. The results are immediately applicable for the more structured systems, e.g., for the sets with a topology or a uniform topology, a metric space, etc.

1. GENERAL CONCEPT OF STABILITY

The essential idea in the stability concept in general is the following. Let $\hat{d}$ and $\hat{e}$ represent the cause and effect of a phenomenon, respectively; i.e., there is assumed a mapping F such that $\hat{e} = F(\hat{d})$. In some other instances, another cause, for example, d, yields then another effect $e = F(d)$. The cause–effect pair $(\hat{e}, \hat{d})$ is stable if small deviations from $\hat{e}$ are caused by small deviations from $\hat{d}$, i.e., for all d close to $\hat{d}$, the corresponding effect $e = F(d)$ is close to $\hat{e}$. Intuitively, then, small perturbations around $\hat{d}$ will not change the effect appreciably.

To formalize the notion of stability, it is apparently necessary to have a way to express "closeness." In general, this can be done simply by using a specified family of subsets, otherwise arbitrary. Let V be an arbitrary set, $\Pi(V)$ the family of all subsets of V, and θ_V a given family of subsets of V, i.e., $\theta_V \subset \Pi(V)$. For any point $v \in V$, the neighborhood system $\bar{N}(v)$ of v relative to θ_V is then defined by

$$\bar{N}(v) = \{\alpha : \alpha \in \theta_V \ \& \ v \in \alpha\}$$

A general concept of stability is then the following definition.

Definition 1.1. Let $F: D \to E$ be a given mapping, θ_D and θ_E given families of subsets of D and E, respectively, and $(\hat{d}, \hat{e}) \in D \times E$, such that $\hat{e} = F(\hat{d})$. $(\hat{d}, \hat{e})$ is stable relative to θ_D and θ_E if and only if

$$(\forall \alpha \in \bar{N}(\hat{e}))(\exists \beta \in \bar{N}(\hat{d}))(\forall d)(d \in \beta \to F(d) \in \alpha)$$

where $\bar{N}(\hat{e}) \subset \theta_E$ and $\bar{N}(\hat{d}) \subset \theta_D$ are the neighborhood systems of $\hat{e}$ and $\hat{d}$ in the sense of θ_E and θ_D, respectively.

(a) Stability of Response

The stability notion from Definition 1.1 can be used to derive a variety of system stability concepts. For example, let D be the initial (or global) state object C, E the output objects Y, and let the initial (global)-response function of a system ρ_0 for a given input $\hat{x} \in X$ be taken for F. We have then the following definition.

Definition 1.2. Let θ_C and θ_Y be given families of subsets of C and Y, respectively. The response $\hat{y} = \rho_o(\hat{c}_o, \hat{x})$ is then stable (relative to θ_C, θ_Y and for the given $\hat{x}$) if and only if

$$(\forall \alpha \in \bar{N}(\hat{y}))(\exists \beta \in \bar{N}(\hat{c}_o))(\forall c_o)(c_o \in \beta \to \rho_o(c_o, \hat{x}) \in \alpha)$$

where $\bar{N}(\hat{y})$ and $\bar{N}(\hat{c}_o)$ are the neighborhood systems of $\hat{y}$ and $\hat{c}_o$, respectively.

When S is a dynamical system, the question of stability is traditionally explored in terms of the state-transition family. This is certainly completely satisfactory when the system has a canonical representation $(\bar{\phi}, \bar{\lambda})$ since the output function λ is a static map and has no effect on stability.

Actually, in such an approach [13] the only information on the system necessary for the stability considerations is the succession of states. All required information of the system can then be presented in a compact form in terms of an ordering in the state space, $\Psi \subset C \times C$. Let us derive such an ordering from the state-transition family $\bar{\phi}$.

Let S be a time-invariant dynamical system $(\bar{\phi}, \bar{\lambda})$ with the state space C. Let $Z = C^T$. For any $x \in X$, the system S defines a relation $\Psi \subset C \times C$ such that

$$(c, c') \in \psi \leftrightarrow (\exists x_{tt'})(c' = \phi_{tt'}(c, x_{tt'})) \tag{9.1}$$

The relation ψ satisfies the following properties: For any c, c', and c'' in C

(i) $(c, c) \in \Psi$,
(ii) $(c, c') \in \Psi \;\&\; (c', c'') \in \Psi \to (c, c'') \in \Psi$.

Condition (i) follows from the convention adopted for the state transition that $\phi_{tt}(c, x_{tt}) = c$, while condition (ii) follows from the time invariancy, the input completeness, and the composition property.†

Let D and E of Definition 1.1 be C and $\Pi(C)$, respectively. Let $F: C \to \Pi(C)$ be such that $F(c) = \psi(c) = \{c' : (c, c') \in \Psi\}$. Let $\theta_D = \theta_E = \theta \subset \Pi(C)$. Then Definition 1.1 yields the following definition.

Definition 1.3. A point $c \in C$ is stable relative to ψ and θ if and only if

$$(\forall \alpha \in \bar{N}(\psi(c)))(\exists \beta \in \bar{N}(c))(\psi(\beta) \subset \alpha)$$

† As a matter of fact, let $(c, c') \in \Psi$ and $(c', c'') \in \Psi$ or $c' = \phi_{tt'}(c, x_{tt'})$ and $c'' = \phi_{ss'}(c', \hat{x}_{ss'})$ for some $x_{tt'}$ and $\hat{x}_{ss'}$; let $t'' = (s' - s) + t'$. (Notice that since S is time invariant, the time set is an Abelian group with a linear ordering.) Then $c'' = \phi_{t't''}(c', F^{t'-s}(\hat{x}_{ss'}))$ because S is time invariant; since

$$x_{tt''} = x_{tt'} \cdot F^{t'-s}(\hat{x}_{ss'}) \in X_{tt''}$$

and since

$$c'' = \phi_{t't''}(c', F^{t'-s}(\hat{x}_{ss'})) = \phi_{t't''}(\phi_{tt'}(c, x_{tt'}), F^{t'-s}(\hat{x}_{ss'})) = \phi_{tt''}(c, x_{tt'} \cdot F^{t'-s}(\hat{x}_{ss'}))$$

$(c, c'') \in \Psi$ holds.

where $\psi(\beta) = \bigcup_{c \in \beta} \psi(c)$ and the neighborhood of a set $\psi(c)$ is defined in the usual manner; i.e., $\bar{N}(\psi(c)) = \{\alpha : \alpha \in \theta \;\&\; \psi(c) \subset \alpha\}$.

It should be pointed out, again, that in Definition 1.3 the only reference to the given system is in terms of the relation ψ, which satisfies (i) and (ii); i.e., it is a preordering relation. Therefore, when using Definition 1.3 we shall talk about stability of the preordering ψ, with the understanding that the relation ψ is defined by the state-transition family of a system.

As an illustration of the application of Definition 1.3, let us consider a dynamical system defined on an Euclidean space with a norm topology, where θ corresponds to that topology. Suppose the dynamical system has no input, i.e., $\phi_{ot}: C \to C$ and $\hat{c} \in C$ is an equilibrium point, $\phi_{ot}(\hat{c}) = \hat{c}$ for every t. Definition 1.3 then can be interpreted as

$$(\forall \varepsilon)(\exists \delta)(\forall c)(\|c - \hat{c}\| < \delta \to (\forall t)(\|\phi_{ot}(c) - \hat{c}\| < \varepsilon))$$

This represents a Lyapunov-type stability, and hence Definition 1.3 is a generalization of Lyapunov-type stability.†

Definition 1.3 can be extended to the stability of a set rather than a point.

Definition 1.4. A set $C' \subset C$ is stable relative to ψ and θ if and only if

$$(\forall \alpha \in \bar{N}(\psi(C')))(\exists \beta \in \bar{N}(C'))(\psi(\beta) \subset \alpha)$$

(b) Stability of an Isolated Trajectory†

The Lyapunov-type stability of a given element $\psi(\hat{c})$, as considered in the preceding section, is defined in relation to other elements $\psi(\beta)$. But there is another type of stability where the stability of $\psi(\hat{c})$ is defined in terms of $\psi(\hat{c})$ itself.

A typical example of this kind is the Poisson-type stability. Let $\hat{z}_c(t) = \phi_{ot}(c, \hat{x}^t)$ for a dynamical system with a fixed input $\hat{x}$; a trajectory $\hat{z}_c: T \to C$ is positively Poisson stable if and only if

$$(\exists \hat{t})(\forall \alpha \in \bar{N}(\hat{z}_c(\hat{t})))(\forall t)(\exists t')(t \le t' \to \hat{z}_c(t') \in \alpha)$$

For instance, if $\hat{z}_c(t) = \sin ct$ for $c \in E^1$, $\hat{z}_c$ is stable in the above sense, where $\hat{t}$ can be any positive number.

As another illustration, let a dynamical system S be defined on a Euclidean space with a norm topology, and furthermore, let S have a unique input $\hat{x}$. Let $\hat{z}(t) = \phi_{ot}(c, \hat{x})$. A classical concept of stability is then expressed as: A trajectory $\hat{z}: T \to C$ is stable if and only if

$$(\forall \varepsilon)(\exists \hat{t})(\forall t)(\forall t')(\hat{t} \le t \;\&\; \hat{t} \le t' \to \|\hat{z}(t) - \hat{z}(t')\| < \varepsilon) \tag{9.2}$$

† See Nemytskii and Stepanov [14].

In relation (9.2) the stability is defined by the trajectory $\hat{z}$ itself. However, the classical stability is, in general, not a Poisson-type stability.

Consider the trajectory of $\hat{z}(t) = 1/(1 + t)$. $\hat{z}$ is stable in the sense of (9.2), but not stable in the sense of Poisson. Furthermore, $z_c(t) = \sin ct$ is not stable in the sense of (9.2). Under some conditions, the classical stability becomes a Lyapunov-type stability, i.e., an asymptotical stability.

(c) Structural Stability†

The concept of structural stability is related with the "morphology" of a system. Let $S \subset X \times Y$ be a general system, which is "parameterized" by a set D in the sense that for any $d \in D$ the system is in a certain "mode." Furthermore, let the system's behavior be classified in terms of various "forms" of behavior, i.e., there is given a function

$$P : \bar{S} \to E$$

such that $e = P(S)$ denotes that $S \in \bar{S}$ has the form e, where $\bar{S} = \{S \subset X \cdot Y\}$. Let $R : D \to \bar{S}$. Let F be the composition of P and R, i.e., $F : D \to E$ such that

$$e = F(d) = P(R(d))$$

Usually, the neighborhood system of $e \in E$ is defined as $\bar{N}(e) = \{\{e\}\}$. Then the form $\hat{e} \in E$ is structurally stable if and only if for each $\hat{d} \in D$, where $\hat{e} = F(\hat{d})$,

$$(\exists \beta \in \bar{N}(\hat{d}))(\forall d)(d \in \beta \to F(d) = \hat{e})$$

As a simple example, let us consider the following dynamic system

$$(d^2 z/dt^2) + 2d(dz/dt) + z = 0$$

Usually, according to the value of the parameter d, the forms of dynamic behavior of the above system are classified as "overdamped" ($= e_1$), "critically damped" ($= e_2$), "underdamped" ($= e_3$), "steady-state oscillation" ($= e_4$), and "unstable" ($= e_5$). Then we have that

$$F(d) = \begin{cases} e_1, & \text{if } d > 1 \\ e_2, & \text{if } d = 1 \\ e_3, & \text{if } 0 < d < 1 \\ e_4, & \text{if } d = 0 \\ e_5, & \text{if } d < 0 \end{cases}$$

† See Thom [15].

If θ_D and θ_E are taken as the usual topology and discrete topology, respectively, it is easy to see that the forms e_1, e_3, and e_5 are structurally stable, while the others are not.

In conclusion, the following remark is of conceptual importance.

The stability concept in this chapter is defined *relative* to a given family of subsets θ. *If no restrictions are imposed, every system can be viewed as stable by an appropriate selection of* θ. However, most often, the family θ is given, and the question is to determine the stability of the system relative to that θ. We shall consider that problem extensively for the stability of sets as introduced in Definition 1.4.

2. STABILITY OF SETS FOR GENERAL SYSTEMS

Stability can be characterized in terms of the properties of the function F. Specifically, we shall derive in this section necessary and sufficient conditions for stability of sets as given in Definition 1.4, and therefore the term "stability" in this section will be referred henceforth solely to that concept.

(a) Preliminaries

As mentioned in Section 1, a family of subsets $\theta \subset \Pi(C)$ is introduced in order to evaluate the changes in C. This notion will be strengthened by introducing the concept of a generalized distance function. This will enable the introduction of a Lyapunov-type function and characterization of stability in terms of the existence of such a function.

Definition 2.1. Given a set C and a family of subsets of C, $\theta \subset \Pi(C)$. Suppose there is given a complete lattice W with the least element o and for each c a subset $W_c \subset W$ such that there is given a function $\rho: C \times C \to W$ which satisfies the following:

(i) $\rho(c, c') \geq \text{o}$ and $\rho(c, c) = \text{o}$
(ii) $\rho(c, c') = \rho(c', c)$
(iii) $\rho(c, c'') \leq \rho(c, c') \vee \rho(c, c'')$
(iv) Let $S(c, w) = \{c' : \rho(c, c') \leq w\}$. Then $\{S(c, w) : w \in W_c\}$ is a base of $c \in C$ in the sense
  (α) $(\forall w \in W_c)(S(c, w) \in \theta)$
  (β) $(\forall \alpha \in \theta)(c \in \alpha \to (\exists w \in W_c)(c \in S(c, w) \subset \alpha))$

Then ρ is called a generalized pseudodistance function with respect to θ. If W_c is a sublattice for each c, ρ is called a generalized distance function.

The following is a main result.

Proposition 2.1. Let C be an arbitrary set and θ a family of subsets of C. (C, θ) has a generalized pseudodistance function ρ.

PROOF. Let $\theta = \{\alpha_i : i \in I\}$, where I is an index set for θ. Let $W = \{0, 1\}^I = \{w : I \to \{0, 1\}\}$. If an ordering $\le$ is defined on W as $w \le w' \leftrightarrow (\forall i)(w(i) \le w'(i))$, where $0 < 1$, then W is a complete lattice whose greatest element is 1, where $1(i) = 1$ for every $i \in I$, and whose least element is 0, where $0(i) = 0$ for every $i \in I$. Let $\rho : C \times C \to W$ be such that

$$\rho(c, c') = w \leftrightarrow w(i) = \begin{cases} 1, & \text{if} \quad (c \in \alpha_i \;\&\; c' \notin \alpha_i) \vee (c \notin \alpha_i \;\&\; c' \in \alpha_i) \\ 0, & \text{otherwise} \end{cases}$$

Let $w_i \in W$ be such that

$$w_i(j) = \begin{cases} 0, & \text{if} \quad i = j \\ 1, & \text{otherwise} \end{cases}$$

Let $W_c = \{w_i : c \in \alpha_i \;\&\; i \in I\}$. Then we can show that conditions (i), (ii), (iii), and (iv) are satisfied.

Condition (i). Apparently, $\rho(c, c') \ge 0$ and since $c \in \alpha_i \;\&\; c \notin \alpha_i$ is impossible, $\rho(c, c) = 0$.

Condition (ii). Let $\rho(c, c') = w$ and $\rho(c', c) = w'$. Then for any $i \in I$,

$$w(i) = 1 \leftrightarrow (c \in \alpha_i \;\&\; c' \notin \alpha_i) \vee (c \notin \alpha_i \;\&\; c' \in \alpha_i) \leftrightarrow w'(i) = 1$$

Hence, $\rho(c, c') = \rho(c', c)$.

Condition (iii). Let $\rho(c, c') = w$, $\rho(c', c'') = w'$, and $\rho(c, c'') = w''$. Then for any $i \in I$,

$$w''(i) = 1 \to (c \in \alpha_i \;\&\; c'' \notin \alpha_i) \vee (c \notin \alpha_i \;\&\; c'' \in \alpha_i) \to w(i) = 1 \vee w'(i) = 1$$

Hence, $\rho(c, c'') \le \rho(c, c') \vee \rho(c', c'')$.

Condition (iv). (α) Let $w_i \in W_c$ be arbitrary. Then we shall show that $\rho(c, c') \le w_i \leftrightarrow c' \in \alpha_i$. Let $\rho(c, c') = w'$. If $c' \in \alpha_i$, then $w'(i) = 0$. Hence $w' \le w_i$. If $c' \notin \alpha_i$, then $w'(i) = 1$. Hence $w' \not\le w_i$. Consequently, $S(c, w_i) = \alpha_i \in \theta$. ($\beta$) Let $\alpha_i \in \theta$ be arbitrary. When $c \in \alpha_i$ holds, $w_i \in W_c$. Hence, $c \in S(c, w_i) = \alpha_i$. Q.E.D.

It is clear that the generalized distance function introduced in the proof of Proposition 2.1 is a formalization of the intuition that the change from c to c' is smaller than that from c to c'' if $(\forall \alpha \in \theta)(c \in \alpha \;\&\; c'' \in \alpha \to c' \in \alpha)$.

We shall consider a characterization of the stability by using a generalized distance function. Let C' be a subset of C and θ a subset of $\Pi(C)$. Let $\{\alpha : \alpha \in \theta \;\&\; C' \subset \alpha\}$ be denoted by $\bar{N}(C')$. Then we have the following corollary.

Corollary 2.1. Let C' be a subset of C and θ a subset of $\Pi(C)$. Let $\rho : C \times C \to W$ be a generalized pseudodistance function given in the proof of Proposition

2.1. Let $\rho(C', c) = \inf_{c' \in C'} \rho(c', c)$. Then there exists $W_{c'} \subset W$ for C' such that the following relations hold:

(a) $\rho(C', c) \geq 0$ and if $c \in C'$, then $\rho(C', c) = 0$
(b) $\rho(C', c'') \leq \rho(C', c') \vee \rho(c', c'')$
(c) $(\forall w)(w \in W_{c'} \rightarrow S(C', w) \in \overline{N}(C'))$
(d) $(\forall \alpha \in \overline{N}(C'))(\exists w \in W_{c'})(C' \subset S(C', w) \subset \alpha)$

PROOF. Let $W_{c'} = \bigcap_{c \in C'} W_c$. We shall show that conditions (a), (b), (c), and (d) are satisfied.

Condition (a). $\rho(C', c) \geq 0$ is clear. If $c \in C'$, $\rho(C', c) \leq \rho(c, c) = 0$. Hence, $\rho(C', c) = 0$.

Condition (b). $\rho(c, c'') \leq \rho(c, c') \vee \rho(c', c'')$ holds for every $c \in C'$. Hence,

$$\rho(C', c'') \leq \rho(c, c'') \leq \rho(c, c') \vee \rho(c', c'')$$

holds for every $c \in C'$, which implies that $\rho(C', c'') \leq \rho(C', c') \vee \rho(c', c'')$.

Condition (c). If $w \in W_{c'}$, then $w(i)$ takes the value of 1 for all $i \in I$ except one point where $w(i) = 0$ and $\alpha_i \in \overline{N}(C')$. Hence, $S(C', w) = \{c : \rho(C', c) \leq w\} = \alpha_i \in \overline{N}(C')$, where $w(i) = 0$.

Condition (d). If $\alpha_i \in \overline{N}(C')$, then $w_i \in W_{c'}$ and hence $C' \subset S(C', w_i) = \alpha_i$.

Q.E.D.

(b) Main Theorem†

Without introducing any more structure, we can give necessary and sufficient conditions for stability by using Corollary 2.1.

Let ψ be a preordering relation in a set C. Let N be an arbitrary partially ordered set and N^+ a fixed subset of N. We can introduce then the following definition [13].

Definition 2.2. A function $f: C \rightarrow N$ is a Lyapunov-type function for a subset $C^* \subset C$ if and only if

(i) $(\forall c)(\forall c')[(c, c') \in \psi \rightarrow f(c) \geq f(c')]$
(ii) $(\forall n)(\exists \alpha)(\forall c)[n \in N^+ \;\&\; C^* \subset \alpha \;\&\; c \in \alpha \rightarrow f(c) \leq n]$
(iii) $(\forall \alpha)(\exists n)(\forall c)[n \in N^+ \;\&\; \psi(C^*) \subset \alpha \;\&\; f(c) \leq n \rightarrow c \in \alpha]$

Theorem 2.1. Let C be an abstract set, ψ a preorder in C, and θ an arbitrary family of subsets of C. Then a subset C^* is stable relative to θ and ψ if and only if there exists a Lyapunov-type function $f: C \rightarrow N$.

PROOF. Let us consider the *if* part first. Let $\alpha \in \overline{N}(\psi(C^*))$ be arbitrary. Then it follows from condition (iii) that for some $\hat{n} \in N^+$, $\{c : f(c) \leq \hat{n}\} \subset \alpha$. By applying condition (ii) to $\hat{n}$, we have that there exists $\beta \in \overline{N}(C^*)$ such that

† See Yoshii [42].

$\beta \subset \{c : f(c) \leq \hat{n}\} \subset \alpha$. Furthermore, conditions (i) implies that if $f(c) \leq \hat{n}$ and if $(c, c') \in \psi$, $f(c') \leq \hat{n}$. Hence, $\psi(\beta) \subset \{c : f(c) \leq \hat{n}\} \subset \alpha$. Consequently, C^* is stable. Let us consider the *only if* part next. Let $N = W$, $N^+ = W_{\hat{c}}$, where $\hat{C} = \psi(C^*)$ and $f(c) = \sup_{c' \in \psi(c)} \rho(\hat{C}, c')$, where W, $W_{\hat{c}}$, and ρ are given in Corollary 2.1. We shall show that the three conditions (i), (ii), and (iii) are satisfied.

Condition (i). Since $f(c) = \sup_{c' \in \psi(c)} \rho(\hat{C}, c')$ and since ψ is transitive, condition (i) is obviously satisfied.

Condition (ii). Let $n \in N^+$ be arbitrary. Since $\hat{C} \subset S(\hat{C}, n) \in \overline{N}(\hat{C})$ from Corollary 2.1 and since C^* is stable, there exists $\alpha \in \overline{N}(C^*)$ such that $\psi(\alpha) \subset S(\hat{C}, n)$; that is,

$$(\forall c)(\forall c')(c \in \alpha \,\&\, (c, c') \in \psi \rightarrow \rho(\hat{C}, c') \leq n)$$

Condition (iii). Let $\alpha \in \overline{N}(\hat{C})$ be arbitrary. Then there exists $n \in N^+$ from Corollary 2.1 such that $\hat{C} \subset S(\hat{C}, n) \subset \alpha$. If $f(c) \leq n$, then

$$n \geq f(c) = \sup_{c' \in \psi(c)} \rho(\hat{C}, c') \geq \rho(\hat{C}, c)$$

which implies $c \in S(\hat{C}, n) \subset \alpha$. Q.E.D.

In many cases, the Lyapunov stability is investigated in reference to an invariant set C^*; i.e., $C^* = \psi(C^*)$. When C^* is invariant, Definition 1.4 can be restated as follows.

An invariant set $C^* \subset C$ is stable relative to ϕ and θ if and only if

$$(\forall \alpha \in \overline{N}(C^*))(\exists \beta \in \overline{N}(C^*))(\psi(\beta) \subset \alpha)$$

In other words, when C^* is not an invariant set, α is taken as a neighborhood of $\psi(C^*)$, but when C^* is invariant, α is taken as a neighborhood of C^*. Consequently, in the case of an invariant set, the definition of a Lyapunov-type function becomes the following definition.

Definition 2.3. A function $f : C \rightarrow N$ is a Lyapunov-type function for an invariant set $C^* \subset C$ if and only if

(i) $(\forall c)(\forall c')((c, c') \in \psi \rightarrow f(c) \geq f(c'))$
(ii) $(\forall n)(\exists \alpha)(\forall c)(n \in N^+ \,\&\, C^* \subset \alpha \,\&\, c \in \alpha \rightarrow f(c) \leq n)$
(iii) $(\forall \alpha)(\exists n)(\forall c)(n \in N^+ \,\&\, C^* \subset \alpha \,\&\, f(c) \leq n \rightarrow c \in \alpha)$

The difference between Definition 2.2 and Definition 2.3 is that in condition (iii) of Definition 2.2, α is a superset of $\psi(C^*)$, while in Definition 2.3 α is a superset of C^*. Apparently, Theorem 2.1 holds for Definition 2.3 also.

Corollary 2.2. Let C be an abstract set, ψ a preordering in C, and θ an arbitrary family of subsets of C. Then an invariant set $C^* \subset C$ is stable relative to θ and ψ if and only if there exists a Lyapunov-type function $f : C \rightarrow N$ as specified in Definition 2.3.

(c) Application of the Main Theorem

It should be noticed that the set C so far has no additional structure except that implied by the arbitrary family of subsets θ. Theorem 2.1 can be applied to a large number of specialized cases by adding more structure to basic sets and functions involved. It is interesting to note that the basic structure of a Lyapunov-type function and the necessary and sufficient conditions are captured by Definition 2.2 and Theorem 2.1. The theorem applies directly to more specific systems, and to show the validity of necessary and sufficient conditions for any particular case, one only has to check that the additional structure is not introduced in a way that violates the basic conditions. Apparently, Theorem 2.1 is true when θ is a general topology; i.e., it satisfies the conditions:

(i) $C \in \theta$ and the empty set $\phi \in \theta$.
(ii) The union of any number of members of θ is again a member of θ.
(iii) The intersection of any finite number of members of θ is again a member of θ.

We have therefore the following corollary.

Corollary 2.3. Let C be a general topological space, $C^* \subset C$, and ψ a preorder in C. C^* is then stable if and only if there exists a Lyapunov-type function $f: C \to N$.

Finally, we shall consider the stability in a metric space [16]. A metric space is interesting on two accounts. First, a metric space can illustrate a basic idea of the generalized distance function. Since the actual construction of a Lyapunov-type function is based on the generalized distance function, the idea of a Lyapunov-type function can be explored clearly in a metric space. Second, the range of a Lyapunov-type function is, in general, much more complicated than that of conventional scalar function. However, when C is a metric space, a Lyapunov-type function whose range is the real line (exactly speaking, nonnegative real line) can be constructed.

Let C be a metric space whose metric is $\rho: C \times C \to R$ and whose topology θ is the metric topology. For a subset $C' \subset C$, let

$$\rho(C', c) = \inf_{c' \in C'} \rho(c', c)$$

$$S(C', \varepsilon) = \{c : \rho(C', c) < \varepsilon\}$$

Then if $W_{c'} = R^+$, where R^+ is the set of positive real numbers, the following facts are apparently true:

(i) $\rho(C', c) \geq 0$ and if $c \in C'$, then $\rho(C', c) = 0$
(ii) $(\forall w)(w \in W_{c'} \to S(C', w) \in \bar{N}(C'))$

where $\bar{N}(C')$ is the whole family of open sets including C'. Furthermore, if C' is a compact set, the following also holds:

(iii) $(\forall \alpha \in \bar{N}(C'))(\exists w \in W_{c'})(C' \subset S(C', w) \subset \alpha)$

As a matter of fact, if (iii) is not true, then for any n ($n = 1, 2, 3, \ldots$), there exists $c_n \in S(C', 1/n) \cap (C \setminus \alpha)$. Since $c_n \in S(C', 1/n)$, there exists $\hat{c}_n \in C'$ such that $\rho(c_n, \hat{c}_n) < 2/n$; since C' is compact, the series $\{\hat{c}_n\}$ has an accumulating point $\hat{c}_0 \in C'$. For the sake of simplicity, suppose $\{\hat{c}_n\}$ converges to $\hat{c}_0$; then

$$\rho(c_n, \hat{c}_0) \leq \rho(c_n, \hat{c}_n) + \rho(\hat{c}_n, \hat{c}_0) \to 0 \qquad \text{as} \quad n \to \infty$$

This contradicts the fact that $c_n \in C \setminus \alpha$.

As the proof of Theorem 2.1 shows, the existence of a Lyapunov-type function for the stability of C^* where $C' = \psi(C^*)$ depends on the above three properties, and the suggested Lyapunov-type function $f: C \to R^+ \cup \{0\}$ is such that

$$f(c) = \sup_{c' \in \psi(c)} \rho(\psi(C^*), c')$$

or if C^* is an invariant set,

$$f(c) = \sup_{c' \in \psi(c)} \rho(C^*, c')$$

Formally, we have the following theorem.

Theorem 2.2. Let (C, θ) be a metric space, C^* a compact subset of C, and ψ a preorder on C. Then C^* is stable relative to θ and ψ if and only if there exists a Lyapunov-type function $f: C \to R^+ \cup \{0\}$, where R^+ is the set of positive real numbers and $N^+ = R^+$.

PROOF. The *if* part is obvious. In the proof of the *only if* part, we may have to be careful of the proof of condition (iii). In the present case, $n \geq \rho(\hat{C}, c)$ does not imply that $c \in S(\hat{C}, n)$. However, since there exists $0 < n' < n$ such that $\hat{C} \subset S(\hat{C}, n') \subset S(\hat{C}, n) \subset \alpha$, we have that $n' \geq \rho(\hat{C}, c)$ implies $c \in S(\hat{C}, n) \subset \alpha$.

Q.E.D.

Chapter X

INTERCONNECTIONS OF SUBSYSTEMS, DECOMPOSITION, AND DECOUPLING

An important application of general systems theory is in the large-scale systems area. Essential in this application is the ability to deal with a system as a family of explicitly recognized and interconnecting subsystems. Actually, this is how the so-called "systems approach" is defined in many instances in practice, and the entire systems theory concerned with a system as an indivisible entity is viewed as nothing but a necessary prerequisite for the consideration of the large-scale and complex problems of real importance.

In this chapter, we shall be concerned with various questions related to interactions between subsystems that are interconnected and constitute a given system; the objective is to provide a foundation for new developments in systems theory aimed at large-scale systems applications.

To demonstrate the breadth of possible applications, specific problems in two different areas are considered. Conditions for decoupling of a multivariable system by means of a feedback are given in terms of the functional controllability of the system; both general and linear systems cases are considered. Conditions for decomposition of a finite discrete time system into a special arrangement of simpler subsystems are given.

1. CONNECTING OPERATORS

Interconnection of two or more systems is an extremely simple operation. All one needs to do is to connect an output of a system with the input of another system or to apply the same input to two systems; in practice, e.g., that might simply mean "connecting the wires" as indicated. Unfortunately, formalization of that simple notion is rather cumbersome primarily because of the "bookkeeping" difficulties, i.e., a need to express precisely what is connectable with what and indeed what is actually being connected. To avoid being bogged down by unduly pedantic definitions, we shall introduce first a class of connectable systems and then define various connecting operations in that class.

For any object $V_i = V_{i1} \times \cdots \times V_{in}$, we shall denote by $\bar{V}_i$ the family of component sets of V_i, $\bar{V}_i = \{V_{i1}, \ldots, V_{in}\}$. Let $S_i \subset X_i \times Y_i$ be a general system with the objects

$$X_i = \times\{X_{ij} : j \in I_{x_i}\}, \qquad Y_i = \times\{Y_{ij} : j \in I_{y_i}\}$$

In general, some, but not all, of the component sets of X_i are available for connection. Let Z_{x_i} denote the Cartesian product of such component sets, i.e., $X_{ij} \in \bar{Z}_{x_i}$ means that X_{ij} is a component set of X_i and is available for connection. Denote by $\bar{X}_i^*$ the family of all component sets not included in Z_{x_i}, $\bar{X}_i^* = \{X_{ij} : X_{ij} \notin \bar{Z}_{x_i}\}$, and by X_i^* the Cartesian product of the members from $\bar{X}_i^*$, $X_i^* = \times\{X_{ij} : X_{ij} \in \bar{X}_i^*\}$. The input object can then be represented as the product of two composite components $X_i = X_i^* \times Z_{x_i}$. Analogously, let Z_{y_i} be the Cartesian product of the output components available for connection. Given a system $S_i \subset X_i \times Y_i$, there can be defined, in general, many "different" connectable systems $S_{iz} \subset (X_i^* \times Z_{x_i}) \times (Y_i^* \times Z_{y_i})$, depending upon the selection of Z_{x_i} and Z_{y_i}. The relationship between S_i as defined on X_i, Y_i and S_{iz} as defined on X_i^*, Z_{x_i}, Y_i^*, and Z_{y_i} is obvious. In both cases, we have essentially the same system except for an explicit recognition of the availability of some components for connection.

We can now define a class of connectable systems

$$\bar{S}_z = \{S_{iz} : S_{iz} \subset (X_i^* \times Z_{x_i}) \times (Y_i^* \times Z_{y_i})\}$$

and the connection operations within that class.

Definition 1.1. Let $\circ : \bar{S}_z \times \bar{S}_z \to \bar{S}_z$ be such that $S_1 \circ S_2 = S_3$, where

$$S_1 \subset X_1 \times (Y_1^* \times Z_{x_1}), \qquad S_2 \subset (X_2^* \times Z_{y_2}) \times Y_2$$

$$S_3 \subset (X_1 \times X_2^*) \times (Y_1^* \times Y_2), \qquad Z_{x_1} = Z_{y_2} = Z$$

and

$$((x_1, x_2), (y_1, y_2)) \in S_3 \leftrightarrow (\exists z)((x_1, (y_1, z)) \in S_1 \;\&\; ((x_2, z), y_2) \in S_2)$$

$\circ$ is termed the cascade (connecting) operation.

Definition 1.2. Let $+ : \bar{S}_z \times \bar{S}_z \to \bar{S}_z$ be such that $S_1 + S_2 = S_3$, where

$$S_1 \subset (X_1{}^* \times Z_{x_1}) \times Y_1, \qquad S_2 \subset (X_2{}^* \times Z_{x_2}) \times Y_2$$

$$S_3 \subset (X_1{}^* \times X_2{}^* \times Z) \times (Y_1 \times Y_2), \qquad Z_{x_1} = Z_{x_2} = Z$$

and

$$((x_1, x_2, z), (y_1, y_2)) \in S_3 \leftrightarrow ((x_1, z), y_1) \in S_1 \;\&\; ((x_2, z), y_2) \in S_2$$

$+$ is termed the parallel (connecting) operation.

Definition 1.3. Let $\mathscr{F}$ be the mapping $\mathscr{F} : \bar{S}_z \to \bar{S}_z$ such that $\mathscr{F}(S_1) = S_2$, where $S_1 \subset (X^* \times Z_x) \times (Y^* \times Z_y)$, $S_2 \subset X^* \times Y^*$, $Z_x = Z_y = Z$, and

$$(x, y) \in S_2 \leftrightarrow (\exists z)(((x, z), (y, z)) \in S_1)$$

$\mathscr{F}$ is termed the feedback (connecting) operation.

Examples of the application of the connection operators are given in Fig. 1.1. It should be noticed that the connection operators could be also defined in alternative ways. For example, rather than defining a feedback using a single system and connecting an output with an input as shown in Fig. 1.1, another subsystem could have been assumed in the feedback path as shown in Fig. 1.2. However, the three basic operations as given in Definitions 1.1 to 1.3 cover in combination most of the cases of interest, and in this sense they can be considered as the primary connecting operators. For example, the connection for the case in Fig. 1.2 is given by $\mathscr{F}(S_1 \circ S_2)$, as shown in Fig. 1.3.

Notice that the operations $\circ$, $+$, and $\mathscr{F}$ are defined as partial functions. Although it is possible to make these functions total, we shall not do that for the sake of simplicity.

When $(S_1 \circ S_2) \circ S_3$ is defined, we have $(S_1 \circ S_2) \circ S_3 = S_1 \circ (S_2 \circ S_3)$. As a matter of fact, if S_1, S_2, and S_3 are defined as

$$S_1 \subset X_1 \times (Y_1{}^* \times Z_{y_1})$$

$$S_2 \subset (X_2{}^* \times X_{x_2}) \times (Y_2{}^* \times Z_{y_2})$$

$$S_3 \subset (X_3{}^* \times Z_{x_3}) \times Y_3$$

$$Z_{y_1} = Z_{x_2} = Z, \qquad \text{and} \qquad Z_{y_2} = Z_{x_3} = Z'$$

then

$$(x_1, x_2, x_3, y_1, y_2, y_3) \in (S_1 \circ S_2) \circ S_3$$

$$\leftrightarrow (\exists z)(\exists z')((x_1, y_1, z) \in S_1 \;\&\; (x_2, z, y_2, z') \in S_2 \;\&\; (x_3, z', y_3) \in S_3)$$

$$\leftrightarrow (x_1, x_2, x_3, y_1, y_2, y_3) \in S_1 \circ (S_2 \circ S_3)$$

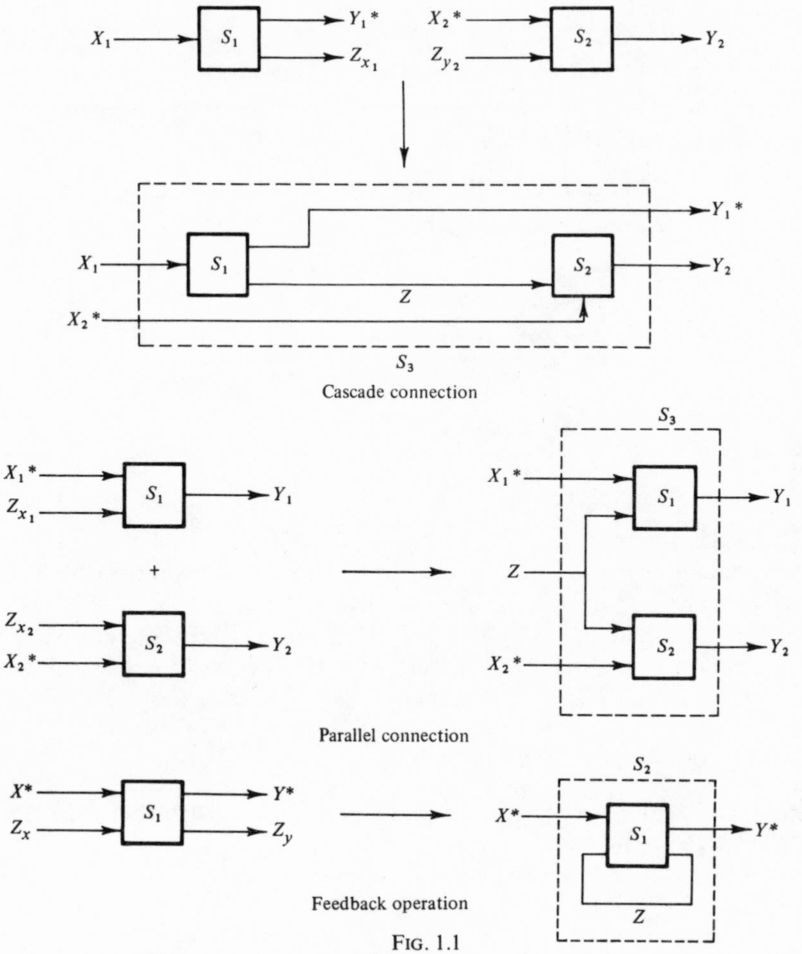
X_1
S_1
$Y_1{}^*$
Z_{x_1}
$X_2{}^*$
Z_{y_2}
S_2
Y_2
Z
S_3
Cascade connection
$X_1{}^*$
Z_{x_1}
Y_1
+
Z_{x_2}
$X_2{}^*$
Y_2
Z
Parallel connection
X^*
Z_x
Y^*
Z_y
Feedback operation

FIG. 1.1

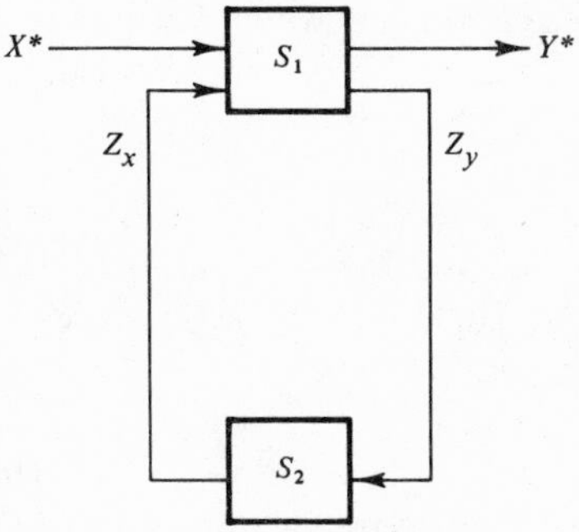
X^*
S_1
Y^*
Z_x
Z_y
S_2

FIG. 1.2

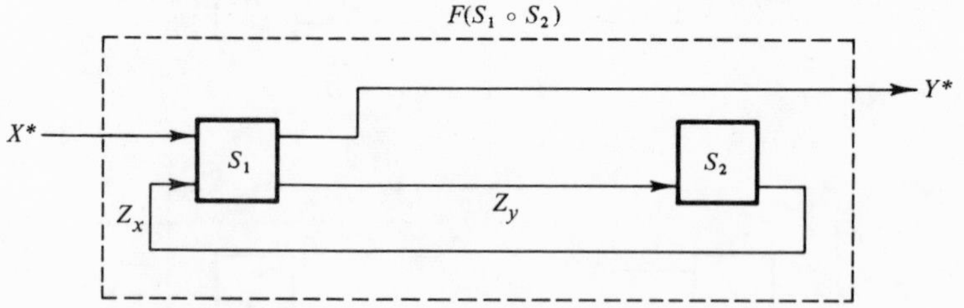

FIG. 1.3

Similarly, we have

$$(S_1 + S_2) + S_3 = S_1 + (S_2 + S_3)$$

if both sides are defined.

The operation $\circ$ does not have identity element. However, in both the input and the output objects, one can define identities $I_x \subset X \times X$, $I_x = \{(x, x): x \in X\}$, and $I_y \subset Y \times Y$, $I_y = \{(y, y): y \in Y\}$. For a given system $S \subset X \times Y$, it is possible then to define the left inverse ${}^{-1}S \subset Y \times X$ and the right inverse $S^{-1} \subset Y \times X$ such that

$${}^{-1}S \circ S = I_y, \qquad \text{and} \qquad S \circ S^{-1} = I_x$$

A function f defined on a family of subsets X_i in X such that $f(X_i) \in X_i$ is called a choice function.

We have then the following immediate propositions.

Proposition 1.1. A system $S \subset X \times Y$ has a right inverse $S^{-1} \subset Y \times X$ if and only if there exists a choice function $f: \{S(x): x \in \mathscr{D}(S)\} \to \mathscr{R}(S)$ such that $\{f(S(x))\} \cap S(x') = \varnothing$ for any $x' \neq x$, where $S(x) = \{y: (x, y) \in S\}$.

Proposition 1.2. Suppose $Y = \mathscr{R}(S)$. Then a system $S \subset X \times Y$ has a left inverse ${}^{-1}S \subset Y \times X$ if and only if there exists a choice function $f: \{(y)S: y \in \mathscr{R}(S)\} \to \mathscr{D}(S)$ such that $\{f((y)S)\} \cap (y')S = \varnothing$ for any $y' \neq y$ where $(y)S = \{x: (x, y) \in S\}$.

For the operation $+$, the empty system $\varnothing$ is the identity. The three operations are interrelated; for instance, $\mathscr{F}(S_1 \circ S_2) = \mathscr{F}(S_2 \circ S_1)$ holds if both sides are defined.

We shall now consider how some properties of the subsystems are affected by interconnections. First, we shall look into the question of nonanticipation. In this respect, we have the following propositions.

Proposition 1.3. If $S_1 \subset X_1 \times (Y_1 \times Z)$ and $S_2 \subset (X_2 \times Z) \times Y_2$ are nonanticipatory, so is $S_3 = S_1 \circ S_2$.

PROOF. Let $(y_1, z) = (\rho_{1y}(c_1, x_1), \rho_{1z}(c_1, x_1))$ and $y_2 = \rho_2(c_2, (x_2, z))$ be nonanticipatory global state representations. Suppose $(x_1, x_2) \mid \bar{T}^t = (x_1', x_2') \mid \bar{T}^t$. Then $y_1 \mid \bar{T}^t = y_1' \mid \bar{T}^t$ and $z \mid \bar{T}^t = z' \mid \bar{T}^t$, where $(y_1', z') = (\rho_{1y}(c_1, x_1'), \rho_{1z}(c_1, x_1'))$, and consequently, $y_2 \mid \bar{T}^t = y_2' \mid \bar{T}^t$, where $y_2' = \rho_2(c_2, (x_2', z'))$. Hence $S_1 \circ S_2$ is nonanticipatory. Q.E.D.

Proposition 1.4. If $S_1 \subset (X_1 \times Z) \times Y$ and $S_2 \subset (X_2 \times Z) \times Y$ are nonanticipatory systems, so is $S_3 = S_1 + S_2$.

Proposition 1.5. Suppose $S \subset (X \times Z) \times (Y \times Z)$ is nonanticipatory such that $(y, z') = (\rho_y(c, x, z), \rho_z(c, x, z))$ is a nonanticipatory systems-response function. If $z \mid \bar{T}^t = \rho_z(c, x, z) \mid \bar{T}^t$ has a unique solution $z \mid \bar{T}^t$ for each (c, x) and t, then $\mathscr{F}(S)$ is also nonanticipatory.

PROOF. It follows from the definition of $\mathscr{F}(S)$ that

$$(x, y) \in \mathscr{F}(S) \leftrightarrow (\exists z)(((x, z), (y, z)) \in S)$$

$$\leftrightarrow (\exists z)(\exists c)(y = \rho_y(c, x, z) \; \& \; z = \rho_z(c, x, z))$$

Since $z = \rho_z(c, x, z)$ has a unique solution for each (c, x), let $z = \phi(c, x) \leftrightarrow z = \rho_z(c, x, z)$. Then

$$(x, y) \in \mathscr{F}(S) \leftrightarrow (\exists c)(y = \rho_y(c, x, \phi(c, x)))$$

Let $\rho(c, x) = \rho_y(c, x, \phi(c, x))$. If $\rho(c, x)$ is nonanticipatory, the desired result follows immediately. Let $\hat{x} \mid \bar{T}^t = \hat{x}' \mid \bar{T}^t$, $\hat{z} = \rho_z(c, \hat{x}, \hat{z})$, and $\hat{z}' = \rho_z(c, \hat{x}', \hat{z}')$. Since ρ_z is nonanticipatory, $\rho_z(c, \hat{x}', \hat{z}) \mid \bar{T}^t = \rho_z(c, \hat{x}, \hat{z}) \mid \bar{T}^t$. Hence, $\phi(c, \hat{x}') \mid \bar{T}^t = \phi(c, \hat{x}) \mid \bar{T}^t$. Since ρ_y is nonanticipatory, we have $\rho(c, \hat{x}) \mid \bar{T}^t = \rho(c, \hat{x}') \mid \bar{T}^t$. Q.E.D.

As an example of the application of Proposition 1.5, consider the system given in Fig. 1.4 and defined by

$$((x, z), (y, z')) \in S \leftrightarrow (\exists y(0))\left(y(t) = z'(t) = y(0) + \int_0^t (x + z)\,dt\right)$$

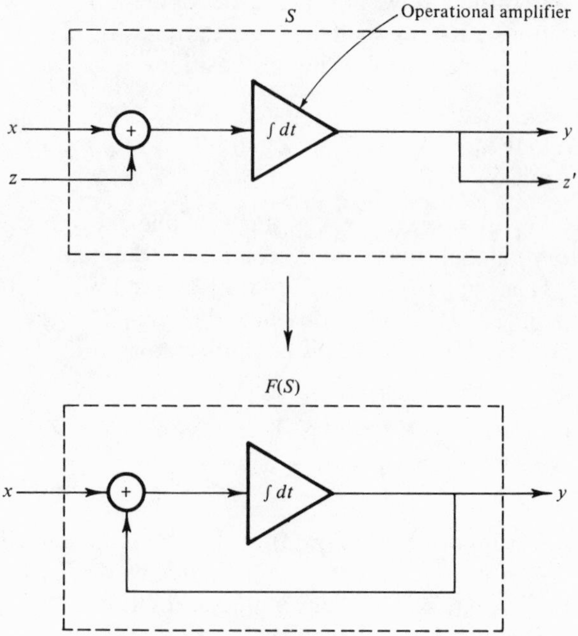

FIG. 1.4

The global-state response function for z is then given by

$$\rho_z(c, x, z) = z' \leftrightarrow z'(t) = c + \int_0^t (x + z)\, dt$$

that is, $z \mid \bar{T}^t = \rho_z(c, x, z) \mid \bar{T}^t$ has a unique solution $z \mid \bar{T}^t$ for each (c, x) and t.

Therefore, by Proposition 1.5 the system is nonanticipatory even after the feedback is closed. This can easily be verified by observing that the system after the feedback is closed is defined by

$$\begin{aligned}(x, y) \in \mathscr{F}(S) &\leftrightarrow (\exists z)(((x, z), (y, z)) \in S) \\ &\leftrightarrow (\exists z)(\exists y(0))\left(y(t) = z(t) = y(0) + \int_0^t (x + z)\, dt\right) \\ &\leftrightarrow dy/dt = x + y\end{aligned}$$

The linearity of a system is preserved by the three interconnecting operations. Recall that a linear system is defined on a linear space.

Proposition 1.6. Suppose S_1 and S_2 are linear systems. Then $S_1 \circ S_2$, $S_1 + S_2$, and $\mathscr{F}(S_1)$ are linear when $S_1 \circ S_2$, $S_1 + S_2$, and $\mathscr{F}(S_1)$ are defined.

PROOF. We shall consider only the case of feedback. Let $S_1 \subset (X \times Z) \times (Y \times Z)$. If $(x, y) \in \mathscr{F}(S_1)$ and $(x', y') \in \mathscr{F}(S_1)$, then there exist $z \in Z$ and $z' \in Z$ such that (x, z, y, z)† $\in S_1$ and $(x', z', y', z') \in S_1$. Since S_1 is linear, $(x + x', z + z', y + y', z + z') \in S_1$. Hence, $(x + x', y + y') \in \mathscr{F}(S_1)$. Similarly, $(\alpha x, \alpha y) \in \mathscr{F}(S_1)$, where α is a scalar multiplier. Q.E.D.

In subsequent sections we shall consider functional systems. Recall that:

(i) a system $S \subset X \times Y$ is functional if and only if

$$(x, y) \in S \;\&\; (x, y') \in S \rightarrow y = y'$$

(ii) a system $S \subset X \times Y$ is one-to-one functional if and only if S is functional and

$$(x, y) \in S \;\&\; (x', y) \in S \rightarrow x = x'$$

Proposition 1.7. If S_1 and S_2 are functional, then $S_1 \circ S_2$ and $S_1 + S_2$ are functional if they are defined. Furthermore, the one-to-one functionality is also preserved by the cascade and parallel operations.

In general, the functionality is not preserved by the feedback operation.

Proposition 1.8. Suppose $S \subset (X \times Z) \times (Y \times Z)$ is functional. Let

$$S(x) = \{z : (\exists y)((x, z, y, z) \in S)\}$$

$$S(x, y) = \{z ; (\exists z')((x, z, y, z') \in S)\}$$

Then $\mathscr{F}(S)$ is functional if and only if for each $x \in X$

(C1) $$(\exists y)(S(x) \subset S(x, y))$$

holds. In particular, if S satisfies the relation

$$(x, z, y, z) \in S \;\&\; (x, z', y', z') \in S \rightarrow z = z'$$

$\mathscr{F}(S)$ is also functional.

PROOF. Suppose condition (C1) holds. Suppose $(x, y) \in \mathscr{F}(S)$ and $(x, y') \in \mathscr{F}(S)$. Then there exist z and z' such that $(x, z, y, z) \in S$ and $(x, z', y', z') \in S$ hold. Therefore, $z \in S(x)$ and $z' \in S(x)$. It follows from condition (C1) that there exists $\hat{y}$ such that $z \in S(x, \hat{y})$ and $z' \in S(x, \hat{y})$ hold; that is, $(x, z, \hat{y}, \hat{z}) \in S$ and $(x, z', \hat{y}, \hat{z}') \in S$ hold for some $\hat{z}$ and $\hat{z}'$. Since S is functional from assumption, $(x, z, y, z) \in S$ and $(x, z, \hat{y}, \hat{z}) \in S$ imply that $y = \hat{y}$. Similarly, we have $y' = \hat{y}$. Consequently, $y = y'$.

† For notational convenience, we shall write (x, z, y, z) instead of $((x, z), (y, z))$.

Conversely, suppose $\mathscr{F}(S)$ is functional. Let $\hat{x} \in X$ be arbitrary. If there is no such $\hat{y}$ as $(\hat{x}, \hat{y}) \in \mathscr{F}(S)$, then $S(\hat{x}) = \varnothing$. Then $S(\hat{x}) \subset S(\hat{x}, y)$ trivially holds. Suppose $(\hat{x}, \hat{y}) \in \mathscr{F}(S)$. Let $z \in S(\hat{x})$ be arbitrary. Then there exists y such that $(\hat{x}, z, y, z) \in S$. Since $\mathscr{F}(S)$ is functional, $y = \hat{y}$ holds. Consequently, $z \in S(\hat{x}, \hat{y})$.
Q.E.D.

2. SUBSYSTEMS, COMPONENTS, AND DECOMPOSITION

There are many ways in which the subunits of a system can be defined. We shall introduce only some of the notions that seem to be among the most interesting in application.

Definition 2.1. Let $S \subset X \times Y$ be a general system. A subsystem of S is a subset $S' \subset S$, $S' \subset X \times Y$. A component of S is a system S^* such that S can be obtained from S^* (possibly in conjunction with some other systems) by the application of the basic connecting operators.

It should be noticed that in application, the term "subsystem" is used for the whole range of different concepts, including the notion of a component as given in Definition 2.1. This should be borne in mind in the interpretation of the statements from this chapter.

We shall first consider the component–system relationship.

Given two systems $S_1 \subset X_1 \times Y_1$ and $S_2 \subset X_2 \times Y_2$, let Π_1 and Π_2 be the projection operators,

$$\Pi_1 : (x_1 \times x_2) \times (Y_1 \times Y_2) \to (X_1 \times Y_1)$$

and

$$\Pi_2 : (X_1 \times X_2) \times (Y_1 \times Y_2) \to (X_2 \times Y_2)$$

such that

$$\Pi_1(x_1, x_2, y_1, y_2) = (x_1, y_1), \qquad \text{and} \qquad \Pi_2(x_1, x_2, y_1, y_2) = (x_2, y_2)$$

Let $S \subset (X_1 \times X_2) \times (Y_1 \times Y_2)$.

Definition 2.2. Two systems $S_1 = \Pi_1(S)$ and $S_2 = \Pi_2(S)$ are considered noninteracting (relative to S) if and only if $S = S_1 + S_2$, and $(\Pi_1(S), \Pi_2(S))$ is called a noninteractive decomposition of S.

Definition 2.3. A noninteractive decomposition $(S_1, \ldots, S_n)$, where $S_1 + \cdots + S_n = S$ is called a maximal noninteractive decomposition if and only if no S_i has a (nontrivial) noninteractive decomposition.

It is an interesting question whether or not a system has a maximal noninteractive decomposition. It is obvious that if a system has finitely many components, i.e., $X = X_1 \times \cdots \times X_n$ and $Y = Y_1 \times \cdots \times Y_n$ for some n and m, it has a maximal noninteractive decomposition. This problem will be considered in relation to the feedback in Section 4.

Let us consider basic decompositions by components.

Proposition 2.1. Every system $S \subset (X_1 \times X_2) \times (Y_1 \times Y_2)$ can be decomposed as $S = \mathscr{F}(S_1 \circ S_2)$, that is, in cascade and feedback components, where $S_1 \subset (X_1 \times Z_1) \times (Y_1 \times Z_2)$ and $S_2 \subset (X_2 \times Z_2) \times (Y_2 \times Z_1)$, and Z_1 and Z_2 are auxiliary sets. (See Fig. 2.1.)

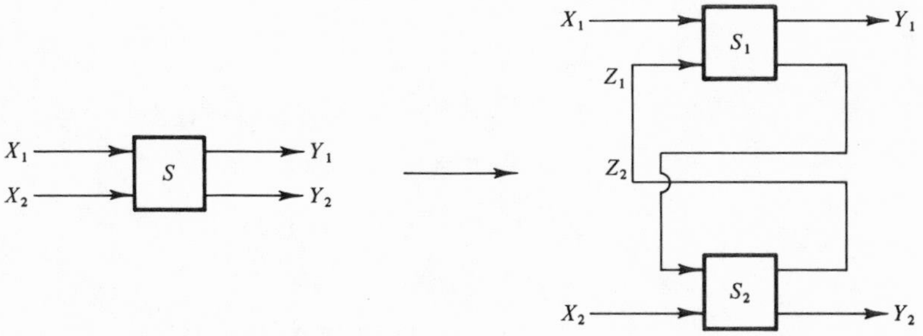

FIG. 2.1

PROOF. Let $Z_1 = X_2 \times Y_2$ and $Z_2 = X_1 \times Y_1$. Let

$$((x_1, z_1), (y_1, z_2)) \in S_1 \leftrightarrow (x_1, x_2, y_1, y_2) \in S \;\&\; z_2 = (x_1, y_1)$$

where $z_1 = (x_2, y_2)$. Let

$$((x_2, z_2), (y_2, z_1)) \in S_2 \leftrightarrow (x_1, x_2, y_1, y_2) \in S \;\&\; z_1 = (x_2, y_2)$$

where $z_2 = (x_1, y_1)$. Now, let $S' = \mathscr{F}(S_1 \circ S_2)$. Then

$$\begin{aligned}(x_1, x_2, y_1, y_2) \in S' &\leftrightarrow (\exists z_1)(\exists z_2)((x_1, z_1, y_1, z_2) \in S_1 \;\&\; (x_2, z_2, y_2, z_1) \in S_2)) \\ &\leftrightarrow (\exists (x_2', y_2'))(\exists (x_1', y_1'))((x_1, x_2', y_1, y_2') \in S \;\&\; (x_1', y_1') \\ &\qquad = (x_1, y_1) \;\&\; (x_1', x_2, y_1', y_2) \in S \;\&\; (x_2', y_2') = (x_2, y_2)) \\ &\leftrightarrow (x_1, x_2, y_1, y_2) \in S\end{aligned}$$

Q.E.D.

Proposition 2.2. Every system $S \subset (X_1 \times X_2) \times (Y_1 \times Y_2)$ can be decomposed into cascade components as shown in Fig. 2.2.

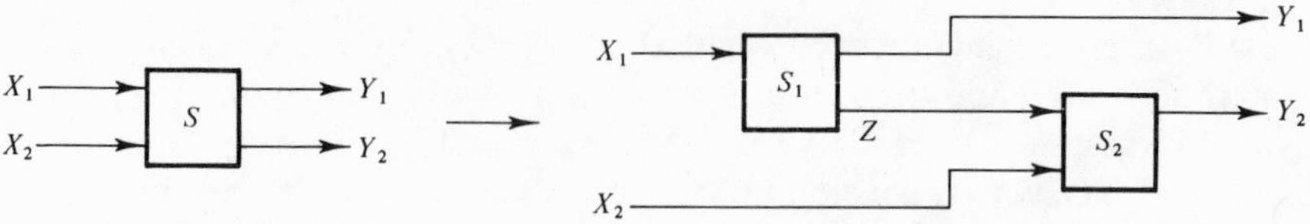

FIG. 2.2

PROOF. Let $Z = X_1 \times Y_1$. Let $S_1 \subset X_1 \times (Y_1 \times Z_1)$ be

$$(x_1, (y_1, z)) \in S_1 \leftrightarrow (\exists(x_2, y_2))((x_1, x_2, y_1, y_2) \in S \;\&\; z = (x_1, y_1))$$

Let $S_2 \subset (X_2 \times Z) \times Y_2$ be: For $z = (x_1, y_1)$

$$((x_2, z), y_2) \in S_2 \leftrightarrow (x_1, x_2, y_1, y_2) \in S$$

Then

$$\begin{aligned}(x_1, x_2, y_1, y_2) \in S_1 \circ S_2 &\leftrightarrow (\exists z)((x_1, y_1, z) \in S_1 \;\&\; (x_2, z, y_2) \in S_2)\\ &\leftrightarrow (\exists z)(\exists(x_2', y_2'))((x_1, x_2', y_1, y_2') \in S\\ &\qquad \& \; z = (x_1, y_1) \;\&\; (x_2, z, y_2) \in S_2)\\ &\leftrightarrow (\exists(x_2', y_2'))((x_1, x_2', y_1, y_2') \in S\\ &\qquad \& \; (x_1, x_2, y_1, y_2) \in S)\\ &\leftrightarrow (x_1, x_2, y_1, y_2) \in S\end{aligned}$$

Q.E.D.

Proposition 2.3. Let $S \subset (X_1 \times X_2) \times (Y_1 \times Y_2)$ and $S(x) = \{y : (x, y) \in S\}$, where $X = X_1 \times X_2$ and $Y = Y_1 \times Y_2$. Let $\Pi_1(y_1, y_2) = y_1$ and $\Pi_2(y_1, y_2) = y_2$. Then, if and only if $S(x) = \Pi_1(S(x)) \times \Pi_2(S(x))$ for every $x \in \mathscr{D}(S)$, S can be decomposed by some S_1 and S_2 as shown in Fig. 2.3.

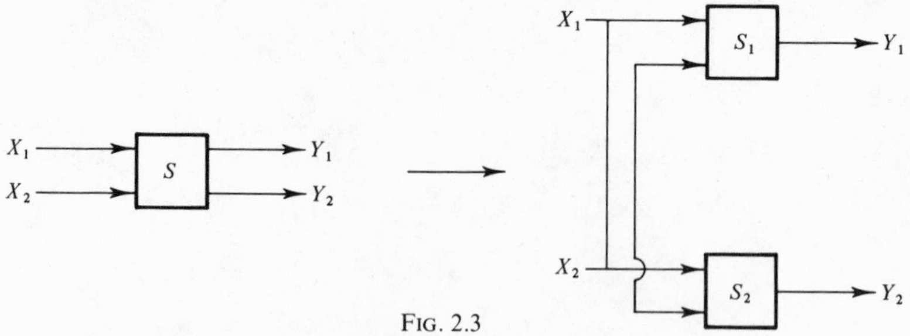

FIG. 2.3

PROOF. Suppose $S(x) = \Pi_1(S(x)) \times \Pi_2(S(x))$ for every $x \in \mathscr{D}(S)$. Let $S_1 \subset X \times Y_1$ be such that

$$(x, y_1) \in S_1 \leftrightarrow (\exists y_2)((x, y_1, y_2) \in S)$$

and $S_2 \subset X \times Y_2$ be such that

$$(x, y_2) \in S_2 \leftrightarrow (\exists y_1)((x, y_1, y_2) \in S)$$

Apparently

$$(x, y_1, y_2) \in S \rightarrow (x, y_1) \in S_1 \;\&\; (x, y_2) \in S_2$$

Conversely, suppose $(x, y_1) \in S_1 \;\&\; (x, y_2) \in S_2$. Then $y_1 \in \Pi_1 S(x)$ and $y_2 \in \Pi_2 S(x)$; that is, $(y_1, y_2) \in S(x)$ because $S(x) = \Pi_1(S(x)) \times \Pi_2(S(x))$. Consequently, $(x, y_1, y_2) \in S$.

Suppose there exist $S_1 \subset X \times Y_1$ and $S_2 \subset X \times Y_2$ such that the above decomposition is realized. Apparently, $S(x) \subset \Pi_1(S(x)) \times \Pi_2(S(x))$. Let $y_1 \in \Pi_1(S(x))$ and $y_2 \in \Pi_2(S(x))$ be arbitrary. Then $(x, y_1) \in S_1$ and $(x, y_2) \in S_2$ should hold. Consequently, $(x, y_1, y_2) \in S$. Q.E.D.

Proposition 2.4. Every system $S \subset X \times Y$ can be decomposed by some S' as shown in Fig. 2.4, where

$$(x, x') \in E_x \leftrightarrow S(x) = S(x')$$

$$(y, y') \in E_y \leftrightarrow (y)S = (y')S$$

$S(x)$ and $(y)S$ are defined in Propositions 1.1 and 1.2, and $\eta_x : X \rightarrow X/E_x$, $\eta_y : Y \rightarrow Y/E_y$ are the natural mappings; that is, $([y], y) \in \eta_y^{-1} \leftrightarrow [y] = \eta_y(y)$.

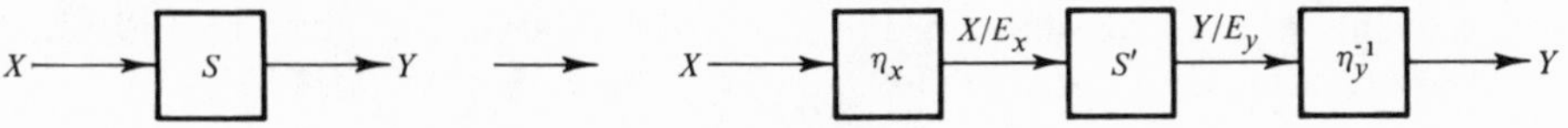

FIG. 2.4

PROOF. Let $S' \subset (X/E_x) \times (Y/E_y)$ be

$$([x], [y]) \in S' \leftrightarrow (\exists \hat{x})(\exists \hat{y})(\hat{x} \in [x] \;\&\; \hat{y} \in [y] \;\&\; (\hat{x}, \hat{y}) \in S)$$

We shall show that

$$(\forall x')(\forall y')(x' \in [x] \;\&\; y' \in [y] \;\&\; ([x], [y]) \in S' \rightarrow (x', y') \in S)$$

Suppose $\hat{x} \in [x]$ and $\hat{y} \in [y]$ such that $(\hat{x}, \hat{y}) \in S$. Such elements exist for $([x], [y]) \in S'$ by definition. Since $x' \in [x]$, we have $S(\hat{x}) = S(x')$. Furthermore, $\hat{y} \in S(\hat{x})$ implies $(x', \hat{y}) \in S$. Consequently, $x' \in (\hat{y})S$. On the other hand, $y' \in [y]$ and $\hat{y} \in [y]$ imply $(y')S = (\hat{y})S$. Hence, we have $x' \in (y')S$, that is, $(x', y') \in S$.

Q.E.D.

In conclusion, we shall briefly consider the subsystem–system relationship.

Definition 2.4. Let S_i $(i \in I)$ be a functional subsystem of $S \subset X \times Y$; i.e., $S_i \subset S$ and $S_i : \mathscr{D}(S) \rightarrow Y$. If $\bigcup_{i \in I} S_i = S$, then $\{S_i : i \in I\}$ is called a functional system decomposition.

A functional system decomposition is apparently equivalent to a global-state representation. Such a decomposition is, therefore, always possible and is not unique.

Definition 2.5. Let $\bar{S}_\alpha = \{S_{\alpha i} : i \in I_\alpha\}$ and $\bar{S}_\beta = \{S_{\beta i} : i \in I_\beta\}$ be two functional system decompositions of $S \subset X \times Y$. Let $\bar{S}_\alpha \leq \bar{S}_\beta$ if and only if $\bar{S}_\alpha \subseteq \bar{S}_\beta$. Then if a decomposition $\bar{S}_o$ satisfies the relation that for any functional system decomposition $\bar{S}_\alpha$

$$\bar{S}_\alpha \leq \bar{S}_o \rightarrow \bar{S}_o = \bar{S}_\alpha$$

$\bar{S}_o$ is called a minimal functional system decomposition of S.

It can be shown quite readily that every finite system has a minimal functional system decomposition.

3. FEEDBACK CONNECTION OF COMPONENTS

In this section, we shall consider the feedback system as given in Fig. 3.1. The overall system is defined as

$$S \subset (X \times Z_x) \times (Y \times Z_y)$$

and the feedback component itself is

$$S_f \subset Z_y \times Z_x$$

Denote by S_B the set of all input–output pairs of the system S in $X \times Y$, i.e.,

$$S_B = \{(x, y) : (\exists(z, z'))(((x, z), (y, z')) \in S)\}$$

and by $\bar{S}_f$ the class of all possible feedback components

$$\bar{S}_f = \{S_f : S_f \subset Z_y \times Z_x\}$$

Let $\mathscr{F}_s$ be the map $\mathscr{F}_s : \bar{S}_f \rightarrow \bar{S}$ such that

$$\mathscr{F}_s(S_f) = \mathscr{F}(S \circ S_f)$$

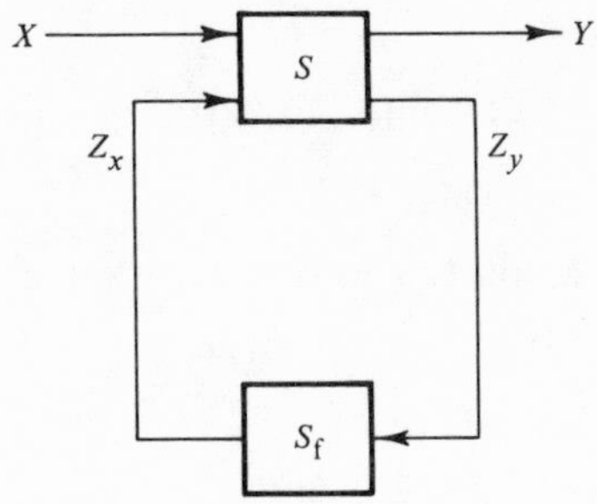

FIG. 3.1

For any given feedback component, $\mathscr{F}_s$ gives the resulting overall system. $\mathscr{F}_s$ therefore indicates the consequences of applying a given feedback. Some conceptually important properties of $\mathscr{F}_s$ are given by the following propositions.

Proposition 3.1. For every $S_f \in \bar{S}_f$, $\mathscr{F}_s(S_f) \subset S_B$.

PROOF.

$$\begin{aligned}(x, y) \in \mathscr{F}_s(S_f) &\leftrightarrow (x, y) \in \mathscr{F}(S \circ S_f)\\ &\leftrightarrow (\exists z)(((x, z), (y, z)) \in S \circ S_f)\\ &\leftrightarrow (\exists z)(\exists z')(((x, z), (y, z')) \in S \;\&\; (z', z) \in S_f)\\ &\rightarrow (x, y) \in S_B\end{aligned}$$

Q.E.D.

Proposition 3.2. Let $\mathscr{G}_s : \Pi(S_B) \rightarrow \bar{S}_f$ be such that

$$(z', z) \in \mathscr{G}_s(S') \leftrightarrow (\exists (x, y) \in S')(((x, z), (y, z')) \in S)$$

Then $S' \subset \mathscr{F}_s(\mathscr{G}_s(S'))$ holds. If S satisfies the relation

(C1) $$(x, z, y, z') \in S \;\&\; (\hat{x}, z, \hat{y}, z') \in S \rightarrow (x, y) = (\hat{x}, \hat{y})$$

then

$$\mathscr{F}_s(\mathscr{G}_s(S')) = S' \leftrightarrow S' \in \mathscr{R}(\mathscr{F}_s)$$

PROOF. In general, since $S' \subset S_B$,

$$(x, y) \in S' \to (\exists(z, z'))((x, z, y, z') \in S \,\&\, (z', z) \in \mathcal{G}_s(S'))$$
$$\to (x, y) \in \mathcal{F}_s(\mathcal{G}_s(S'))$$

Suppose condition (C1) holds. Obviously, $\mathcal{F}_s(\mathcal{G}_s(S')) = S' \to S' \in \mathcal{R}(\mathcal{F}_s)$. Conversely, we have that

$$(x, y) \in \mathcal{F}_s(\mathcal{G}_s(S')) \to (\exists(z', z))((z', z) \in \mathcal{G}_s(S') \,\&\, (x, z, y, z') \in S)$$
$$\to (\exists(z', z))(\exists(\hat{x}, \hat{y}))((\hat{x}, \hat{y}) \in S' \,\&\, (\hat{x}, z, \hat{y}, z') \in S$$
$$\& \, (x, z, y, z') \in S)$$
$$\to (x, y) = (\hat{x}, \hat{y}) \in S'$$

(because of the condition stated in the proposition).

Consequently, since $S' \subset \mathcal{F}_s(\mathcal{G}_s(s'))$ generally holds, we have $S' = \mathcal{F}_s(\mathcal{G}_s(S'))$.

Q.E.D.

Let $S_{fB} = \{(z', z) : (\exists(x, y))((x, z, y, z') \in S)\} \subset Z_y \times Z_x$. Let

$$\bar{S}_f^* = \{S_f^* : S_f^* \subset S_{fB}\}$$

Proposition 3.3. Suppose S satisfies the relation

$$(x, z, y, z') \in S \,\&\, (x, \hat{z}, y, \hat{z}') \in S \to (z, z') = (\hat{z}, \hat{z}')$$

The restriction of $\mathcal{F}_s$ to $\bar{S}_f^*$, $\mathcal{F}_s \mid \bar{S}_f^*$ is then a one-to-one mapping, and if $S' = \mathcal{F}_s(S_f^*)$, then $S_f^* = \mathcal{G}_s(S')$.

PROOF. Notice that if $S' = \mathcal{F}_s(S_f^*)$ holds, we have

$$(z', z) \in \mathcal{G}_s(S') \leftrightarrow (\exists(x, y))((x, y) \in S' \,\&\, (x, z, y, z') \in S)$$
$$\leftrightarrow (\exists(x, y))(\exists(\hat{z}', \hat{z}))((\hat{z}', \hat{z}) \in S_f^* \,\&\, (x, \hat{z}, y, \hat{z}') \in S \,\&\, (x, z, y, z') \in S)$$
$$\leftrightarrow (z', z) = (\hat{z}', \hat{z}) \in S_f^*$$

Q.E.D.

Consider now conceptual interpretation of Propositions 3.1, 3.2, and 3.3. Important properties of a feedback and the consequences of applying feedback to a given system (e.g., for the purpose of decoupling) can be studied in reference to the properties of $\mathcal{F}_s$. Proposition 3.1 indicates that the application of any feedback results in a restriction in S_B. To illustrate the importance of Proposition 3.2, consider a feedback connection defined on a linear space as given in Fig. 3.2, i.e., $S \subset (X \times Z_x) \times (Y \times Z_y)$, $Z_x = X$, $Z_y = Y$, and

$$(x, z, y, z') \in S \leftrightarrow z' = y \,\&\, (x + z, y) \in S_1$$

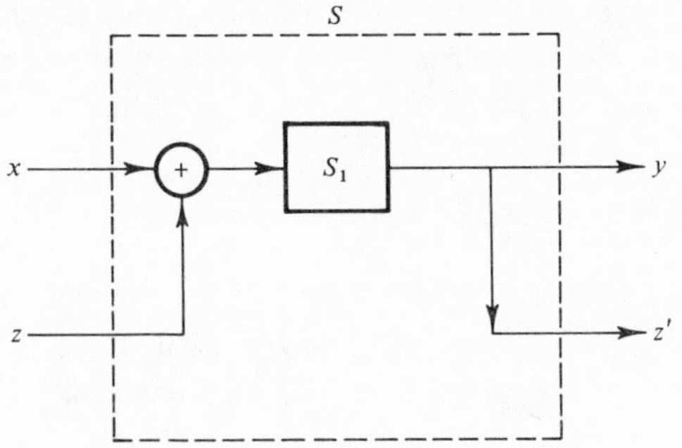

FIG. 3.2

Since X is linear, the addition operation is defined in X. Assume that S is functional (e.g., an initial state is fixed) and the input space is reduced such that S_1 is a one-to-one mapping. The conditions from Propositions 3.2 and 3.3 are then satisfied. Suppose S' is an arbitrary subsystem of S_B, $S' \subset S_B$. Proposition 3.2 gives, then, the conditions when S' can be synthesized by a feedback. S' can be obtained from S by applying feedback if and only if $\mathscr{F}_s(\mathscr{G}_s(S')) = S'$; furthermore, if S' can be synthesized, the required feedback is given by $\mathscr{G}_s(S')$. This will be studied further in Section 4.

Additional properties of $\mathscr{F}_s$ are given in the following propositions.

Proposition 3.4. $\mathscr{F}_s$ is an order homomorphism, i.e., $S_f \subseteq S_f' \rightarrow \mathscr{F}_s(S_f) \subseteq \mathscr{F}_s(S_f')$.

Proposition 3.5. $\mathscr{G}_s$ is an order homomorphism, i.e., $S' \subseteq \hat{S}' \rightarrow \mathscr{G}_s(S') \subseteq \mathscr{G}_s(\hat{S}')$.

Proposition 3.6. Suppose S satisfies the condition of Proposition 3.3. Then the class of the systems that can be synthesized by feedbacks forms a closure system on S_B. $\mathscr{F}_s \circ \mathscr{G}_s$ is the closure operator for that closure system.

PROOF. Let $J = \mathscr{F}_s \circ \mathscr{G}_s : \Pi(S_B) \rightarrow \Pi(S_B)$. It follows from Propositions 3.4 and 3.5 that if $S' \subseteq \hat{S}' \subseteq S_B$, then $J(S') \subseteq J(\hat{S}')$. Proposition 3.2 shows that $S' \subseteq J(S')$ for any $S' \subseteq S_B$. Furthermore, since $\mathscr{F}_s(\mathscr{G}_s(S')) \in \mathscr{R}(\mathscr{F}_s)$ for any S', it follows from Proposition 3.2 also that $JJ(S') = J(S')$ for any $S' \subseteq S_B$. Since $J(S') = S' \leftrightarrow S' \in \mathscr{R}(\mathscr{F}_s)$, $\mathscr{R}(\mathscr{F}_s)$ forms a closure system on S_B.

Q.E.D.

4. DECOUPLING AND FUNCTIONAL CONTROLLABILITY

The objective of this section is to give necessary and sufficient conditions for the decoupling of a system via feedback. It represents an illustration of the use of the general systems framework developed in this chapter. Many other systems problems that are essentially structural or algebraic in nature can be treated in a similar way.

(a) Decoupling of Functional Systems

We shall consider a system $S \subset (X \times Z_x) \times (Y \times Z_y)$ with a feedback component $S_f \subset Z_y \times Z_x$ connected as shown in Fig. 4.1. Furthermore, it will always be assumed that the following condition is satisfied:

(P1) $$(x, z_x, y, z_y) \in S \rightarrow y = z_y$$

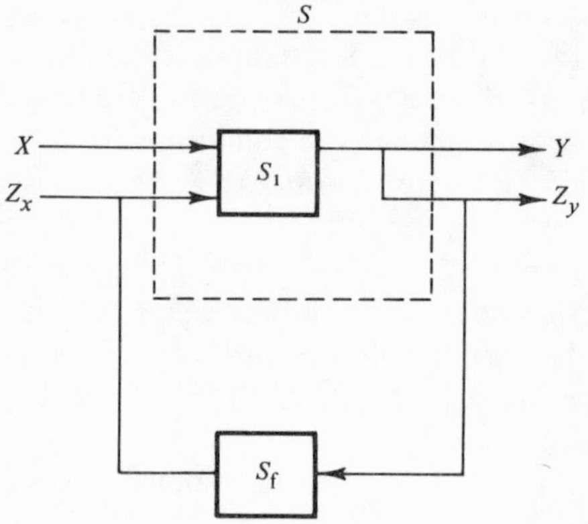

FIG. 4.1

i.e., $Z_y = Y$. We shall consider, therefore, the feedback system as defined on $(X \times Z_x) \times Y$ rather than on $(X \times Z_x) \times (Y \times Z_y)$. The notation $\mathscr{F}(S \circ S_f)$, where $S_f \subset Y \times Z_x$, is then not quite consistent with the previous definitions. However, this inconsistency should be allowed here for simplification of the notation.

Before considering the question of decoupling explicitly, some additional properties of the feedback connection need to be established.

Proposition 4.1. Consider a feedback system $S \subset (X \times Z_x) \times Y$ whose feedback component is $S_f \subset Y \times Z_x$. Suppose

(i) S_f and $\mathscr{F}_s(S_f)$ are functional, i.e., $S_f:(Y) \rightarrow Z_x$, $\mathscr{F}_s(S_f):(X) \rightarrow Y$.
(ii) S satisfies the condition

(P2) $$(x, z, y) \in S \;\&\; (x', z, y) \in S \rightarrow x = x'$$

Then $\mathscr{F}_s(S_f):(X) \rightarrow Y$ is a one-to-one functional.

PROOF. Suppose $\mathscr{F}_s(S_f)(x) = y$ and $\mathscr{F}_s(S_f)(x') = y$. Then

$$(\exists z)((x, z, y) \in S \ \& \ z = S_f(y))$$

and

$$(\exists z')((x', z', y) \in S \ \& \ z' = S_f(y))$$

Consequently, we have $z = z'$. Hence, (P2) implies $x = x'$. Q.E.D.

Using Proposition 4.1 we shall show a basic result.

Proposition 4.2. Suppose

(i) S is functional, $S: \mathscr{D}(S) \to Y$;
(ii) S satisfies (P2);
(iii) S satisfies the condition

(P3) $$(x, z, y) \in S \ \& \ (x, z', y) \in S \to z = z'$$

Let $\bar{S}_f \subset \Pi(Y \times Z_x)$ be the set of all functional feedback components defined in $Y \times Z_x$, i.e., $S_f: \mathscr{D}(S_f) \to Z_x$, such that $\mathscr{F}(S \cdot S_f)$ is functional. Let $\hat{S}$ be an arbitrary system $\hat{S} \subset X \times Y$, and $K(X, Y)$ is a family of subsystems of $\hat{S}$ such that

$$K(X, Y) = \{S' : S' \subset \hat{S} \ \& \ S' \text{ is one-to-one functional}\}$$

Then

$$\mathscr{F}_s(\bar{S}_f) = K(X, Y) \qquad \text{if and only if} \quad \hat{S} = S_B$$

where

$$S_B = \{(x, y) : (\exists z)((x, z, y) \in S)\}$$

PROOF. We consider the *only if* part first. Suppose $\mathscr{F}_s(\bar{S}_f) = K(X, Y)$ holds. Let $(x, y) \in \hat{S}$ be arbitrary. Since $\mathscr{F}_s(\bar{S}_f) = K(X, Y)$ and since $\bigcup K(X, Y) = \hat{S}$, there exists $S_f \in \bar{S}_f$ such that $(x, y) \in \mathscr{F}_s(S_f)$; that is, $(\exists z)((y, z) \in S_f \ \& (x, z, y) \in S)$. Hence, $(x, y) \in S_B$. Let $(\hat{x}, \hat{y}) \in S_B$ be arbitrary such that $(\exists \hat{z})((\hat{x}, \hat{z}, \hat{y}) \in S)$. Let $S_f = \{(\hat{y}, \hat{z})\}$. We shall show that $S_f \in \bar{S}_f$. Notice that if $(x, y) \in \mathscr{F}(S \cdot S_f)$ and $(x, y') \in \mathscr{F}(S \cdot S_f)$, then $y = \hat{y} = y'$ holds; that is, $\mathscr{F}(S \cdot S_f)$ is trivially functional. Furthermore, S_f is functional. Therefore, $S_f \in \bar{S}_f$. Naturally, $(\hat{x}, \hat{y}) \in \mathscr{F}_s(S_f)$. Since $\mathscr{F}_s(\bar{S}_f) = K(X, Y)$, $(\hat{x}, \hat{y}) \in \hat{S}$.

Next, we consider the *if* part. Suppose $\hat{S} = S_B$ holds. Let $S' \in \mathscr{F}_s(\bar{S}_f)$ be arbitrary. Then there exists $S_f \in \bar{S}_f$ such that $S' = \mathscr{F}_s(S_f)$. S' is functional due to the assumption of $\bar{S}_f$ and satisfies the relation $S' \subset S_B = \hat{S}$ because $S' = \mathscr{F}(S \cdot S_f)$. Since S' is one to one from Proposition 4.1, S' is an element of $K(X, Y)$.

Let $S' \in K(X, Y)$ be arbitrary. Since $\hat{S} = S_B$, we have $S' \subset S_B$. Let $S_f = \mathscr{G}_s(S')$; that is,

$$S_f = \{(y, z) : (\exists x)((x, y) \in S' \,\&\, (x, z, y) \in S)\}$$

Suppose $(y, z) \in S_f$ and $(y, z') \in S_f$. Then

$$(\exists x)(\exists x')((x, y) \in S' \,\&\, (x', y) \in S' \,\&\, (x, z, y) \in S \,\&\, (x', z', y) \in S)$$

Since S' is one to one, we have $x = x'$. Consequently, (P3) implies that $z = z'$. Therefore, S_f is functional. We can show $\mathscr{F}(S \cdot S_f) = S'$ because

$$\begin{aligned}\mathscr{F}(S \cdot S_f) \ni (x, y) &\leftrightarrow (\exists z)((x, z, y) \in S \,\&\, (y, z) \in S_f)\\ &\leftrightarrow (\exists z)(y = S(x, z) \,\&\, (\exists \hat{x})(y = S'(\hat{x}) \,\&\, y = S(\hat{x}, z)))\\ &\leftrightarrow (\exists z)(\exists \hat{x})(y = S(x, z) \,\&\, y = S'(\hat{x}) \,\&\, y = S(\hat{x}, z))\\ &\leftrightarrow (\exists z)(y = S(x, z) \,\&\, y = S'(x))\end{aligned}$$

because (P2) implies $x = \hat{x} \leftrightarrow (x, y) \in S'$, and because $S' \subset S_B$. Therefore, $\mathscr{F}(S \cdot S_f) = S'$, where S' is functional by definition. Consequently, $S_f \in \bar{S}_f$, which implies $S' \in \mathscr{F}_s(\bar{S}_f)$. Q.E.D.

We shall apply Proposition 4.2 to decoupling of multivariable systems. We shall first introduce the formal definitions of decoupling by a feedback and functional controllability.

Definition 4.1. A general system $S \subset X \times Y$ is functionally controllable if and only if

$$(\forall y \in Y)(\exists x \in X)((x, y) \in S)$$

Definition 4.2. A multivariable general system $S \subset (X_1 \times \cdots \times X_n) \times Z_x \times (Y_1 \times \cdots \times Y_n)$ is called decoupled by feedback if and only if there exists a feedback system $S_f \subset (Y_1 \times \cdots \times Y_n) \times Z_x$ such that

$$\mathscr{F}(S \circ S_f) = S_1 + \cdots + S_n$$

where $S_i \subset X_i \times Y_i$ $(i = 1, \ldots, n)$ is functionally controllable.

Decoupling as defined by Definition 4.2 means that after the appropriate feedback S_f is applied, each of the output components, for example, y_i, can be changed by changing solely the corresponding input x_i, while no other output is affected by such changes of x_i. The functional controllability, on the other hand, means that any output $(y_1, \ldots, y_n) \in Y$ can be reproduced by an appropriate selection of the input x.

Definitions 4.1 and 4.2 are given for general systems. If a general system is functional as considered in this section, the necessary modifications are obvious. In particular, if S is a multivariable functional system, i.e.,

$$S:(X_1 \times \cdots \times X_n) \times Z_x \to (Y_1 \times \cdots \times Y_n)$$

a feedback system S_f associated with S is always assumed functional; i.e., $S_f: Y_1 \times \cdots \times Y_n \to Z_x$. Relations between the two concepts are given by the following propositions.

Proposition 4.3. Let S be a multivariable functional system $S:(X \times Z_x) \to Y$, where $X = X_1 \times \cdots \times X_n$, $Z_x = Z_{x_1} \times \cdots \times Z_{x_n}$, and $Y = Y_1 \times \cdots \times Y_n$. If S can be decoupled by a feedback S_f, then S is functionally controllable, i.e.,

$$(\forall y \in Y)(\exists (x, z) \in X \times Z_x)(y = S(x, z))$$

PROOF. Let $S_f: Y \to Z_x$ be the feedback that yields the decoupling. Let $\hat{y} = (\hat{y}_1, \ldots, \hat{y}_n) \in Y$ be arbitrary. Let $\hat{x}_i \in X_i$ be such that $\hat{y}_i = S_i(\hat{x}_i)$. This is possible because S_i is functionally controllable. Then $\mathscr{F}(S \circ S_f)(\hat{x}) = \hat{y}$, where $\hat{x} = (\hat{x}_1, \ldots, \hat{x}_n)$. This is also true because $\mathscr{F}(S \circ S_f) = S_1 + \cdots + S_n$. Consequently, it follows from the definitions of the feedback that there exists $\hat{z} \in Z_x$ such that $\hat{y} = S(\hat{x}, \hat{z})$ and $\hat{z} = S_f(\hat{y})$ hold. Q.E.D.

In application, the problem of decoupling is of interest in a more specialized setting. Namely, there is given a system $S_o: X_o \to Y$, and the question is asked whether a feedback input can be combined, in a given sense, with the external input so that the resulting feedback system is decoupled. To provide for the possibility to combine the external inputs and with feedback inputs, an additional component H is given, resulting in the overall system as shown in Fig. 4.2. H will be referred to as the input port, $H: X \times Z_x \to X_o$. A

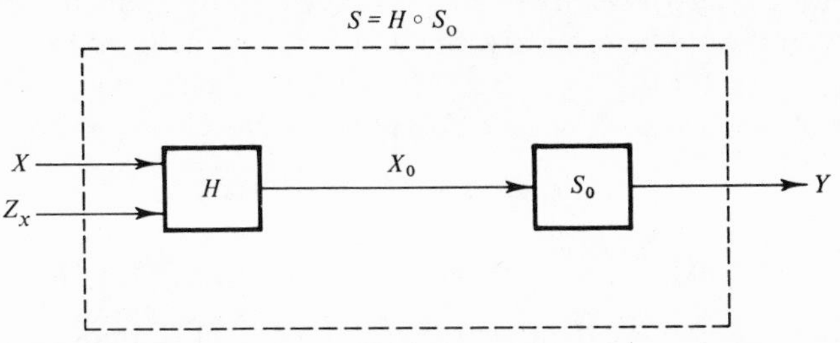

FIG. 4.2

problem of decoupling is then the following. Given a system S_o, an input port of a given type H, and a class of feedbacks $\bar{S}_f$, what conditions S_o should be satisfied so that there exists a feedback $S_f \in \bar{S}_f$ such that $\mathscr{F}_s(S_f)$ is decoupled, where $S = H \circ S_o$?

The answer to this question is given by the following definition.

Definition 4.3. An input port $H: X \times Z_x \to X_0$ is termed complete if and only if it satisfies the following condition:

$$(\forall x \in X)(\forall x_0 \in X_0)(\exists z \in Z_x)(x_0 = H(x, z))$$

Proposition 4.4. Let $X = X_1 \times \cdots \times X_n$, $Y = Y_1 \times \cdots \times Y_n$, and $Z_x = Z_{x_1} \times \cdots \times Z_{x_n}$. Suppose

(i) $S_0: X_0 \to Y$ is one to one, i.e., reduced with respect to the input;
(ii) H is complete and furthermore satisfies the conditions

$$H(x, z) = H(x', z) \to x = x', \qquad H(x, z) = H(x, z') \to z = z'$$

(iii) $\bar{S}_f \subset \Pi(Y \times Z_x)$ is the set of all functional systems S_f such that $\mathscr{F}((H \circ S_0) \circ S_f)$ is functional;
(iv) the cardinality of Y_i is equal to that of X_i for each i.

Then S_0 is functionally controllable if and only if $S = H \circ S_0$ can be decoupled by a feedback from $\bar{S}_f$.

PROOF. We consider the *if* part first. Suppose $H \circ S_0$ can be decoupled by a feedback from $\bar{S}_f$. Then Proposition 4.3 implies that $H \circ S_0$ is functionally controllable, i.e.,

$$(\forall \hat{y})(\exists(\hat{x}, \hat{z}))(\hat{y} = (H \circ S_0)(\hat{x}, \hat{z}))$$

Let $\hat{x}_0 = H(\hat{x}, \hat{z})$. Then the above condition implies

$$(\forall \hat{y})(\exists \hat{x}_0)(\hat{y} = S_0(\hat{x}_0))$$

that is, S_0 is functionally controllable.

Next, we consider the *only if* part. Suppose S_0 is functionally controllable. Let $\hat{x} \in X$ and $\hat{y} \in Y$ be arbitrary. Since S_0 is functionally controllable, there exists $\hat{x}_0 \in X$ such that $\hat{y} = S_0(\hat{x}_0)$. Furthermore, since H is complete, there exists $\hat{z} \in Z_x$ such that $\hat{x}_0 = H(\hat{x}, \hat{z})$. Therefore, we have

$$(\forall(x, y))(\exists z)((x, z, y) \in H \circ S_0)$$

Let

$$\hat{S} = S_B = \{(x, y) : (\exists z)((x, z, y) \in H \circ S_0)\} = X \times Y$$

Proposition 4.2 implies, then, that $\mathscr{F}_s(\bar{S}_f) = K(X, Y)$. Suppose $S_i: X_i \to Y_i$ is one-to-one functional and functionally controllable for each $i \le n$. Since the cardinality of Y_i is equal to that of X_i, such S_i exists. Then $S' = S_1 + \cdots + S_n \subset \hat{S}$ because of the structure of $\hat{S}$. Furthermore, S' is one-to-one functional. Consequently, $\mathscr{F}_s(S_f) = S_1 + \cdots + S_n$ for some feedback $S_f \in \bar{S}_f$. Q.E.D.

We shall consider now linear systems. The counterpart of Proposition 4.2 for linear systems is given by the following proposition.

Proposition 4.5. Suppose S satisfies conditions (i), (ii), and (iii) from Proposition 4.2 and furthermore

(iv) S is a linear system.

Let $\bar{S}_f^L$ be a subset of $\bar{S}_f$ defined in Proposition 4.2, $\bar{S}_f^L \subset \bar{S}_f$ such that $S_f \in \bar{S}_f^L$ if and only if S_f is linear. Let $\hat{S} \subset X \times Y$ be an arbitrary linear system and $L(X, Y)$ a subset of the family of subsystems defined in Proposition 4.2, $L(X, Y) \subset K(X, Y)$ such that $S' \in L(X, Y)$ if and only if S' is linear.

Then $\mathscr{F}_s(\bar{S}_f^L) = L(X, Y)$ if $\hat{S} = S_B$ where

$$S_B = (\exists z)((x, z, y) \in S))$$

The proof of Proposition 4.5 is almost the same as that of Proposition 4.2 except that when we define a feedback component S_f in the proof, we have to show that it is linear. The proof is left as an exercise.

For the linear systems, the following mapping $H: X \times Z_x \to X_0$ is conventionally used for an input port: $H(x, z) = x + z$. It is easy to see that H is complete and that condition (ii) of Proposition 4.4 is apparently satisfied. Then the counterpart of Proposition 4.4 for linear systems is given by the following proposition.

Proposition 4.6. Let $X = X_1 \times \cdots \times X_n$, $Y = Y_1 \times \cdots \times Y_n$, and $Z_x = Z_{x_1} \times \cdots \times Z_{x_n}$. Suppose

(i) $S_0: X_0 \to Y$ is linear and reduced with respect to the input;

(ii) $\bar{S}_f^L \subset \Pi(Y \times Z_x)$ is the set of all linear functional systems S_f such that $\mathscr{F}((H \circ S_0) \circ S_f)$ is functional;

(iii) for each pair (X_i, Y_i), there exists a linear one-to-one correspondence $S_i: X_i \to Y_i$.

Then S_0 is functionally controllable if and only if $S = H \circ S_0$ can be decoupled by a feedback from $\bar{S}_f^L$.

We shall now consider time systems. Our first concern is with nonanticipation. Recall that a functional time system $S_0: X_0 \to Y$ is nonanticipatory if and only if

$$(\forall t)(\forall x)(\forall \hat{x})(x \mid \bar{T}^t = \hat{x} \mid \bar{T}^t \to S_0(x) \mid \bar{T}^t = S_0(\hat{x}) \mid \bar{T}^t)$$

Consequently, a functional time system S_0 is nonanticipatory if and only if $S_0 \mid \bar{T}^t$ is functional for each t. We have then the following proposition.

Proposition 4.7. Let $S \subset (X \times Z_x) \times Y$. Let $S_f \subset Y \times Z_x$ and $\mathscr{F}(S \cdot S_f) = \mathscr{F}_s(S_f)$ be functional and nonanticipatory. If S satisfies the condition that:

for each $t \in T$

$$(x, z, y) \in S \,\&\, (\hat{x}, \hat{z}, \hat{y}) \in S \,\&\, (z, y) \mid \bar{T}^t = (\hat{z}, \hat{y}) \mid \bar{T}^t \rightarrow x \mid \bar{T}^t = \hat{x} \mid \bar{T}^t$$

then $\mathscr{F}_s(S_f) \mid \bar{T}^t$ is one-to-one functional.

PROOF. Since $\mathscr{F}_s(S_f)$ is functional and nonanticipatory, $\mathscr{F}_s(S_f) \mid \bar{T}^t$ is functional. Suppose $y = \mathscr{F}_s(S_f)(x)$ and $\hat{y} = \mathscr{F}_s(S_f)(\hat{x})$. Suppose $y \mid \bar{T}^t = \hat{y} \mid \bar{T}^t$. We shall show that $x \mid \bar{T}^t = \hat{x} \mid \bar{T}^t$. It follows from the definition of $\mathscr{F}_s(S_f)$ that

$$(\exists z)(\exists \hat{z})((x, z, y) \in S \,\&\, (\hat{x}, \hat{z}, \hat{y}) \in S \,\&\, z = S_f(y) \,\&\, \hat{z} = S_f(\hat{y}))$$

Since $y \mid \bar{T}^t = \hat{y} \mid \bar{T}^t$ and since S_f is nonanticipatory, we have $z \mid \bar{T}^t = \hat{z} \mid \bar{T}^t$. Then, since $(z, y) \mid \bar{T}^t = (\hat{z}, \hat{y}) \mid \bar{T}^t$, we have $x \mid \bar{T}^t = \hat{x} \mid \bar{T}^t$. Hence, $\mathscr{F}_s(S_f) \mid \bar{T}^t$ is one-to-one functional. Q.E.D.

The following is the counterpart of Proposition 4.2 for nonanticipatory linear time system.

Proposition 4.8. Suppose

(i) $S \subset (X \times Z_x) \times Y$ is linear, nonanticipatory, functional, and furthermore satisfies the conditions that for each $t \in T$

$$S(x, z) \mid \bar{T}^t = S(\hat{x}, \hat{z}) \mid \bar{T}^t \,\&\, z \mid \bar{T}^t = \hat{z} \mid \bar{T}^t \rightarrow x \mid \bar{T}^t = \hat{x} \mid \bar{T}^t$$

and

$$S(x, z) \mid \bar{T}^t = S(\hat{x}, \hat{z}) \mid \bar{T}^t \,\&\, x \mid \bar{T}^t = \hat{x} \mid \bar{T}^t \rightarrow z \mid \bar{T}^t = \hat{z} \mid \bar{T}^t$$

(ii) $\bar{S}_f \subset \Pi(Y \times Z_x)$ is the family of linear nonanticipatory functional systems such that if $S_f \in \bar{S}_f$, then $\mathscr{F}(S \circ S_f)$ is functional and nonanticipatory.

Let $\hat{S} = S_B = \{(x, y) : (\exists z)((x, z, y) \in S)\}$ and let $L(X, Y)$ be the family of all linear subsystems S' of $\hat{S}$ such that $S' \mid \bar{T}^t$ is one-to-one functional for each t.

Then $\mathscr{F}_s(\bar{S}_f) = L(X, Y)$.

PROOF. Let $S' \in \mathscr{F}_s(\bar{S}_f)$ be arbitrary. Then there exists $S_f \in \bar{S}_f$ such that $S' = \mathscr{F}_s(S_f) = \mathscr{F}(S \circ S_f)$. Since S and S_f are linear, S' is linear and functional due to condition (ii). Furthermore, since $S' = \mathscr{F}(S \circ S_f)$, we have $S' \subset S_B = \hat{S}$. Since $S' \mid \bar{T}^t$ is one-to-one functional due to Proposition 4.7, S' is an element of $L(X, Y)$. Let $S' \in L(X, Y)$ be arbitrary. Since $\hat{S} = S_B$, we have $S' \subset S_B$. Let

$$S_f = \{(y, z) : (\exists x)((x, y) \in S' \,\&\, (x, z, y) \in S\} = \mathscr{G}_s(S')$$

S_f is apparently linear. Suppose $(y, z) \in S_f$ and $(\hat{y}, \hat{z}) \in S_f$. Then

$$(\exists x)(\exists \hat{x})(y = S'(x) \,\&\, \hat{y} = S'(\hat{x}) \,\&\, y = S(x, z) \,\&\, \hat{y} = S(\hat{x}, \hat{z}))$$

Suppose $y \mid \bar{T}^t = \hat{y} \mid \bar{T}^t$. Since $S' \mid \bar{T}^t$ is one-to-one functional, $x \mid \bar{T}^t = \hat{x} \mid \bar{T}^t$ holds. Therefore, $z \mid \bar{T}^t = \hat{z} \mid \bar{T}^t$. Hence, S_f is linear, nonanticipatory, and functional. Since $\mathscr{F}(S \circ S_f) = S'$ holds due to Proposition 4.2, $\mathscr{F}(S \circ S_f)$ is functional and nonanticipatory. Consequently, $S_f \in \bar{S}_f$. Q.E.D.

The relation between the functional controllability and the decoupling for linear nonanticipatory time systems is given in the following, which is the counterpart of Proposition 4.4. Let $H: X \times Z_x \to X_o$ be the same as used for Proposition 4.6. Then we have the following proposition.

Proposition 4.9. Let $X = X_1 \times \cdots \times X_n$, $Y = Y_1 \times \cdots \times Y_n$, and $Z_x = Z_{x_1} \times \cdots \times Z_{x_n}$. Suppose

(i) $S_o: X_o \to Y$ is linear, nonanticipatory, and satisfies the following condition:

$$(\forall t)(\forall x)(S_o(x) \mid \bar{T}^t = 0 \to x \mid \bar{T}^t = 0)$$

(ii) $\bar{S}_f \subset \Pi(Y \times Z_x)$ is the family of linear nonanticipatory functional systems such that if $S_f \in \bar{S}_f$, then $\mathscr{F}((H \circ S_o) \circ S_f)$ is functional and nonanticipatory;

(iii) for each pair (X_i, Y_i), there exists a linear mapping $S_i: X_i \to Y_i$ such that $S_i \mid \bar{T}^t$ is a one-to-one correspondence for each $t \in T$.

Then S_o is functionally controllable if and only if $S = H \circ S_o$ can be decoupled by a feedback from $\bar{S}_f$.

If we relax the conditions of Definition 4.1 (for instance, condition (iii) is omitted or the domain of S_i is allowed to be a proper subset of X_i), condition (iv) of Proposition 4.4, condition (iii) of Proposition 4.6, and condition (iii) of Proposition 4.9 can be relaxed.

Propositions 4.8 and 4.9 are presented for linear nonanticipatory time systems. However, it is apparent that they also hold for general nonanticipatory time systems with slight modifications.

(b) Decoupling of General Systems

Following the reasoning in subsection (a), we shall investigate in the sequel decoupling of general systems. Figure 4.3 is a schematic diagram of systems considered in this section, where $S \subset X_o \times Y$ is a general system, $H: X \times Z \to X_o$ is an input part, and $S_f: Y \to Z$ is a feedback functional system.

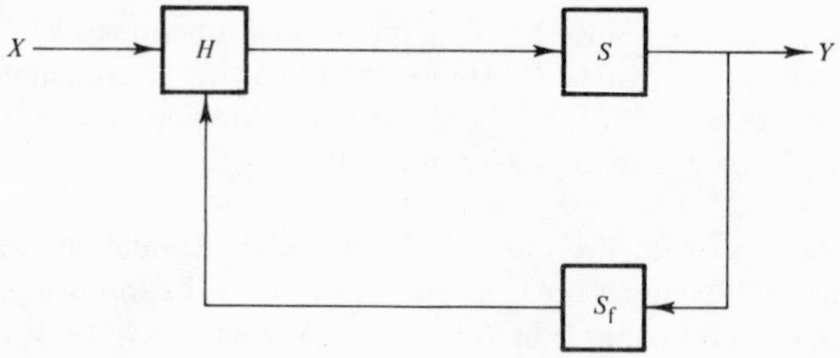

FIG. 4.3

The following two propositions are of conceptual importance.

Proposition 4.10. Let $S_1 \subset X_1 \times Y_1$, $S_2 \subset X_2 \times Y_2$, and $S = S_1 + S_2 \subset (X_1 \times X_2) \times (Y_1 \times Y_2)$. Then S is functionally controllable if and only if both S_1 and S_2 are functionally controllable.

Proposition 4.11. Let $S \subset X_0 \times Y$ and $\hat{S} \subset X \times Y$ be systems and let $H: X \times Z \to X_0$ be an input port. Suppose there exists $S_f: Y \to Z$ such that $\hat{S} = \mathscr{F}(H \circ S \circ S_f)$. Then S is functionally controllable if $\hat{S}$ is so.

PROOF. The assertion follows from $\mathscr{R}(\hat{S}) \subset \mathscr{R}(S)$. Q.E.D.

Proposition 4.12. Let $S \subset X_0 \times Y$ be a functionally controllable system and let $H: X \times Z \to X_0$ be an input port. Then, given a functionally controllable system $\hat{S} \subset X \times Y$, there exists a feedback component $S_f: Y \to Z$ such that $\mathscr{F}(H \circ S \circ S_f) = \hat{S}$ if and only if for any $y \in Y$ there exists $z \in Z$ such that

(i) $H((y)\hat{S}, z) \subset (y)S$;
(ii) $H(x, z) \in (y)S \to x \in (y)\hat{S}$;

where $(y)S = \{x : (x, y) \in S\}$ and $(y)\hat{S} = \{x : (x, y) \in \hat{S}\}$ are defined in Section 1.

PROOF. We shall first prove the *if* part. Let $V(y)$ be defined by

$$z \in V(y) \leftrightarrow (H((y)\hat{S}, z) \subset (y)S$$

and

$$H(x, z) \in (y)S \to x \in (y)\hat{S})$$

By the premise $V(y) \neq \phi$. Using the correspondence $y \mapsto V(y)$ and the axiom of choice, we can construct a function $S_f: Y \to Z$. We claim $\hat{S} = \mathscr{F}(H \circ S \circ S_f)$. $\hat{S} \subset \mathscr{F}(H \circ S \circ S_f)$ is clear. To prove $\hat{S} \supset \mathscr{F}(H \circ S \circ S_f)$, let $(x, y) \in \mathscr{F}(H \circ S \circ S_f)$. Then there exist $z \in Z$, $x_0 \in X_0$ such that $H(x, z) = x_0$, $(x_0, y) \in S$, and $S_f(y) = z$. Hence, $H(x, z) \in (y)S$, which implies $x \in (y)\hat{S}$. Thus $(x, y) \in \hat{S}$.

To prove the *only if* part, suppose $\hat{S} = \mathscr{F}(H \circ S \circ S_f)$. Then $(y)\hat{S} = \{x : H(x, S_f(y)) \in (y)S\}$. Therefore, $H((y)\hat{S}, S_f(y)) \subset (y)S$. Let $H(x, S_f(y)) \in (y)S$. Then $(x, y) \in \mathscr{F}(H \circ S \circ S_f) = \hat{S}$. Thus, $x \in (y)\hat{S}$. Q.E.D.

Proposition 4.13. Let $S \subset X_o \times Y$ be a system and let $H: X \times Z \to X_o$ be an input port. Suppose that $S_1 \subset X_1 \times Y_1$ and $S_2 \subset X_2 \times Y_2$ are controllable systems such that there exists a system $\hat{S} \subset X \times Y$ which satisfies the following conditions:

(i) $\hat{S} = S_1 + S_2$;
(ii) for any $y \in \mathscr{R}(S)$, there exists $z \in Z$ such that (α) $H((y)\hat{S}, z) \subset (y)S$, and ($\beta$) $H(x, z) \in (y)S \to x \in (y)\hat{S}$.

Then S can be decoupled into S_1 and S_2 by a functional feedback if and only if S is functionally controllable.

PROOF. The *only if* part follows from Proposition 4.11 since S_1 and S_2 are functionally controllable (hence so is $S_1 + S_2$ by Proposition 4.10). The *if* part follows from Proposition 4.12. Q.E.D.

Propositions 4.12 and 4.13 are also true for linear systems as follows.

Proposition 4.14. Let $S \subset X_o \times Y$ be a functionally controllable linear system and let $H: X \times Z \to X_o$ be a linear input port.

Then, given a functionally controllable linear system $\hat{S} \subset X \times Y$, there exists a linear functional feedback component $S_f: Y \to Z$ such that $\mathscr{F}(H \circ S \circ S_f) = \hat{S}$ if and only if for any $y \in Y$, there exists $z \in Z$ such that

(i) $H((y)\hat{S}, z) \subset (y)S$; and
(ii) $H(x, z) \in (y)S \to x \in (y)\hat{S}$.

PROOF. The *only if* part can be proved by the same reasoning as in the proof of Proposition 4.12. To prove the *if* part, it is enough to show that we can construct a linear $S_f: Y \to Z$. Let $V(y)$ be as in the proof of Proposition 4.12. We claim

$$z \in V(y), \qquad \text{and} \qquad \hat{z} \in V(\hat{y}) \to \alpha z + \beta \hat{z} \in V(\alpha y + \beta \hat{y})$$

for scalars α and β.

To see this, assume that $z \in V(y)$, $\hat{z} \in V(\hat{y})$. First, we shall prove

$$H(x, \alpha z + \beta \hat{z}) \in (\alpha y + \beta \hat{y})S \to x \in (\alpha y + \beta \hat{y})\hat{S}$$

Let $x_1 \in X$ be chosen so that $H(x_1, z) \in (y)S$. Let $x_2 = (1/\beta)(x - \alpha x_1)$, where we assume that $\beta \neq 0$, because if $\beta = 0$, then the statement is clear. Then

$x = \alpha x_1 + \beta x_2$. Hence,

$$H(x, \alpha z + \beta \hat{z}) = \alpha H(x_1, z) + \beta H(x_2, \hat{z})$$

By the assumption,

$$(\alpha H(x_1, z) + \beta H(x_2, \hat{z}), \alpha y + \beta \hat{y}) \in S$$
$$(\alpha H(x_1, z), \alpha y) \in S$$

Hence, $(\beta H(x_2, \hat{z}), \beta \hat{y}) \in S$. Therefore, $H(x_2, \hat{z}) \in (\hat{y})S$. By the assumption, we have $x_1 \in (y)\hat{S}$ and $x_2 \in (\hat{y})\hat{S}$. Thus, $x \in (\alpha y + \beta \hat{y})\hat{S}$.

To prove that

$$H((\alpha y + \beta \hat{y})\hat{S}, \quad \alpha z + \beta \hat{z}) \subset (\alpha y + \beta \hat{y})S$$

let $x \in (\alpha y + \beta \hat{y})\hat{S}$. Since $\hat{S}$ is functionally controllable, there exists $\hat{x} \in X$ such that $(\hat{x}, \hat{y}) \in \hat{S}$. Hence, $(x - \beta \hat{x}, \alpha y) \in \hat{S}$. Thus,

$$H(x, \alpha z + \beta \hat{z}) = \alpha H((1/\alpha)(x - \beta \hat{x}), z) + \beta H(\hat{x}, \hat{z}) \in (\alpha y + \beta \hat{y})S$$

proving that

$$H((\alpha y + \beta \hat{y})\hat{S}, \alpha z + \beta \hat{z}) \subset (\alpha y + \beta \hat{y})S$$

where we assume $\alpha \neq 0$, because if $\alpha = 0$, the assertion is clear.

Let $\hat{S}_f \subset Y \times Z$ be defined by

$$(y, z) \in \hat{S}_f \leftrightarrow z \in V(y)$$

Then $\hat{S}_f$ is a linear system. Therefore, by using Zorn's lemma we can show that there exists a linear map $S_f: Y \to Z$ such that $\{(y, S_f(y)) : y \in Y\} \subset \hat{S}_f$. S_f is a desired linear map. Q.E.D.

Proposition 4.15. Let $S \subset X_o \times Y$ be a linear system and let $H: X \times Z \to X_o$ be a linear input port. Suppose that $S_1 \subset X_1 \times Y_1$ and $S_2 \subset X_2 \times Y_2$ are functionally controllable linear systems such that there exists a system $\hat{S} \subset X \times Y$ which satisfies the following conditions:

(i) $\hat{S} = S_1 + S_2$;
(ii) for any $y \in \mathcal{R}(S)$, there exists $z \in Z$ such that (α) $H((y)\hat{S}, z) \subset (y)S$, and ($\beta$) $H(x, z) \in (y)S \to x \in (y)\hat{S}$.

Then S can be decoupled into S_1 and S_2 by linear functional feedback if and only if S is functionally controllable.

PROOF. The *only if* part follows from Proposition 4.11 since S_1 and S_2 are functionally controllable (hence so is $S_1 + S_2$ by Proposition 4.10). The *if* part follows from Proposition 4.14. Q.E.D.

As an example, let us consider a system $S \subset X_o \times Y$ defined by $\dot{y} = Ay + Bx_o$, where

$$X_o = L_2(o, \infty), \qquad A: n \times n \text{ matrix}, \qquad B: n \times n \text{ nonsingular matrix}$$

Then S is functionally controllable. Let us examine the feedback system in Fig. 4.4, where $X = L_2(o, \infty)$. Let $S_i \subset X_i \times Y_i$ $(i = 1, \ldots, n)$ be defined by $\dot{y}_i = \alpha_i y_i + \beta_i x_i$, where x_i and y_i are scalar functions of t, and α_i, β_i are constant reals. We assume $\beta_i \neq o$ and $x_i \in L_2(o, \infty)$. Then

$$\hat{S} = S_1 + S_2 + \cdots + S_n \subset X \times Y$$

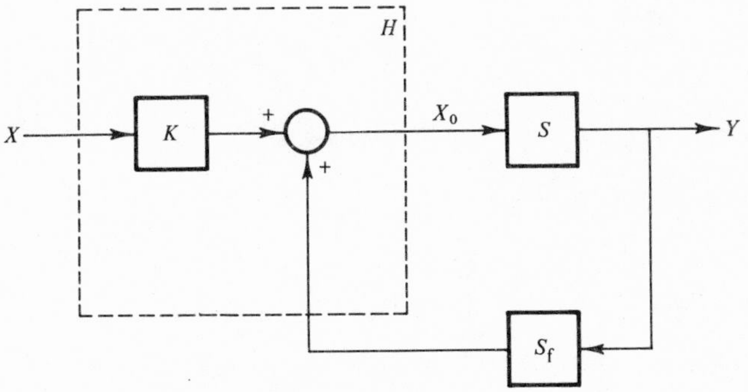

FIG. 4.4

is functionally controllable. Suppose that S is decoupled into $S_1, \ldots, S_n$. Then for any y there exists z such that $H((y)\hat{S}, z) \subset (y)S$. Therefore, if x_i and y_i satisfy $\dot{y}_i = \alpha_i y_i + \beta_i x_i$, then

$$\frac{d}{dt}\begin{bmatrix} o \\ \vdots \\ y_i \\ o \\ \vdots \\ o \end{bmatrix} = A\begin{bmatrix} o \\ \vdots \\ y_i \\ o \\ \vdots \\ o \end{bmatrix} + B\left(K\begin{bmatrix} o \\ \vdots \\ x_i \\ o \\ \vdots \\ o \end{bmatrix} + z_i\right)$$

Hence

$$B^{-1}(\alpha_i I - A)\begin{bmatrix} o \\ \vdots \\ y_i \\ \vdots \\ o \end{bmatrix} + (\beta_i B^{-1} - K)\begin{bmatrix} o \\ \vdots \\ x_i \\ \vdots \\ o \end{bmatrix} = z_i$$

z_i must be independent of x_i. Therefore,

$$(\beta_i B^{-1} - K)\begin{bmatrix} \mathrm{o} \\ \vdots \\ 1 \\ \vdots \\ \mathrm{o} \end{bmatrix} = \mathrm{o}$$

for each i. Thus

$$z_i = B^{-1}(\alpha_i I - A)\begin{bmatrix} \mathrm{o} \\ \vdots \\ y_i \\ \vdots \\ \mathrm{o} \end{bmatrix}$$

Hence we have $K = B^{-1}\beta$, $S_{\mathrm{f}} = B^{-1}(\alpha - A)$, where

$$\beta = \begin{bmatrix} \beta_1 \mathrm{o} & \cdots & \mathrm{o} \\ \mathrm{o} & & \mathrm{o} \\ \vdots & & \\ \mathrm{o} & \cdots & \mathrm{o}\beta_n \end{bmatrix} \qquad \alpha = \begin{bmatrix} \alpha_1 \mathrm{o} & \cdots & \mathrm{o} \\ \mathrm{o} & & \vdots \\ \vdots & & \mathrm{o} \\ \mathrm{o} & \cdots & \mathrm{o}\alpha_n \end{bmatrix}$$

In fact, S can be decoupled into $S_1, \ldots, S_n$ if $K = B^{-1}\beta$ and $S_{\mathrm{f}} = B^{-1}(\alpha - A)$. Furthermore, if rank $B = n$ (where B is an $n \times m$ matrix), then the analogous argument can be applied.

Proposition 4.12 is applicable, with additional conditions, to nonanticipatory systems as follows.

Proposition 4.16. Let $S \subset X_{\mathrm{o}} \times Y$ be a functionally controllable system with a nonanticipatory systems-response function ρ and let $H: X \times Z \to X_{\mathrm{o}}$ be an input port. Let $\hat{S} \subset X \times Y$ be a given functionally controllable nonanticipatory system with a nonanticipatory systems-response function $\hat{\rho}$. Assume that ρ, $\hat{\rho}$, and H satisfy the following conditions, respectively:

(1a) $\rho(c, x) \mid \bar{T}^t = \rho(c, \hat{x}) \mid \bar{T}^t \to x \mid \bar{T}^t = \hat{x} \mid \bar{T}^t$
(1b) $(\forall c)(\forall y \in \mathscr{R}(S))(\exists x)(\rho(c, x) = y)$
(2a) $\hat{\rho}(c, x) \mid \bar{T}^t = \hat{\rho}(c, \hat{x}) \mid \bar{T}^t \to x \mid \bar{T}^t = \hat{x} \mid \bar{T}^t$
(2b) $(\forall c)(\forall y \in \mathscr{R}(S))(\exists x)(\hat{\rho}(c, x) = y)$
(3) $H(x, z) \mid \bar{T}^t = H(\hat{x}, \hat{z}) \mid \bar{T}^t \;\&\; x \mid \bar{T}^t = \hat{x} \mid \bar{T}^t \to z \mid \bar{T}^t = \hat{z} \mid \bar{T}^t$

Then there exists a nonanticipatory feedback component $S_f: Y \to Z$ such that $\mathscr{F}(H \circ S \circ S_f) = \hat{S}$ if and only if for any $y \in Y$, there exists $z \in Z$ such that

(i) $H((y)\hat{S}, z) \subset (y)S$
(ii) $H(x, z) \in (y)S \to x \in (y)\hat{S}$

PROOF. The *only if* part is clear. We know that if (i) and (ii) hold, there exists a feedback component $S_f: Y \to Z$ such that $\mathscr{F}(H \circ S \circ S_f) = \hat{S}$. Therefore, to prove the *if* part, it is enough to show that if $\mathscr{F}(H \circ S \circ S_f) = \hat{S}$ and if the given conditions hold, then S_f is nonanticipatory.

Suppose $y, \hat{y} \in \mathscr{R}(\hat{S})$. Then there exist c, d, x, and $\hat{x}$ such that

$$y = \hat{\rho}(d, x) = \rho(c, H(x, S_f(y))), \qquad \text{and} \qquad \hat{y} = \hat{\rho}(d, \hat{x}) = \rho(c, H(\hat{x}, S_f(\hat{y})))$$

Suppose $y \mid \bar{T}^t = \hat{y} \mid \bar{T}^t$. Then by the assumptions, we have $x \mid \bar{T}^t = \hat{x} \mid \bar{T}^t$ and

$$H(x, S_f(y)) \mid \bar{T}^t = H(\hat{x}, S_f(\hat{y})) \mid \bar{T}^t$$

Hence, $S_f(y) \mid \bar{T}^t = S_f(\hat{y}) \mid \bar{T}^t$. Therefore, S_f is nonanticipatory. Q.E.D.

5. ABSTRACT POLE ASSIGNABILITY

The problem of pole assignability has been considered for the class of linear time-invariant ordinary differential equation systems, and the equivalence between pole assignability and state controllability is a well-known result for that class of systems. This is at least conceptually interesting and worthy of being studied in our abstract framework. However, the result is deeply rooted into specific structures of differential equations; in particular, the matrix description of a system is directly related to the definition of pole assignability. For abstract developments, therefore, we shall first modify the concept of pole assignability in a way suitable for our abstract framework.

The type of feedback system considered in this section is shown in Fig. 5.1, where

(i) S is a strongly nonanticipatory time-invariant linear dynamical system $(\bar{\rho}, \bar{\phi})$.

(ii) The state space is of finite dimension and is taken for the output of S, i.e., $B = C$ and $Y \subset C^T$.

(iii) The input alphabet A is equal to the state space C, i.e., $A = C$.

(iv) The feedback S_f is a linear static functional system, i.e., $S_f: C \to C(=A)$ and $S_f(y)(t) = S_f(y(t))$. The class of S_f in consideration in this section is denoted by $\bar{S}_f$.

(v) The input port H is defined by $H(x, z) = x + z$.

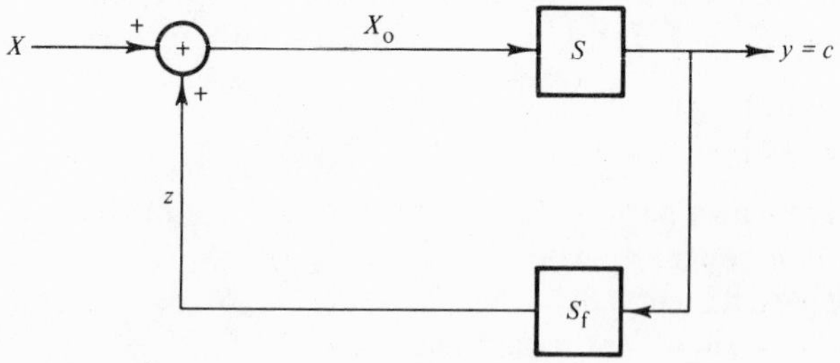

FIG. 5.1

(vi) A system S in consideration is assumed to satisfy the following conditions; for any $S_f \in \bar{S}_f$, $c \in C$, $x \in X$, and $t \in T$

(α) $\rho_o(c, x + S_f(y)) \mid T_t = y \mid T_t$ has a unique solution y;
(β) $\rho_o(c, x + S_2(y)) = y \,\&\, \rho_o(c, \hat{x} + S_2(\hat{y})) = \hat{y} \,\&\, x^t = \hat{x}^t \to \bar{y}^t = \bar{\hat{y}}^t$.

Due to assumption (vi) (α), a function $\hat{\phi}_{ot}: C \times X^t \to C$ can be defined as follows:

$$\hat{\phi}_{ot}(c, x^t) = \phi_{ot}(c, x^t + (S_f(y)) \mid T^t)$$

where y is the solution of $\rho_o(c, x^t \cdot x_t + S_f(y)) = y$, and x_t is arbitrary.

Furthermore, it is not difficult to show that $\{\hat{\phi}_{ot} : t \in T\}$ satisfies the composition property, i.e.,

$$\hat{\phi}_{ot}(c, x^t \cdot x_{tt'}) = \hat{\phi}_{o\tau}(\hat{\phi}_{ot}(c, x^t), F^{-t}(x_{tt'}))$$

where $\tau = t' - t$.

This can be proven by the time invariance of $\bar{\phi}$ and assumptions (iv) and (vi). [Notice that S_f is commutative with F^t, i.e., $F^t(S_f(y)) = S_f(F^t(y))$.]

$\hat{\phi}_{ot}$ will be referred to as the *state-transition function of the feedback system.* As usual, we shall write

$$\hat{\phi}_{ot}(c, x^t) = \hat{\phi}_{10t}(c) + \hat{\phi}_{20t}(x^t)$$

Under the above assumptions, the pole assignability is defined in the abstract framework as follows.

Definition 5.1. Let S be a linear dynamical system specified as above. Let $f(w) = w^n + \alpha_{n-1}w^{n-1} + \cdots + \alpha_o$ be an arbitrary monic polynomial with the indeterminate w, where the coefficients are elements of the field $\mathscr{A}$ (over which S is defined), α_o is not zero, and n is equal to the dimension of C. The system S is then said to be pole assignable if and only if for any $t > 0$ there exists a feedback $S_f \in \bar{S}_f$ such that

$$f(\hat{\phi}_{10t}) = (\hat{\phi}_{10t})^n + \alpha_{n-1}(\hat{\phi}_{10t})^{n-1} + \cdots + \alpha_o I = o$$

holds, where pointwise addition, pointwise scalar multiplication, and multiplication by composition are assumed for $\hat{\phi}_{10t}$, and $I:C \to C$ is the identity mapping.

The relationship between pole assignability and state controllability is given by the following proposition.

Proposition 5.1. Let S be a linear dynamical system specified as above whose state space C is of dimension n. Assume that the field $\mathscr{A}$ has more than $(n + 1)$ elements and that the state response $\phi_{10t}:C \to C$ of S is an isomorphism. Then, if S is pole assignable, the system S is controllable from zero.

PROOF. Let $\hat{c} \in C$ be arbitrary. Since ϕ_{10t} is an isomorphism, there exists $\alpha_o \neq o$ such that $(\phi_{10t})^n + \alpha_o I \equiv L_t$ is an isomorphism. Choose S_f such that $\hat{\phi}_{10t}$ satisfies $(\hat{\phi}_{10t})^n + \alpha_o I = 0$. Then we have $(\hat{\phi}_{10t})^n - (\phi_{10t})^n = -L_t$. Since L_t is an isomorphism, there exists $c \in C$ such that $((\hat{\phi}_{10t})^n - (\phi_{10t})^n)(c) = \hat{c}$; that is, $\hat{\phi}_{10\hat{t}}(c) - \phi_{10\hat{t}}(c) = \hat{c}$ (due to the state composition property), where $\hat{t} = nt$. Furthermore, we have from the definition of $\hat{\phi}_{10t}$ that for any $c \in C$ and $S_f \in \bar{S}_f$,

$$(\hat{\phi}_{10\hat{t}} - \phi_{10\hat{t}})(c) = \phi_{o\hat{t}}(o, S_f(y)^{\hat{t}})$$

holds, where y satisfies the relation $\rho_o(c, S_f(y)) = y$. Consequently,

$$\phi_{o\hat{t}}(o, S_f(y)^{\hat{t}}) = \hat{c}$$

Q.E.D.

In the above proposition, ϕ_{10t} is assumed to be an isomorphism. As Proposition 4.3, Chapter VII, shows, this holds under rather mild conditions; in particular, this is true for a linear time-invariant differential equation system.

6. SIMPLIFICATION THROUGH DECOMPOSITION OF DISCRETE-TIME DYNAMICAL SYSTEMS†

In the preceding sections, we have considered the decomposition of a system, primarily into two subsystems of a rather general kind. In the present section, we shall consider decomposition of a system into a family of subsystems of a prescribed, much simpler type.

We shall consider time-invariant discrete-time dynamical systems defined by the maps

$$\rho_t:C \times X_t \to Y_t, \qquad \phi_{tt'}:C \times X_{tt'} \to C$$

† See Hartmanis and Stearns [12].

where $T = \{0, 1, 2, \ldots\}$.

To simplify the developments, the following will be assumed:

(i) If the system is (strongly) nonanticipatory, which is the most interesting case in practice, the dynamic behavior of the system is completely represented by $\bar{\phi} = \{\phi_{tt'} : t, t' \in T\}$. In this section, we shall consider, therefore, only $\bar{\phi}$, refer to it as a state-transition system, and be concerned with the decomposition of $\bar{\phi}$.

(ii) Since the system is time invariant and discrete time, $\bar{\phi}$ is characterized by a single mapping, namely, $\phi_{01} : C \times A \to C$; ϕ_{0t}, or in general $\phi_{tt'}$, can be determined uniquely from ϕ_{01} by using the composition property of the state-transition function. We shall denote ϕ_{01} by ϕ and refer to ϕ itself as a state-transition system.

(iii) Let $X = A^T$ and $\bar{X} = \bigcup_{t \in T} X^t$. If a composition operation $\circ$ is defined on $\bar{X}$ as $\circ : \bar{X} \times \bar{X} \to \bar{X}$ such that $x^t \circ \hat{x}^{t'} = x^t \cdot F^t(\hat{x}^{t'})$, where F^t is the shift operator, $\bar{X}$ is essentially a free monoid on A; the null element of that monoid will be denoted by x^0. In considering the structure of $\bar{X}$ later on in this section, we shall treat $\bar{X}$ as a monoid in the above sense.

For the sake of notational convenience and in order to follow the conventional literature, we shall define a special form of the cascade connection for the state-transition systems.

Definition 6.1. Let $\phi_1 : C_1 \times A \to C_1$ and $\phi_2 : C_2 \times (C_1 \times A) \to C_2$ be two state-transition systems, where A is the input alphabet for ϕ_1, while $C_1 \times A$ is the input alphabet for ϕ_2. The cascade connection $\hat{\phi}$ of ϕ_1 and ϕ_2 is then

$$\hat{\phi} : (C_1 \times C_2) \times A \to C_1 \times C_2$$

such that

$$\hat{\phi}((c_1, c_2), a) = (\phi_1(c_1, a), \phi_2(c_2, (c_1, a)))$$

Definition 6.2. A state-transition system $\phi : C \times A \to C$ is decomposed into a cascade connection of two state-transition systems $\phi_1 : C_1 \times A \to C_1$ and $\phi_2 : C_2 \times (C_1 \times A) \to C_2$ if and only if there exists a mapping $R : C_1 \times C_2 \to C$ such that R is onto and the following diagram is commutative:

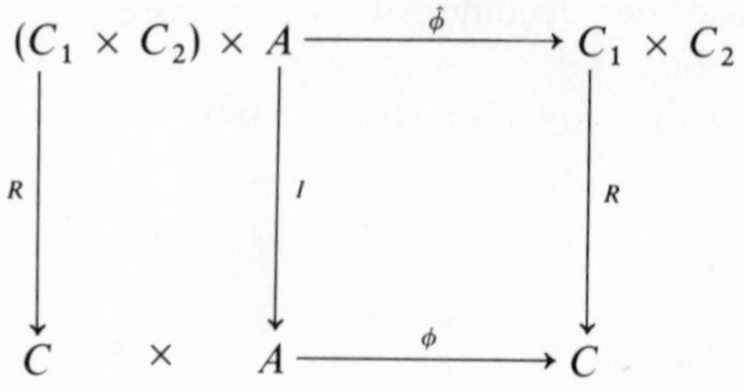

Intuitively, ϕ is decomposed into the cascade connection of ϕ_1 and ϕ_2 if and only if $\phi = \phi_1 \cdot \phi_2 \cdot R = \hat{\phi} \cdot R$ holds. (Strictly speaking, some notational modifications of ϕ_1 and ϕ_2 are necessary for $\phi_1 \cdot \phi_2 \cdot R$ to be compatible with the cascade operation introduced in Section 1.)

The following is a basic method to decompose a state-transition system.

Proposition 6.1. Let $\phi: C \times A \to C$ be a state-transition system. Given a class of subsets $\{C_\alpha : \alpha \in C_1\}$ of C such that

(i) $C = \bigcup_{\alpha \in C_1} C_\alpha$
(ii) $(\forall C_\alpha)(\forall a)(\exists C_\beta)(\phi(C_\alpha, a) \subset C_\beta)$

Then there exists a set C_2 and two state-transition systems $\phi_1 : C_1 \times A \to C_1$ and $\phi_2 : C_2 \times (C_1 \times A) \to C_2$ such that ϕ is decomposed into the cascade connection of ϕ_1 and ϕ_2.

PROOF. Let $\hat{R} \subset C_1 \times C$ be a general system such that $(\alpha, c) \in \hat{R} \leftrightarrow c \in C_\alpha$. Let a set C_2 be a global state of $\hat{R}$; that is, there exists a mapping $R: C_1 \dot{\times} C_2 \to C$ such that $c \in C_\alpha \leftrightarrow (\alpha, c) \in \hat{R} \leftrightarrow (\exists c_2)(c = R(\alpha, c_2))$. Let $\phi_1 : C_1 \times A \to C_1$ and $\phi_2 : C_2 \times (C_1 \times A) \to C_2$ be such that $\phi_1(\alpha, a) = \beta \to \phi(C_\alpha, a) \subset C_\beta$ and

$$\phi_2(c_2, (\alpha, a)) = c_2' \to \phi(R(\alpha, c_2), a) = R(\phi_1(\alpha, a), c_2')$$

There may be more than one β that satisfies the relation $\phi(C_\alpha, a) \subset C_\beta$. In that case, any β satisfying the condition can be used to define ϕ_1. As for ϕ_2, we shall show that there exists at least one c_2' that satisfies the relation $\phi(R(\alpha, c_2), a) = R(\phi_1(\alpha, a), c_2')$. The definition of R implies $R(\alpha, c_2) \in C_\alpha$. Let $\phi_1(\alpha, a) = \beta$. The definition ϕ_1 then implies $\phi(C_\alpha, a) \subset C_\beta$. Consequently, $\phi(R(\alpha, c_2), a) \in C_\beta$. Therefore, there exists $c_2' \in C_2$ such that $\phi(R(\alpha, c_2), a) = R(\beta, c_2')$. If there are more than one c_2' satisfying the condition, any one of them can be used for ϕ_2. Let

$$\hat{\phi}((c_1, c_2), a) = (\phi_1(c_1, a), \phi_2(c_2, (c_1, a)))$$

We shall show that $R[\hat{\phi}((c_1, c_2), a)] = \phi(R(c_1, c_2), a)$ holds. Let $\phi_2(c_2, (c_1, a)) = c_2'$. The definition of ϕ_2 implies then that $\phi(R(c_1, c_2), a) = R(\phi_1(c_1, a), c_2')$. Therefore, $R[\hat{\phi}((c_1, c_2), a)] = \phi(R(c_1, c_2), a)$ holds. Furthermore, since $\bigcup_{\alpha \in C_1} C_\alpha = C$, R is onto. Q.E.D.

Let us apply Proposition 6.1 and its proof to the case when the state space C of a state-transition system is finite. Let $C_\alpha = C - \{\alpha\}$, where $\alpha \in C$. Then it is easy to show that $\{C_\alpha : \alpha \in C\}$ satisfies the conditions of Proposition 6.1. In other words, a finite state-transition system always has a cascade decomposition. In order to discuss stronger results on decomposition of the finite case, we shall first introduce the following definitions.

Definition 6.3. Let $\phi : C \times A \to C$ be a state-transition system. For each $a \in A$, let $\phi_a : C \to C$ be such that $\phi_a(c) = \phi(c, a)$. If ϕ_a is onto, the input a is called a permutation input. If ϕ_a is a constant function, the input a is called a reset input. If ϕ_a is the identity function, the input a is called an identity input.

Definition 6.4. Let $\phi : C \times A \to C$ be a state-transition system. If all the inputs of ϕ are permutation, ϕ will be referred to as a P system. If all the inputs of ϕ are a reset or an identity, ϕ will be referred to as an R system. If all the inputs of ϕ are either permutation or reset, ϕ will be referred to as a P–R system.

Let us return to the finite case where $C_\alpha = C - \{\alpha\}$. If the input a is a permutation for ϕ, it is also a permutation for the component subsystems ϕ_1 as given in Proposition 6.1. (Refer to the proof of Proposition 6.1.) If the input a is not a permutation, there exists $C_\beta \in \{C_\alpha : \alpha \in C\}$ such that $\phi(C_\alpha, a) \subset C_\beta$ for all $a \in C_1 = C$. Consequently, the input a is a reset for ϕ_1. Therefore, ϕ_1 is a P–R system. Notice that the cardinality of C_2 can be less than that of C_1 or C. If we apply repeatedly the procedure of the cascade decomposition given in Proposition 6.1 to a state-transition system, the system is decomposed finally into the cascade connection of P–R systems, because a state-transition system whose state space has two elements has to be a P–R system. Consequently, we have the following proposition.

Proposition 6.2. A finite state-transition system can be decomposed into a cascade connection of P–R systems.

In the obvious manner, we can extend the domain of a state-transition system $\phi : C \times A \to C$ to $C \times \bar{X}$, where $\bar{X}$ is the free monoid of A; that is, $\phi(c, x^t) = \phi_{0t}(c, x^t)$. Let $\phi_x : C \to C$ be such that $\phi_x(c) = \phi(c, x)$. It is easy to show that $\phi_{x'} \cdot \phi_x = \phi_{x \cdot x'}$, where $\phi_{x'} \cdot \phi_x$ is the usual composition of the two functions ϕ_x and $\phi_{x'}$. Consequently, we can consider $x \in \bar{X}$ itself as a mapping, i.e., $x : C \to C$ such that $x(c) = \phi_x(c)$ and $\Lambda(c) = c$, where $\Lambda = x^0$ is the identity element of the monoid $\bar{X}$. From now on $\bar{X}$ will be considered a monoid of functions in the above sense.

Let a relation $E \subset \bar{X} \times \bar{X}$ be defined by

$$(x, x') \in E \leftrightarrow (\forall c)(x(c) = x'(c))$$

E is apparently an equivalence relation. Moreover, E is a congruence relation with respect to the monoid operation of $\bar{X}$. Let $\bar{X}/E = \{[x] : x \in \bar{X}\}$. A composition operation on $\bar{X}/E$, therefore, can be defined as

$$[x] \cdot [x'] = [x \cdot x']$$

$\bar{X}/E$ is also a monoid with respect to the new operation.

Proposition 6.3. Let $\phi: C \times A \to C$ be a state-transition system. Let C_1 be a group in $\bar{X}/E$, where the group operation is the monoid operation of $\bar{X}/E$. The state-transition system ϕ then can be decomposed into the cascade connection of two state-transition systems $\phi_1: C_1 \times A \to C_1$ and $\phi_2: C \times (C_1 \times A) \to C$ given by

$$\phi_1(c_1, a) = \begin{cases} c_1 \cdot [a], & \text{if } [a] \in C_1 \\ c_1, & \text{otherwise} \end{cases}$$

$$\phi_2(c, (c_1, a)) = \begin{cases} c, & \text{if } [a] \in C_1 \\ c_1^{-1}((c_1 \cdot [a])(c)), & \text{otherwise} \end{cases}$$

where c_1^{-1} is the inverse of c_1 in the group C_1.

PROOF. Let $R: C_1 \times C \to C$ be defined by

$$R(c_1, c) = c_1(c)$$

Then,

$$\begin{aligned} R(\hat{\phi}((c_1, c), a)) &= R(\phi_1(c_1, a), \phi_2(c, (c_1, a))) \\ &= \phi_1(c_1, a)(\phi_2(c, (c_1, a))) \\ &= (c_1 \cdot [a])(c) \\ &= \phi(R(c_1, c), a) \end{aligned}$$

Furthermore, if $c_1 = [\Lambda] \in C_1$, we have $c_1(c) = c$. Hence, R is onto. Q.E.D.

Let us consider a finite state-transition system again. Proposition 6.2 shows that every finite state-transition system can be decomposed into a cascade connection of P–R systems. Let us apply Proposition 6.3 to a P–R system. Let P be the set of all permutation inputs. Let P^* be the free monoid on P. Then it is easy to show that $P^*/E \subset \bar{X}/E$ is a finite group. Furthermore, since every element of P^* is an onto mapping, if an input a is a reset, we have $[a] \notin P^*/E$. Consequently, if a is a reset, $\phi_2(c, (c_1, a))$ of Proposition 6.3 does not depend on c, and if a is a permutation, $\phi_2(\alpha(c_1, a))$ is the identity mapping, where $C_1 = P^*/E$. Therefore, ϕ_2 is an R system.

Proposition 6.4. Every finite state-transition system can be decomposed into a cascade connection of P systems and R systems.

We shall consider ϕ_1 of Proposition 6.3. Let us recall the following definitions.

Definition 6.5. A subgroup H of a group G is called normal subgroup if and only if $aH = Ha$ for every $a \in G$, where $aH = \{x : (\exists y)(y \in H \;\&\; x = a \cdot y)\}$.

Definition 6.6. A group G is a simple group if and only if it has no nontrivial (i.e., different from G itself or the identity) normal subgroup.

Definition 6.7. Let $\phi: C \times A \to C$ be a state-transition system, where C is a group. If C is simple, the system is called simple.

Let $\phi: C \times A \to C$ be a state-transition system whose state space C is a group. Let G be a group and let $\psi: C \to G$ be a group homomorphism. Let $E_\psi \subset C \times C$ be such that

$$(c, c') \in E_\psi \leftrightarrow \psi(c) = \psi(c')$$

Then E_ψ is apparently a congruence relation. Let $H_\psi = \{c : \psi(c) = e\}$, where e is the identity element of G.

Proposition 6.5. Let $\psi: C \to G$ be a group homomorphism such that the condition

$$\psi(c) = \psi(c') \to \psi(\phi(c, a)) = \psi(\phi(c', a)) \qquad \text{for every} \quad a \in A$$

holds. Then $\phi: C \times A \to C$ can be decomposed into the cascade connection of $\phi_1: (C/E_\psi) \times A \to (C/E_\psi)$ and $\phi_2: H_\psi \times ((C/E_\psi) \times A) \to H_\psi$.

PROOF. Let $C_\alpha = [\alpha] \in C/E_\psi$, where $\alpha \in C$. Then $\bigcup_{\alpha \in C} C_\alpha = C$. Furthermore, we can show that for any C_α and $a \in A$, $\phi(C_\alpha, a) \subset C_\beta$, where $\beta = \phi(\alpha, a)$. Let $c \in C_\alpha$ be arbitrary; that is, $\psi(c) = \psi(\alpha)$. Then the property of ψ assumed in Proposition 6.5 implies that $\psi(\phi(c, a)) = \psi(\phi(\alpha, a))$; i.e., $\phi(c, a) \in C_\beta$, where $\beta = \phi(\alpha, a)$. Let $h: C/E_\psi \to C$ be a choice function; i.e., $h([\alpha]) \in C_\alpha$, where $\alpha \in C$. Let $R: (C/E_\psi) \times H_\psi \to C$ be defined by $R(C_\alpha, y) = h(C_\alpha) \cdot y$, where $y \in H_\psi$. Then

$$\begin{aligned} c \in C_\alpha &\leftrightarrow \psi(c) = \psi(\alpha) \\ &\leftrightarrow (\exists y)(y \in H_\psi \,\&\, c = \alpha y) \\ &\leftrightarrow (\exists y)(\exists \hat{y})(y \in H_\psi \,\&\, \hat{y} \in H_\psi \,\&\, c = \alpha\hat{y} \cdot \hat{y}^{-1} y) \qquad \text{where} \quad h([\alpha]) = \alpha\hat{y} \\ &\leftrightarrow (\exists y)(y \in H_\psi \,\&\, c = R(C_\alpha, y)) \end{aligned}$$

The final result can be proven in the same way as Proposition 6.1. Q.E.D.

Let us return to $\phi_1: C_1 \times A \to C_1$ of Proposition 6.3.

Proposition 6.6. Let $\phi_1: C_1 \times A \to C_1$ be the cascade component of Proposition 6.3. Let $\psi: C_1 \to G$ be an arbitrary group homomorphism. Then $\psi(c_1) = \psi(c_1')$ implies that $\psi(\phi_1(c, a)) = \psi(\phi_1(c_1', a))$ for every $\alpha \in A$. Consequently, if C_1 is not a simple group, the state-transition system ϕ_1 can be decomposed into the cascade connection of $\phi_{11}: (C_1/H) \times A \to (C_1/H)$ and $\phi_{12}: H \times ((C_1/H) \times A) \to H$ where $H \subset C_1$ is a normal subgroup.

PROOF. Suppose $\psi(c_1) = \psi(c_1')$. Suppose $[a] \in C_1$. The definition of ϕ_1 implies that

$$\psi(\phi_1(c_1, a)) = \psi(c_1 \cdot [a]) = \psi(c_1) \cdot \psi([a]) = \psi(c_1') \cdot \psi([a]) = \psi(\phi_1(c_1', a))$$

If $[a] \notin C_1$, then

$$\psi(\phi_1(c_1, a)) = \psi(c_1) = \psi(c_1') = \psi(\phi_1(c_1', a))$$

Suppose H is a normal subgroup of C_1. Then there exists another group G and a group homomorphism $\psi : C_1 \to G$ such that $H = H_\psi$ and $\alpha H = [\alpha]$, where $\alpha \in C_1$. The final result follows by Proposition 6.5.   Q.E.D.

As we mentioned in Proposition 6.6, the state-transition system ϕ_1 has a peculiar property; i.e., for every group homomorphism ψ,

$$\psi(c_1) = \psi(c_1') \to \psi(\phi_1(c_1, a)) = \psi(\phi_1(c_1', a))$$

for every $a \in A$. This property is also preserved by ϕ_{11} and ϕ_{12} when ϕ_{11} and ϕ_{12} are defined in the same way as ϕ_1 and ϕ_2 in Proposition 6.1. Notice that if $\phi : C \times A \to C$ is a finite state-transition system, C_1 of ϕ_1 of Proposition 6.3 is finite. Consequently, we have the following proposition.

Proposition 6.7. If $\phi : C \times A \to C$ is a finite state-transition system, the system can be decomposed into a cascade connection of state-transition systems which are either simple or reset.

A reset system has a parallel decomposition. We shall define a parallel decomposition of a state-transition system in the following way.

Definition 6.8. Let $\phi : C \times A \to C$ be a state-transition system. Let $\{\phi_i : C_i \times A \to C_i : i = 1, \ldots, n\}$ be a family of state-transition systems. If there exists a mapping $R : C_1 \times \cdots \times C_n \to C$ such that R is onto and the following diagram is commutative:

$$
\begin{array}{ccc}
(C_1 \times \cdots \times C_n) \times A & \xrightarrow{\bar{\phi}} & C_1 \times \cdots \times C_n \\
\downarrow{\scriptstyle R} \quad \downarrow{\scriptstyle I} & & \downarrow{\scriptstyle R} \\
C \quad \times \quad A & \xrightarrow{\phi} & C
\end{array}
$$

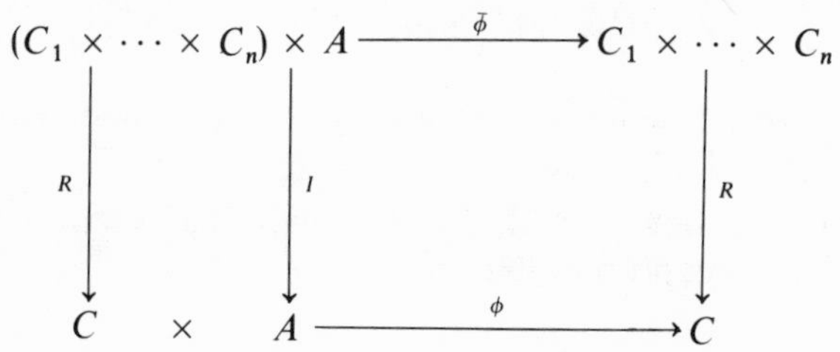

$\bar{\phi} = \{\phi_1, \ldots, \phi_n\}$ is called a parallel decomposition of ϕ, where

$$\bar{\phi}((c_1, \ldots, c_n), a) = (\phi_1(c_1, a), \ldots, \phi_n(c_n, a))$$

Proposition 6.8. Let $\phi: C \times A \rightarrow C$ be a reset system. Given a family of sets $\{C_i : i = 1, \ldots, n\}$ and an onto mapping $R: C_1 \times \cdots \times C_n \rightarrow C$. Then there exists a family of state-transition systems $\{\phi_i : C_i \times A \rightarrow C_i : i = 1, \ldots, n\}$ such that $\bar{\phi} = \{\phi_i : i = 1, \ldots, n\}$ is a parallel decomposition of the reset system ϕ.

PROOF. Let a state-transition system $\phi_i : C_i \times A \rightarrow C_i$ be given by

$$\phi_i(c_i, a) = \begin{cases} c_i, & \text{if } a \text{ is an identity input, i.e., } \phi_a \text{ is the identity} \\ c_{a_i}, & \text{if } a \text{ is a reset input} \end{cases}$$

where

$$\bar{c}_a = (c_{a_1}, \ldots, c_{a_n}) \in C_1 \times \cdots \times C_n$$

such that $R(\bar{c}_a) = \phi(c, a)$ for every $c \in C$. Since R is onto, when a is a reset input, it is clear that there exists $\bar{c}_a \in C_1 \times \cdots \times C_n$ such that $R(\bar{c}_a) = \phi(c, a)$ for every c. If there are more than two $\bar{c}_a$ that satisfy the above condition, any one of them can be used to define ϕ_i. We shall show that R satisfies the commutative diagram of Definition 6.8. Suppose a is an identity input. Then

$$\begin{aligned} R(\bar{\phi}((c_1, \ldots, c_n), a)) &= R(c_1, \ldots, c_n) \\ &= \phi(R(c_1, \ldots, c_n), a) \end{aligned}$$

Suppose a is a reset input. Then

$$\begin{aligned} R(\bar{\phi}((c_1, \ldots, c_n), a)) &= R(\bar{c}_a) \\ &= \phi(c, a) \\ &= \phi(R(c_1, \ldots, c_n), a) \end{aligned}$$

Q.E.D.

Suppose the state space C of a reset system is finite. There exist then a positive integer n and a mapping $R: \{0, 1\}^n \rightarrow C$ such that R is onto. Consequently, it follows from Proposition 6.8 that every finite reset system $\phi: C \times A \rightarrow C$ has a parallel decomposition

$$\{\phi_i : \{0, 1\} \times A \rightarrow \{0, 1\} : i = 1, \ldots, n\}$$

Definition 6.9. If the state space C of a state-transition system ϕ has two elements, ψ is called a two-state transition system.

Combining Definition 6.9 with Proposition 6.7 by using Proposition 6.8, we have the following proposition.

Proposition 6.9 (Krohn–Rhodes [17]).If $\phi: C \times A \rightarrow C$ is a finite state-transition system, the system can be decomposed into a cascade and parallel connection of the state-transition systems which are either simple or two states.

The results of the principal decomposition theorems in this section are shown diagrammatically in Fig. 6.1.

Finite state discrete-time dynamical system (FSDTD)

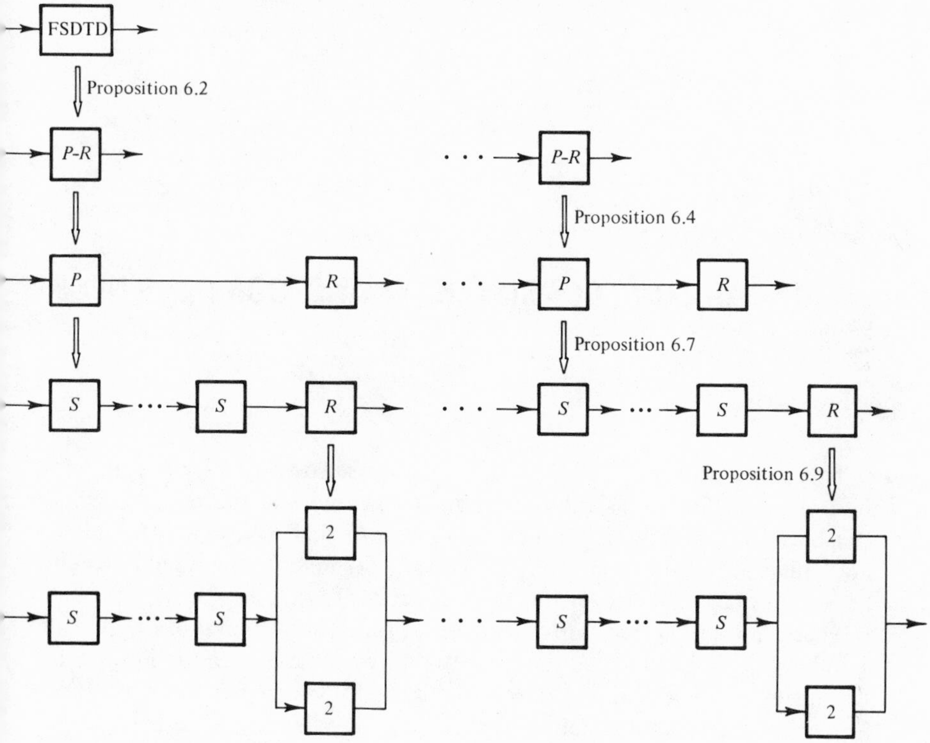

P-R: *P-R* system
P: *P* system
R: *R* system

S: Simple system
2: Two-state system

FIG. 6.1

Chapter XI

COMPUTABILITY, CONSISTENCY, AND COMPLETENESS

There are a number of important discrete-time processes, such as computation, theorem proving, symbol-manipulation processes, and the like, which can be represented by dynamical systems. We shall show in this chapter how this can be done and, using an abstract formulation for such a system, prove some key results regarding the properties of such systems. In particular, we shall consider the questions of consistency and completeness of a formal (or axiomatic logic) system and the computability problems. We shall prove the famous Goedel theorem in the context of the consistency and completeness properties of a general system. The Goedel diagonalization argument will be developed without reference to the specifics of the "dynamics" of a logical system. In this way, we shall reveal the most essential feature of an important result—in the spirit of the general methodology penetrated in the first chapter.

1. COMPUTATION AS DYNAMICAL PROCESS

A computation can be described essentially in the following way. There is given a set of initial data on which there is applied a sequence of transformations according to a preassigned arrangement (schedule or algorithm); when the process stops, the last set of data represents the result of computation. Conditions for ending transformation process are in the rules for applying transformations themselves and depend also on the initial data.

It is quite apparent that a computation represents a dynamical process.

Since the computation process by definition is nonanticipatory, stationary, and discrete in time, the dynamical system used to represent a computation can be specified solely in reference to a state-transition function.

We shall start, therefore, with a class of time-invariant state-transition families $\bar{\phi} = \{\bar{\phi}_i\}$, where $\bar{\phi}_i = \{\phi^i_{0t} : C \to C : t \in T\}$, and an output function $\lambda : C \to B$. Since the state-transition family $\bar{\phi}_i$, which specifies a computation process, is, by definition, time invariant, $\bar{\phi}_i$ is specified by $\{\phi^i_{0t}\}$. Furthermore, since $\bar{\phi}_i$ is discrete in time, that is, $T = \{0, 1, 2, \ldots\}$, $\bar{\phi}_i$ is essentially specified by ϕ^i_{01}. Each $\bar{\phi}_i$ of $\bar{\phi}$ represents one algorithm (schedule), and so the selection of the class $\bar{\phi}$ itself specifies a type of algorithm (for example, the class of Turing machines). Since $\bar{\phi}_i$ is free from an input, $(\bar{\phi}_i, \lambda)$ will be referred to as a *free dynamical system.* An output function $\lambda : C \to B$ represents a "decoding" in the sense that it gives an interpretation of the present state. In this chapter, one fixed output function λ will be assumed.

For any given state-transition family $\bar{\phi}_i \in \bar{\phi}$, a state $c \in C$ is called an *equilibrium state if and only if*

$$(\forall t)(\phi^i_{0t}(c) = c)$$

holds. Let C^e_i be the set of equilibrium states of $\bar{\phi}_i$; i.e.,

$$C^e_i = \{c : (\forall t)(\phi^i_{0t}(c) = c)\}$$

In this chapter, we assume for the class $\bar{\phi}$ that there exists a subset $C^e \subset C$ such that it represents the equilibrium set common to every member $\bar{\phi}_i$ of $\bar{\phi}$; i.e., $C^e_i = C^e$ for every $\bar{\phi}_i$.

Starting from a given initial state (initial data) c and an algorithm $\bar{\phi}_i \in \bar{\phi}$, we note that the changes of the state represent the evolution of the computational process in time, and if the system reaches an equilibrium state in a finite time, the computation process has been completed. Computation, therefore, can be interpreted as the so-called finite time stability, i.e., the ability of a system to reach an equilibrium state in a finite time. A trajectory reaching an equilibrium state in a finite time will be called a *computation.* This leads to the following definition.

Definition 1.1. A function $f : C \to B$ is computable by $\bar{\phi}$ if and only if there exists $\bar{\phi}_i \in \bar{\phi}$ such that

$$(\forall c)(\exists t)(\phi^i_{0t}(c) \in C^e \;\&\; f(c) = \lambda(\phi^i_{0t}(c)))\quad.$$

Before proceeding any further, it will be convenient to simplify the framework still more. Namely, for the basic question of computability of importance is only the fact that the process specified by a given $\bar{\phi}_i$ and c has terminated in an equilibrium state in a finite time. How this is achieved, i.e., the computation process itself, is of no importance. We can, therefore, dispense with the dynamics of each state-transition family and consider solely a static mapping from the state space into the set of computation processes. To that end we shall define a new system

$$\rho : C \times X \to Y$$

such that $X = \bar{\phi}$, $Y = C^T$, and

$$\rho(c, \bar{\phi}_i) = y \leftrightarrow (\forall t)(y(t) = \phi^i_{ot}(c))$$

For notational convenience, we introduce a subset $Y_0 \subset Y$ and a mapping $\text{Res} : Y \to C$ as follows:

$$Y_0 = \{y : (\exists \hat{t})(\forall t)(t \geq \hat{t} \to y(t) = y(\hat{t})\ \&\ y(\hat{t}) \in C^e)\}$$

and

$$\text{Res}\,(y) = \begin{cases} y(\hat{t}), & \text{if } \ y \in Y_0 \\ \text{undefined}, & \text{otherwise} \end{cases}$$

In the present framework, Definition 1.1 can be restated as in the following definition.

Definition 1.2. A function $f : (C) \to B$ is computable by ρ if and only if there exists $\hat{x} \in X$ such that

(i) $c \in (C) \leftrightarrow \rho(c, \hat{x}) \in Y_0$
(ii) $f(c) = \lambda \cdot \text{Res} \cdot \rho(c, \hat{x})$

It should be noticed that in Definition 1.2, the function f can be a partial function when (C) is proper subset of C. The development can proceed now solely by using the mapping ρ. It is conceptually important, however, to have in mind the interpretation of ρ as a dynamic system, i.e., as resulting from a computation process. To help make this bridge, we shall give in the last section some standard interpretations of ρ. We can introduce now the following concepts.

Definition 1.3. A given subset $C' \subset C$ is acceptable by ρ if and only if there exists $\hat{x} \in X$ such that

$$c \in C' \leftrightarrow \rho(c, \hat{x}) \in Y_0$$

i.e., C' is the domain on which computations can exist for a given $\hat{x}$.

Definition 1.4. A given subset $Y' \subset Y_0$ is representable by ρ if there exists $\hat{x} \in X$ such that

$$y \in Y' \leftrightarrow (\exists c)(\rho(c, \hat{x}) = y)$$

Definition 1.5. A subset $C' \subset C$ is decidable if and only if both C' and $C \setminus C'$ are acceptable by ρ.

Definition 1.6. Let W be an arbitrary subset of Y. The system ρ is W-consistent if and only if

$$W \cap Y_0 = \phi$$

and W-complete if and only if

$$W \cup Y_0 = Y$$

2. FUNDAMENTAL DIAGONALIZATION (GOEDEL) THEOREM

We shall now prove a fundamental theorem, which contains a basic argument for the impossibility of solving certain kinds of decidability problems. The theorem is an abstraction of the famous Goedel theorem, and we shall present it in the context of consistency and completeness properties of a system.

Let $g: X \to C$ be an injection, which we shall term a *Goedel mapping*. In reference to the given Goedel mapping, we can now define a *norm* or *diagonalization* y_x for any $x \in X$ by

$$y_x = \rho(g(x), x)$$

Let Q be an arbitrary subset of Y; there is defined, then, a set of elements $X_Q{}^d$ whose norms are in Q, i.e.,

$$x \in X_Q{}^d \leftrightarrow \rho(g(x), x) \in Q$$

Let $C_Q{}^d$ be the image of $X_Q{}^d$ under g; i.e.,

$$C_Q{}^d = g(X_Q{}^d)$$

Theorem 2.1. Let $\rho: C \times X \to Y$ be a given system and $W \subset Y$ a subset of the outputs. If $C_W{}^d$ is an acceptable set, the system is either inconsistent or incomplete.

PROOF. Since $C_W{}^d$ is an acceptable set, there exists $x^* \in X$ such that for any $c \in C$

$$c \in C_W{}^d \leftrightarrow \rho(c, x^*) \in Y_0$$

and, in particular, for $g(x^*) \in C$, we have

$$g(x^*) \in C_W{}^d \leftrightarrow \rho(g(x^*), x^*) \in Y_0 \tag{11.1}$$

where C_W^{d}, by definition, is the set of Goedel images of the norms that are elements of W, i.e.,

$$g(x) \in C_W^{\mathrm{d}} \leftrightarrow \rho(g(x), x) \in W \tag{11.2}$$

Since relation (11.2) holds for any $x \in X$, it follows from (11.1) and (11.2) that, for the given $x^* \in X$,

$$\rho(g(x^*), x^*) \in Y_o \leftrightarrow \rho(g(x^*), x^*) \in W \tag{11.3}$$

Denote by y^* the output of the system such that $y^* = \rho(g(x^*), x^*)$. From (11.3) it follows that either $y^* \in W \cap Y_o$, i.e., the system is W-inconsistent, or $y^* \notin W \cup Y_o$, i.e., the system is W-incomplete. Q.E.D.

3. APPLICATION OF THE FUNDAMENTAL THEOREM TO FORMAL SYSTEMS

We shall now show how the fundamental theorem can be applied to specific systems with more structure. We shall consider the so-called formal or symbol manipulating systems as being the closest to the general framework developed so far. Similar application can be made for other types of systems, as, e.g., logical systems of different varieties. As a representative of a formal system, we shall use the representation system [17] defined as an ordered sextuple

$$K = (E, S, T, R, P, \phi)$$

such that

(1) E is a denumerable set and represents expressions;
(2) $S \subset E$ represents sentences;
(3) $T \subset S$ represents theorems of S;
(4) $R \subset S$ represents refutable sentences;
(5) $P \subset E$ are (unary) predicates;
(6) N denotes the set of integers.

Let two mappings be given by $g: E \to N$, $\phi: E \times N \to E$, such that g is a injection and $\phi(e, n) \in S$ whenever e is a predicate, $e \in P$. Then, for any $e \in E$, $g(e)$ is the *Goedel number* of e.

It is quite easy to construct a general system for K by establishing the following correspondences:

(i) Predicates P are inputs of the system;
(ii) Expressions E are the states;

(iii) Sentences S are outputs;

(iv) The Goedel (restricted) function g of the general system is a restriction of the mapping g of K, $g = g \mid P$, and the state representation $\rho : E \times P \to S$ is

$$\rho(e, p) = \phi(p, g(e))$$

(v) Theorem set $T \subset S$ corresponds to Y_o.

We can now state a Goedel (-like) theorem for the representation system K.

Theorem 3.1. Let R^* be the set of Goedel numbers for all the elements whose norm is in $R \subset S$. If R^* is representable in K, then K is either R-inconsistent or R-incomplete.

PROOF. Let ρ be a general system for K constructed as above. To R^* of K corresponds a set of inputs $C_W{}^{d}$ precisely as defined in Section 2. Since R^* is representable, $C_W{}^{d}$ is an acceptable set. But then, by Theorem 2.1, there exists an input $x \in X$ such that

$$\rho(g(x), x) \in Y_o \leftrightarrow \rho(g(x), x) \in W$$

There exists, then, a corresponding predicate $p \in P$ such that

$$\phi(p, g(p)) \in T \leftrightarrow \phi(p, g(p)) \in R$$

$\phi(p, g(p))$ is an example of a sentence which is either in both T and R or is outside of both T and R. K is, therefore, either inconsistent or incomplete.

Q.E.D.

4. REALIZATION BY TURING MACHINES†

The abstract dynamical system and the concepts introduced in Section 1 can be realized in terms of many specific models used in the computation field.

We shall consider here briefly the realization as a Turing machine. Let I be the set of nonnegative integers; i.e., $I = \{0, 1, 2, \ldots\}$. Let V be a countable set; i.e., $V = \{v_o, v_1, \ldots\}$. Let V^* be the set of finite sequences from V; that is, V^* is the free monoid of V, whose identity element is Λ (empty string).

A Turing machine is a time-invariant free dynamical system such that:

(i) the time set T is the set of nonnegative integers;

(ii) the state space $C = V^* \times I \times V^*$;

† See Davis [18].

(iii) the output object B is V^*;

(iv) for each Turing machine, there are given two positive integers m and n and a mapping $\phi: I_n \times V_m \to \tilde{V}_m \times I_n$, where $I_n = \{0, \ldots, n\}$, $V_m = \{v_o, \ldots, v_m\}$, and $\tilde{V}_m = V_m \cup \{R, L\}$. R and L are special elements, and ϕ satisfies the condition that $\phi(i, v_j) \neq (v_j, i)$ for all $i \in I_n$ and $v_j \in V_m$ except for $i = 1$ where $\phi(i, v_j) = (v_j, i)$ for every $v_j \in V_m$. The state transition ϕ_{01} is then given by

$$\phi_{01}(\alpha v_k, i, v_j\beta) = \begin{cases} (\alpha v_k v_j, i', \beta), & \text{if } \phi(i, v_j) = (R, i') \text{ and } \beta \neq \Lambda \\ (\alpha v_k v_j, i', v_o), & \text{if } \phi(i, v_j) = (R, i') \text{ and } \beta = \Lambda \\ (\alpha, i', v_k v_j \beta), & \text{if } \phi(i, v_j) = (L, i') \text{ and } \alpha v_k \neq \Lambda \\ (\Lambda, i', v_o v_j \beta), & \text{if } \phi(i, v_j) = (L, i') \text{ and } \alpha v_k = \Lambda \\ (\alpha v_k, i', v_e \beta), & \text{if } \phi(i, v_j) = (v_e, i') \\ (\alpha v_k, i, v_j \beta), & \text{otherwise} \end{cases}$$

where α, α', β, and β' are elements of V^*, while i and i' are elements of I. Since the state-transition family is time invariant and since the time set is discrete, the state-transition family is uniquely determined by ϕ_{01}. The pair of integers (m, n) will be referred to as the characteristic numbers;

(v) the output function $\lambda: C \to B$ is given by

$$\lambda(\alpha, i, \beta) = \alpha \cdot \beta$$

Let V_m^* be the free monoid on V_m. Let (m, n) be the characteristic number of a Turing machine S. Then, if an initial state is in $V_m^* \times I_n \times V_m^*$, the state of the system S remains in $V_m^* \times I_n \times V_m^*$ at all $t \in T$. In other words, if an initial state is in $V_m^* \times I_n \times V_m^*$, the set of reachable and interesting equilibrium points is a subset of $V^* \times \{1\} \times V^*$. (The states of the form (α, i, Λ) are also trivial equilibrium states.) Following the custom, we assume that an initial state $c = (\alpha, i, \beta)$ of a Turing machine with the characteristic number (m, n) is selected from $\{\Lambda\} \times \{0\} \times V_m^*$ such that $c = (\Lambda, 0, \beta)$ where $\beta \neq \Lambda$. Consequently, for the class $\bar{\Phi}$ of Turing machines, the set $V^* \times \{1\} \times V^*$ can be taken as C^e. Then $(\bar{\Phi}, \lambda)$ is the entire class of all Turing machines, and a general system ρ is defined by these Turing machines as given in Section 1.

Chapter XII

CATEGORIES OF SYSTEMS AND ASSOCIATED FUNCTORS

In the preceding chapters, we have considered a number of rather general systems properties, giving conditions when a system has some stated properties. Obviously, this classifies the systems in different types in reference to various properties they possess. It is of interest to define these classes more precisely and to establish the relationships between them in a rigorous manner. An appropriate framework for this is provided by the categorical algebra, which is concerned with the classes and relationships between them.†

Various classes of general systems are defined in this chapter as categories with the homomorphisms between the objects of a category representing a modeling relation. Relationships between classes of systems are established in terms of the appropriate functors, using results from the representation and realization theories as derived in the preceding chapters. The natural transformations between these functors are also established. The diagram in Fig. 4.1 displays the relationships between the main concepts.

1. FORMATION OF CATEGORIES OF GENERAL SYSTEMS AND HOMOMORPHIC MODELS

(a) Formation of Categories of General Systems

To construct a category one needs a class of objects and a class of morphisms between them. For the objects we shall simply take systems, i.e., relations on the appropriately defined sets. The choice is less obvious as far

† We shall use the objective concept of a category given in the appendix.

as morphisms are concerned. Actually there are many ways to introduce the necessary morphisms, and we shall present some of the approaches that we have found interesting.

(i) *General Approach*

Since our approach to general systems theory is set-theoretic, the most obvious candidates for morphisms are simply all the functions defined on respective sets. A category of general systems will be defined by all relations $S \subset X \times Y$ and the mappings between them. Other categories are constructed in reference to additional structure of the sets X and Y and the selection of appropriate functions as morphisms.

(ii) *Structure Modeling Approach*

When the structure of a system is of prime concern, a relational homomorphism between systems can be used as a morphism. Let $S \subset X \times Y$, $S' \subset X' \times Y'$ be two general systems, h and k the mappings, $h: X \to X'$, $k: Y \to Y'$. The mapping $h \times k: X \times Y \to X' \times Y'$ is called a relational homomorphism from S to S' if the following holds:

$$(\forall(x, y))((x, y) \in S \to (h(x), k(y)) \in S')$$

A category of general systems will be then defined by all relations $S \subset X \times Y$ and the relational homomorphisms between them.

(iii) *Algebraic Modeling Approach*

If X, X', Y, and Y' are algebras, the mappings h and k of the structure modeling approach can be limited to homomorphisms from X to X' and from Y to Y', respectively, to yield yet another category of general systems.

The categories constructed in this way are somewhat more restrictive than the ones defined by using the structure modeling approach, but they have important advantages too, namely:

(i) Additional structure allows a deeper analysis of various systems properties. The introduction of such additional structure algebraically is most appropriate for the categorical considerations.

(ii) The homomorphisms have rather nice interpretation as "modeling functions." Namely, in various applications, the homomorphisms can be defined as simplifying the structure of the original system while preserving the essential relationships and suppressing the unessential details. The homomorphic image represents then "a simplified" system, i.e., a model of the original system. This argument is also applicable to the structure modeling approach. However, if we require a model of, for instance, a linear system to be linear, the structure modeling approach does not satisfy the requirement.

In this chapter, we shall follow, in principle, the algebraic modeling approach. However, when we treat only the categories of general systems without any specific algebraic structures, the approach can be eventually considered equal to the structure modeling approach. More specifically, when considering general systems, our approach can be viewed as a special case of the algebraic modeling approach, where since no algebraic structure is explicitly assumed to exist in the objects, every function can be a homomorphism.

If a system $S \subset X \times Y$ is a time system, elements of X and Y have a special structure; i.e., $x: T \to A$ and $y: T \to B$, where $X \subset A^T$ and $Y \subset B^T$. Consequently, when we call a function $h: X \to X'$ a homomorphism of inputs of time systems, we always assume that h is constructed by a function (homomorphism) $h_a: A \to A'$ such that $h(x)(t) = h_a(x(t))$ for every $t \in T$ where $X' \subset A'^T$. This construction implies the following property:

$$(\forall x^t)(\forall x_t')(\forall x_t'')(x^t \cdot x_t' \in X \;\&\; x^t \cdot x_t'' \in X \to h(x^t \cdot x_t') \mid T^t = h(x^t \cdot x_t'') \mid T^t)$$

Therefore, we can define an extention of a homomorphism $h: X \to X'$ of a time system as $h: X^t \to X''$ such that $h(x^t) = h(x^t \cdot x_t) \mid T^t$, where x_t is arbitrary if $x^t \cdot x_t \in X$ holds. Similarly, $h: X_{tt'} \to X'_{tt}$ can be defined as

$$h(x_{tt'}) = h(x^t \cdot x_{tt'} \cdot x_{t'}) \mid T_{tt'}$$

In other words, a homomorphism of time objects is required to be at least a homomorphism with respect to the operations of concatenation and restriction.

(b) Homomorphic Models

In general, there are many ways to formulate the concept of models. However, conceptually, the most convincing mathematical formulation is by means of homomorphisms. This is the modeling approach used in this chapter. We shall explore briefly the meaning of such homomorphic models.

Let X, X', Y, and Y' be Ω-algebras, while

$$h_x: X \to X', \qquad \text{and} \qquad h_y: Y \to Y'$$

are homomorphisms; it is easy to show then that

$$h = \langle h_x, h_y \rangle : X \times Y \to X' \times Y'$$

such that

$$h(x, y) = (h_x(x), h_y(y))$$

is also a homomorphism, where the algebraic structures of $X \times Y$ and $X' \times Y'$ are defined in the usual way. The concept of a homomorphic model is then shown in the following definition.

Definition 1.1. Let $S \subset X \times Y$ and $S' \subset X' \times Y'$ be general systems, where $h = \langle h_x, h_y \rangle$ is a homomorphism from $X \times Y$ into $X' \times Y'$ and h_x is onto. Then S' is a model of S if and only if

$$(\forall(x, y))((x, y) \in S \rightarrow h(x, y) \in S')$$

Furthermore, S and S' are called equivalent if and only if h is an isomorphism and

$$(\forall(x', y'))((x', y') \in S' \rightarrow h^{-1}(x', y') \in S)$$

is also satisfied.

Similarly, in the case of dynamical systems, the concept of homomorphic models is given by the following definition.

Definition 1.2. Let

$$S = \{(\rho_t : C_t \times X_t \rightarrow Y_t, \phi_{tt'} : C_t \times X_{tt'} \rightarrow C_{t'}) : t, t' \in T\}$$

and

$$\hat{S} = \{(\hat{\rho}_t : \hat{C}_t \times \hat{X}_t \rightarrow \hat{Y}_t, \hat{\phi}_{tt'} : \hat{C}_t \times \hat{X}_{tt'} \rightarrow \hat{C}_{t'}) : t, t' \in T\}$$

be dynamical systems, where $h = \langle h_x, h_y, h_c \rangle$ is a homomorphism from $X_t \times Y_t \times C_t$ into $\hat{X}_t \times \hat{Y}_t \times \hat{C}_t$ for every $t \in T$ and h_x is onto. Then $\hat{S}$ is a model of S if and only if for every c_t and x_t,

$$h_y(\rho_t(c_t, x_t)) = \hat{\rho}_t(h_c(c_t), h_x(x_t))$$

Furthermore, S and $\hat{S}$ are called equivalent if and only if h is an isomorphism.

Similar definitions can be introduced for other kinds of system representations, such as global-state response, input response, and others.

As mentioned above, a homomorphic model has a convenient property; i.e., it preserves algebraic structures that are of interest and suppresses non-interesting details. This is fully in accord with the intuitive concept of a model.

Let us examine the situation in some detail for the class of dynamical systems.

Given the response function of a dynamical system, $\rho_t : C_t \times X_t \rightarrow Y_t$, and homomorphisms $h = \langle h_x, h_y, h_c \rangle$. Let $\hat{C}_t$, $\hat{X}_t$, and $\hat{Y}_t$ be the homomorphic images of C_t, X_t, and Y_t, respectively, and let $\hat{\rho}_t$ be the homomorphic image of ρ_t. Let $E_x \subset X_t \times X_t$ be such that

$$(x_t, x_t') \in E_x \leftrightarrow h_x(x_t) = h_x(x_t')$$

Naturally, E_x is an equivalence relation. Similarly, let E_y and E_c be the equivalence relations derived from h_y and h_c. Since h_x, h_y, and h_c are homomorphisms, the quotient sets X_t/E_x, Y_t/E_y, and C_t/E_c can have the same algebraic structures as X_t, Y_t, and C_t. However, two different elements x_t and x_t', where $(x_t, x_t') \in E_x$, may not be considered equivalent with respect to another algebraic structure not embodied in the definition of the homomorphism h_x; such structure is completely ignored by h_x. Let

$$\rho_t^M : (C_t/E_c) \times (X_t/E_x) \to (Y_t/E_y)$$

$$\rho_t^M([c_t], [x_t]) = [\rho_t(c_t, x_t)]$$

where $c_t \in [c_t] \in C_t/E_c$ and $x_t \in [x_t] \in X_t/E_x$. The above definition is proper because

$$h_y(\rho_t(c_t, x_t)) = \hat{\rho}_t(h_c(c_t), h_x(x_t))$$

holds. It is easy to show that ρ_t^M and $\hat{\rho}_t$ are equivalent; i.e., an isomorphism can be found between them. We have, therefore, the following commutative diagram, where η_c, η_x, and η_y are the natural mappings:

$$\begin{array}{ccc} C_t \times X_t & \xrightarrow{\rho_t} & Y_t \\ \downarrow (\eta_c, \eta_x) & & \downarrow \eta_y \\ (C_t/E_c) \times (X_t/E_x) & \xrightarrow{\rho_t^M} & (Y_t/E_y) \\ \updownarrow \text{Isomorphism} & & \updownarrow \text{Isomorphism} \\ \hat{C}_t \times \hat{X}_t & \xrightarrow{\hat{\rho}_t} & \hat{Y}_t \end{array}$$

(with $(h_c, h_x): C_t \times X_t \to \hat{C}_t \times \hat{X}_t$ and $h_y: Y_t \to \hat{Y}_t$)

$\hat{\rho}_t$ is a simplified model of ρ_t because of the isomorphic relationship it has with ρ_t^M, which is defined on the quotient sets.

It should be noticed that any homomorphism can generate a model; i.e., a model is determined by a homomorphism, and the selection of a homomorphism depends on the properties of the original system that are taken as important for modeling. Also, notice that only h_x is required to be an onto mapping. Hence, the homomorphic image of the original system is, in general, a proper subset of the model. The model, then, has certain implications not present in the original system, which, again, is compatible with the intuitive notion of a model.

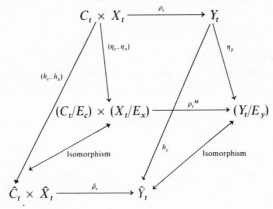

The homomorphism $h = \langle h_x, h_y, h_c \rangle$ in Definition 1.2 is given solely in reference to ρ; it is *not* required, therefore, that the following diagram be satisfied:

$$\begin{array}{ccc} C_t \times X_{tt'} & \xrightarrow{\phi_{tt'}} & C_{t'} \\ {\scriptstyle h_c}\downarrow \quad \downarrow{\scriptstyle h_x} & & \downarrow{\scriptstyle h_c} \\ \hat{C}_t \times \hat{X}_{tt'} & \xrightarrow{\hat{\phi}_{tt'}} & \hat{C}_{t'} \end{array}$$

However, we shall show now that an appropriately defined state-transition function of a model given by Definition 1.2 is related with the state transition of the corresponding original system in a way which is quite natural from the input–output viewpoint, where the concept of the state is a secondary, derived concept.

Let $E_t \subset C_t \times C_t$ be such that

$$(c_t, c_t') \in E_t \leftrightarrow (\forall x_t)(\rho_t(c_t, x_t) = \rho_t(c_t', x_t))$$

Then E_t is an equivalence relation. It should be noticed that if $(c_t, c_t') \in E_t$ holds, c_t need not be distinguished from c_t' since both c_t and c_t' always produce the same output for the same input. So we can use the quotient set C_t/E_t for a state object.

We shall show that E_t is a congruence relation with respect to h_c and $\phi_{tt'}(-, x_{tt'})$, i.e.,

(i) $[c_t] = [c_t'] \rightarrow [h_c c_t] = [h_c c_t']$
(ii) $[c_t] = [c_t'] \rightarrow (\forall x_{tt'})([\phi_{tt'}(c_t, x_{tt'})] = [\phi_{tt'}(c_t', x_{tt'})])$

where, naturally, $[h_c c_t]$ is an equivalence class on $\hat{X}_t/\hat{E}_t$ and $\hat{E}_t \subset \hat{C}_t \times \hat{C}_t$.

(i) $[c_t] = [c_t'] \rightarrow [h_c c_t] = [h_c c_t']$

$$\begin{aligned} [c_t] = [c_t'] &\rightarrow (\forall x_t)(\rho_t(c_t, x_t) = \rho_t(c_t', x_t)) \\ &\rightarrow (\forall x_t)(h_y \rho_t(c_t, x_t) = h_y \rho_t(c_t', x_t)) \\ &\rightarrow (\forall x_t)(\hat{\rho}_t(h_c c_t, h_x x_t) = \hat{\rho}_t(h_c c_t', h_x x_t)) \\ &\rightarrow (\forall \hat{x}_t)(\hat{\rho}_t(h_c c_t, \hat{x}_t) = \hat{\rho}_t(h_c c_t', \hat{x}_t)) \qquad (h_x \text{ is onto!}) \\ &\rightarrow [h_c c_t] = [h_c c_t'] \end{aligned}$$

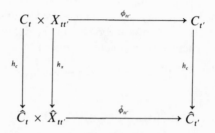

(ii) $[c_t] = [c_t'] \rightarrow [\phi_{tt'}(c_t, x_{tt'})] = [\phi_{tt'}(c_t', x_{tt'})]$ for every $x_{tt'} \in X_{tt'}$

$$[c_t] = [c_t'] \rightarrow (\forall x_t)(\rho_t(c_t, x_t) = \rho_t(c_t', x_t))$$

$$\rightarrow (\forall x_{tt'})(\forall x_{t'})(\rho_{t'}(\phi_{tt'}(c_t, x_{tt'}), x_{t'}) = \rho_{t'}(\phi_{tt'}(c_t', x_{tt'}), x_{t'}))$$

$$\rightarrow (\forall x_{tt'})([\phi_{tt'}(c_t, x_{tt'})] = [\phi_{tt'}(c_t', x_{tt'})])$$

Consequently, we can define extensions of $\phi_{tt'}$ and h_c as follows:

$$\phi_{tt'}:(C_t/E_t) \times X_{tt'} \rightarrow C_{t'}/E_{t'}$$

such that

$$\phi_{tt'}([c_t], x_{tt'}) = [\phi_{tt'}(c_t, x_{tt'})]$$

and

$$h_c: C_t/E_t \rightarrow \hat{C}_{t'}/\hat{E}_{t'}$$

such that

$$h_c([c_t]) = [h_c(c_t)]$$

We have now the following proposition.

Proposition 1.1. The commutative diagram associated with ρ_t as given in Definition 1.2 implies the commutativity of the following diagram:

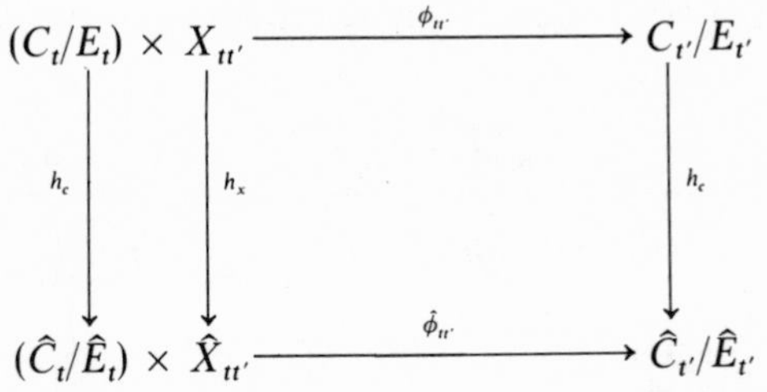

where

$$(c_t, c_t') \in E_t \leftrightarrow (\forall x_t)(\rho_t(c_t, x_t) = \rho_t(c_t', x_t))$$

$$\phi_{tt'}([c_t], x_{tt'}) = [\phi_{tt'}(c_t, x_{tt'})]$$

$$h_c([c_t]) = [h_c(c_t)]$$

PROOF. Let $x_t = x_{tt'} \cdot x_{t'}$ and c_t be arbitrary. Then

$$
\begin{aligned}
h_y\rho_t(c_t, x_t) &= \hat{\rho}_t(h_c c_t, h_x x_t) \\
&\rightarrow h_y\rho_t(c_t, x_t) \mid T_{t'} = \hat{\rho}_t(h_c c_t, h_x x_t) \mid T_{t'} \\
&\rightarrow h_y\rho_{t'}(\phi_{tt'}(c_t, x_{tt'}), x_{t'}) = \hat{\rho}_{t'}(\hat{\phi}_{tt'}(h_c c_t, h_x x_{tt'}), h_x x_{t'}) \\
&\rightarrow \hat{\rho}_{t'}(h_c\phi_{tt'}(c_t, x_{tt'}), h_x x_{t'}) = \hat{\rho}_{t'}(\hat{\phi}_{tt'}(h_c c_t, h_x x_{tt'}), h_x x_{t'}) \\
&\rightarrow [h_c\phi_{tt'}(c_t, x_{tt'})] = [\hat{\phi}_{tt'}(h_c c_t, h_x x_{tt'})] \qquad \text{(because } h_x \text{ is onto)} \\
&\rightarrow h_c\phi_{tt'}([c_t], x_{tt'}) = \hat{\phi}_{tt'}(h_c[c_t], h_x x_{tt'})
\end{aligned}
$$

Q.E.D.

The above proposition means that if the state object C_t is allowed to have redundant, "useless," states, the state transition may not satisfy the commutative diagram; however, the state transition does satisfy the desired commutativity if we consider only the essential aspect of C_t, i.e., if the redundant states are eliminated. It should be noticed again that the redundancy of states is a direct consequence of considering the state objects as a secondary concept.

Before moving to the categorical considerations of general systems, let us illustrate the basic idea, which we shall use for the categorical classifications.

Let A and B in Fig. 1.1 be two categories of systems. The objects in each category (i.e., the systems of a given type) are related by a homomorphic modeling relation, which is a morphism in respective category. The categories of systems are related by functors; specifically, we shall be interested in two types of functors, termed constructive functor and forgetful functor, respectively. The constructive functor maps the systems with less structure into the systems with more structure (i.e., more structure in the respective sets S).

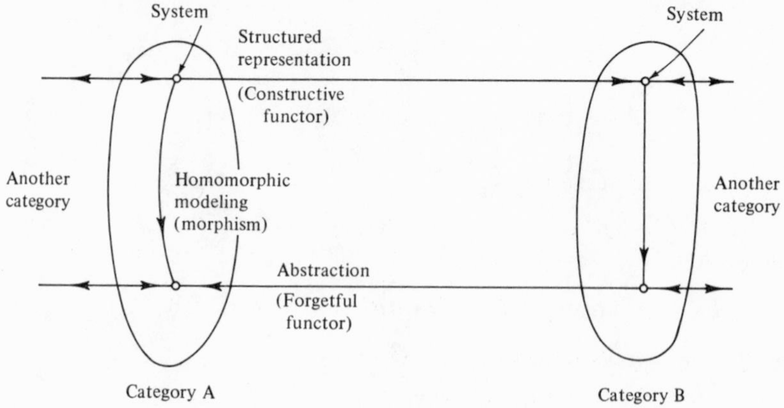

FIG. 1.1

For instance, a constructive functor maps general systems into systems with response functions or time systems into dynamical systems. The forgetful functor maps the systems into opposed direction; e.g., it maps dynamical systems into time systems, etc.

2. CATEGORIES OF GENERAL SYSTEMS

In this section, we shall construct several categories of general systems and establish the existence of some functors and relationships between them. Since it is, in general, necessary to prove that a class of objects and a class of morphisms form a category, we shall present various definitions, including those of categories, in the form of propositions. Where the required conditions are obviously satisfied, the proof will be omitted.

By following the algebraic approach, the most immediate category is given by the following proposition.

Proposition 2.1. Let $\mathrm{Ob}\,\bar{S}$ be the class of all relations, $S \subset X \times Y$, where X and Y are arbitrary sets, $X = \mathscr{D}(S)$, and for each pair, $S, S' \in \mathrm{Ob}\,\bar{S}$, let $\mathrm{Mor}\,(S, S')$ be the set of all pairs of functions $f = \langle h_x, h_y \rangle$, $h_x : X \to X'$ and $h_y : Y \to Y'$ such that h_x is onto and

$$(x, y) \in S \to (h_x(x), h_y(y)) \in S'$$

Define the composition in $\{\mathrm{Mor}\,(S, S') : S, S' \in \mathrm{Ob}\,S\}$ by

$$\langle h_x, h_y \rangle \cdot \langle h_x', h_y' \rangle = \langle h_x \cdot h_x', h_y \cdot h_y' \rangle$$

where $h_x \cdot h_x'$ and $h_y \cdot h_y'$ are usual compositions of functions, i.e.,

$$h_x \cdot h_x'(x') = h_x(h_x'(x'))$$

$(\mathrm{Ob}\,\bar{S}, \{\mathrm{Mor}\,(S, S') : S, S' \in \mathrm{Ob}\,\bar{S}\})$ then is a category, which will be denoted by $\bar{S}$ and termed category of general systems.

The category of general systems as defined in Proposition 2.1 has a difficulty in subsequent theoretical developments. This difficulty is due to the fact that under general functions as used in Proposition 2.1, a function-type relation R which is defined by

$$(\forall (x, y))(\forall (x', y'))((x, y), (x', y') \in R \;\&\; x = x' \to y = y')$$

can be mapped into a nonfunction relation, i.e., a single-valued relation into a many-valued relation. To avoid this, we shall introduce a subcategory of $\bar{S}$.

Proposition 2.2. Let Ob $\bar{S}^f$ be the class as defined in Proposition 2.1 and $\text{Mor}^f(S, S')$ a subclass of functions as defined in Proposition 2.1 such that for any $f \in \text{Mor}^f(S, S')$, if $R \in \text{Ob}\,\bar{S}^f$ is functional, so is $f(R)$.

$$(\text{Ob}\,\bar{S}^f, \{\text{Mor}^f(S, S'): S, S' \in \text{Ob}\,\bar{S}^f\})$$

then is a category, which will be denoted by $\bar{S}^f$ and termed function-preserving category of general systems.

The class of objects for both $\bar{S}$ and $\bar{S}^f$ are the same, but when we want to indicate which category is in question, the class of objects for $\bar{S}^f$ will be denoted by Ob $\bar{S}^f$. Although both categories refer to the same class of systems, the modeling relationships used are different, with the requirements for $\bar{S}^f$ being slightly stronger than for $\bar{S}$. Namely, a system S' can be a model of another system S in $\bar{S}$ even if not so in $\bar{S}^f$; conversely, however, if S' is a model of S in $\bar{S}^f$, it is always so in $\bar{S}$ as well. This fact can be expressed by a functor in the following proposition.

Proposition 2.3. Let $H_1 : \bar{S}^f \to \bar{S}$ be the identity; that is, for every $S \in \text{Ob}\,\bar{S}^f$, $H_1(S) = S$, and for every $\langle h_x, h_y \rangle \in \text{Mor}(S, S')$ in $\bar{S}^f$, $H_1(\langle h_x, h_y \rangle) = \langle h_x, h_y \rangle$. H_1 is then a functor.

We shall now introduce a category for the consideration of systems-response functions.

Proposition 2.4. Let Ob $\bar{S}_g$ be the class of all functions, $\rho: C \times X \to Y$, where C, X, and Y are arbitrary sets. For any $\rho, \rho' \in \text{Ob}\,\bar{S}_g$, let $\text{Mor}(\rho, \rho')$ be the set of all triplets of functions, $f = \langle h_x, h_y, h_c \rangle$, $h_x : X \to X'$, $h_y : Y \to Y'$, $h_c : C \to C'$ such that h_x is onto and the diagram

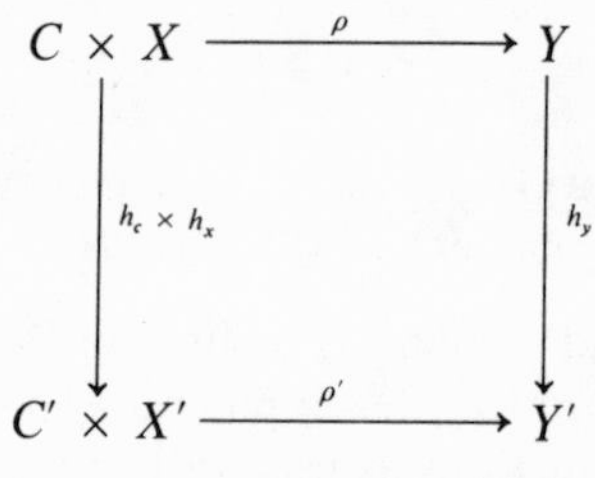

is commutative, i.e.,

$$h_y \cdot \rho(c, x) = \rho'(h_c \cdot c, h_x \cdot x)$$

for every $x \in X$ and $c \in C$. Define the composition in $\{\text{Mor}(\rho, \rho') : \rho, \rho' \in \text{Ob}\,\bar{S}_g\}$ by

$$\langle h_x, h_y, h_c \rangle \cdot \langle h_x', h_y', h_c' \rangle = \langle h_x \cdot h_x', h_y \cdot h_y', h_c \cdot h_c' \rangle$$

$(\text{Ob}\,\bar{S}_g, \{\text{Mor}(\rho, \rho') : \rho, \rho' \in \text{Ob}\,\bar{S}_g\})$ is then a category, which we shall denote by $\bar{S}_g$ and refer to as the category of global-systems responses.

Consider now the relationship among the categories $\bar{S}$, $\bar{S}^f$, and $\bar{S}_g$. A functor has been already established between $\bar{S}$ and $\bar{S}^f$. Relationships of $\bar{S}$ and $\bar{S}^f$ with $\bar{S}_g$ are characterized by the existence of two functors: a forgetful functor $F_1 : \bar{S}_g \to \bar{S}$, which to every global-response function associates the corresponding general system, and the constructive functor $G_1 : \bar{S}^f \to \bar{S}_g$, which to every general system assigns a global response.

For each relation $S \subset X \times Y$, let

$$\text{Fun}(S) = \{f : f : X \to Y \,\&\, f \subseteq S\}$$

Proposition 2.5. Let $G_1 : \bar{S}^f \to \bar{S}_g$ be a mapping such that

(i) for every $S \in \text{Ob}\,\bar{S}^f$,

$$G_1(S) = \rho$$

where $C = \text{Fun}(S) \subseteq Y^X$, $\rho : C \times X \to Y$ such that

$$\rho(c, x) = c(x)$$

for all $(c, x) \in C \times X$;

(ii) for every $\langle h_x, h_y \rangle \in \text{Mor}(S, S')$,

$$G_1(\langle h_x, h_y \rangle) = \langle h_x, h_y, h_c \rangle$$

where

$$h_c(c) = c' \leftrightarrow c'(h_x \cdot x) = h_y \cdot c(x) \qquad \text{for every} \quad x$$

that is, the following diagram is commutative:

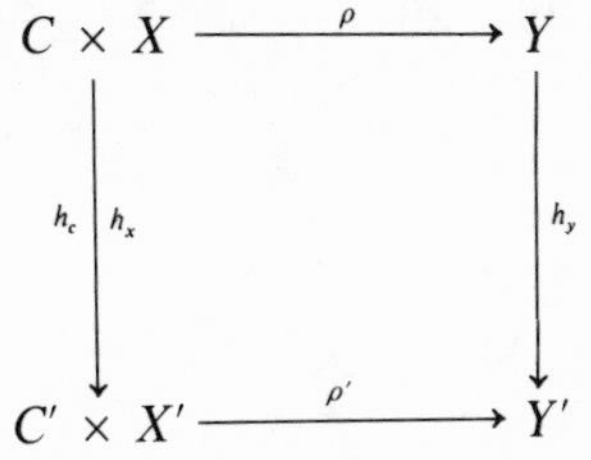

G_1 is then a functor.

PROOF. It should be noticed that h_c is well defined because h_x is onto, which implies $h_x(X) = X'$, and since $R = \{(x, c(x)) : x \in X\}$ is functional,

$$h_x \times h_y(R) = \{(h_x \cdot x, h_y \cdot c(x)) : x \in X\}$$

is also functional, which implies that

$$h_x \cdot x = h_x \cdot \hat{x} \to h_y \cdot c(x) = h_y \cdot c(\hat{x})$$

For $S \in \text{Ob}\, \bar{S}^f$,

$$G_1(I_S) = G_1(\langle I_x, I_y \rangle) = \langle I_x, I_y, h_c \rangle$$

where $h_c(c) = c' \leftrightarrow c'(x) = c(x)$ for every x. Hence, $h_c = I_c$. That is,

$$G_1(I_S) = I_{G_1(S)}$$

For the morphisms $\langle h_x, h_y \rangle$, $\langle h_x', h_y' \rangle$ where

$$X \overset{h_x}{\to} X' \overset{h_x'}{\to} X'', \qquad \text{and} \qquad Y \overset{h_y}{\to} Y' \overset{h_y'}{\to} Y''$$

$$G_1(\langle h_x', h_y' \rangle \cdot \langle h_x, h_y \rangle) = G_1(\langle h_x' \cdot h_x, h_y' \cdot h_y \rangle) = \langle h_x' \cdot h_x, h_y' \cdot h_y, \hat{h}_c \rangle$$

where, by definition, $\hat{h}_c : C \to C''$ is

$$\hat{h}_c(c) = c'' \leftrightarrow c''(h_x' \cdot h_x \cdot x) = h_y' \cdot h_y \cdot c(x) \qquad \text{for every} \quad x \in X$$

On the other hand,

$$G_1(\langle h_x, h_y \rangle) = \langle h_x, h_y, h_c \rangle$$

where

$$h_c(c) = c' \leftrightarrow c'(h_x \cdot x) = h_y \cdot c(x) \qquad \text{for every} \quad x \in X$$

and

$$G_1(\langle h_x', h_y' \rangle) = \langle h_x', h_y', h_c' \rangle$$

where

$$h_c'(c') = c'' \leftrightarrow c''(h_x' \cdot x') = h_y' \cdot c'(x') \qquad \text{for every} \quad x' \in X'$$

Therefore,

$$\begin{aligned} h_c' \cdot h_c(c) = c'' &\leftrightarrow c''(h_x' \cdot x') = h_y'[h_c(c)](x') && \text{for every} \quad x' \in X' \\ &\leftrightarrow c''(h_x' \cdot h_x \cdot x) = h_y'[h_c(c)](h_x \cdot x) && \text{for every} \quad x \in X \\ & && \text{(notice that } h_x \text{ is onto)} \\ &\leftrightarrow c''(h_x' \cdot h_x \cdot x) = h_y' \cdot c'(h_x \cdot x) && \text{for every} \quad x \in X, \\ & && \text{where } c' = h_c(c) \\ &\leftrightarrow c''(h_x' \cdot h_x \cdot x) = h_y' \cdot h_y \cdot c(x) && \text{for every} \quad x \in X \\ &\leftrightarrow \hat{h}_c(c) = c'' \end{aligned}$$

Consequently,

$$G_1(\langle h_x', h_y'\rangle \cdot \langle h_x, h_y\rangle) = \langle h_x', h_y', h_c'\rangle \cdot \langle h_x, h_y, h_c\rangle$$
$$= G_1(\langle h_x', h_y'\rangle) \cdot G_1(\langle h_x, h_y\rangle)$$

Q.E.D.

Proposition 2.6. Let $F_1 : \bar{S}_g \to \bar{S}$ be a mapping such that for every $\rho \in \mathrm{Ob}\,\bar{S}_g$

$$F_1(\rho) = S \subset X \times Y$$

where

$$(x, y) \in S \leftrightarrow (\exists c)(y = \rho(c, x))$$

and for every $\langle h_x, h_y, h_c\rangle$,

$$F_1(\langle h_x, h_y, h_c\rangle) = \langle h_x, h_y\rangle$$

F_1 is then a functor.

The considerations so far have established the diagram in Fig. 2.1. To

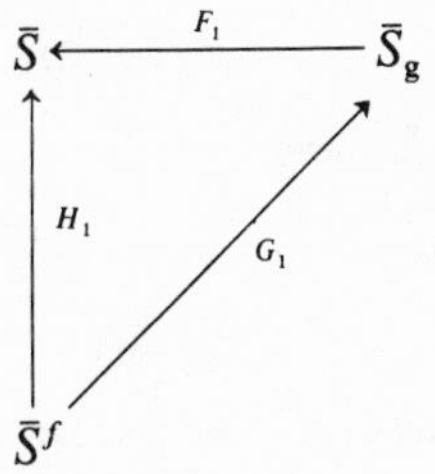

FIG. 2.1

make the relationship implied by Fig. 2.1 more symmetrical, we shall introduce the concept of the function-preserving category of global-systems response.

Proposition 2.7. Let $\mathrm{Ob}\,\bar{S}_g^f$ be the class of objects defined in Proposition 2.4; i.e., $\mathrm{Ob}\,\bar{S}_g^f = \mathrm{Ob}\,\bar{S}_g$ and $\mathrm{Mor}^f(\rho, \rho')$ a subset of morphisms defined in Proposition 2.4 such that the diagram

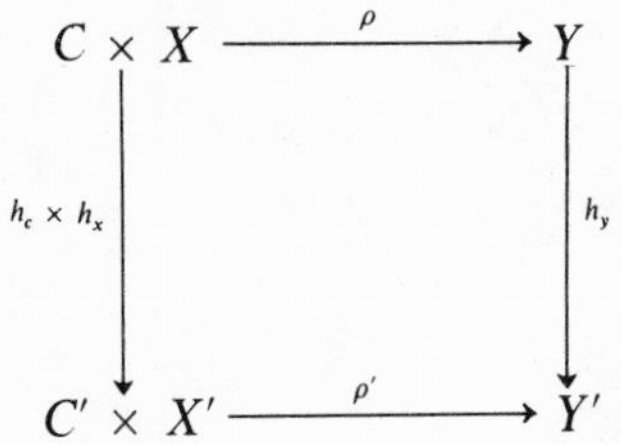

is commutative, and $\langle h_x, h_y\rangle$ is function preserving for any functional relation $R \subset F_1(\rho)$; that is $(h_x \times h_y)(R)$ is a function-type relation.

$(\text{Ob}\,\bar{S}_g{}^f, \{\text{Mor}^f(\rho, \rho'):\rho, \rho' \in \text{Ob}\,\bar{S}_g{}^f\})$ is then a category, which we shall denote by $\bar{S}_g{}^f$ and refer to as the function-preserving category of global-system response.

PROOF. Let $R \subset F_1(\rho)$ be a function-type relation. Let

$$\langle h_x, h_y, h_c\rangle \in \text{Mor}^f(\rho, \rho')$$

Then

$$\begin{aligned}(x, y) \in R &\to (x, y) \in F_1(\rho)\\ &\to (\exists c)(y = \rho(c, x))\\ &\to (\exists c')(h_y \cdot y = \rho'(c', h_x \cdot x))\\ &\to (h_x \cdot x, h_y \cdot y) \in F_1(\rho')\\ &\to (h_x \times h_y)(x, y) \in F_1(\rho')\end{aligned}$$

Therefore, $(h_x \times h_y)(R) \subset F_1(\rho')$. Consequently, if $\langle h_x, h_y, h_c\rangle \in \text{Mor}^f(\rho, \rho')$ and $\langle h_x', h_y', h_c'\rangle \in \text{Mor}^f(\rho', \rho'')$, then

$$\langle h_x', h_y', h_c'\rangle \cdot \langle h_x, h_y, h_c\rangle \in \text{Mor}^f(\rho, \rho'')$$

Q.E.D.

The categories $\bar{S}_g^f$ and $\bar{S}_g$ are obviously related by an identity functor

$$H_2 : \bar{S}_g^f \to \bar{S}_g$$

i.e., $H_2(\rho) = \rho$ and $H_2(\langle h_x, h_y, h_c\rangle) = \langle h_x, h_y, h_c\rangle$. Furthermore, $\bar{S}^f$ and $\bar{S}_g^f$ are related by the functor G_1 as defined in Proposition 2.5 in one direction and by the restriction F_1^f of the functor F_1 in another direction. The diagram is now as shown in Fig. 2.2.

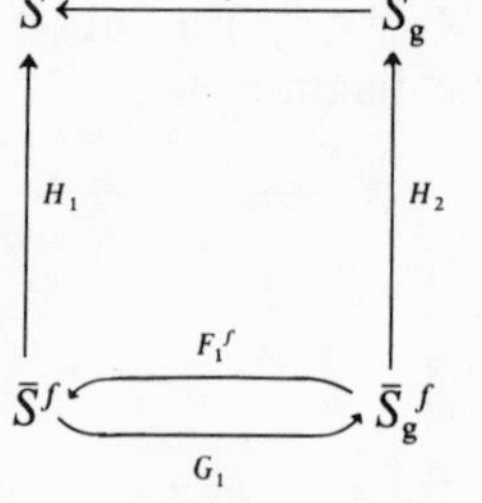

FIG. 2.2

The functors G_1 and F_1^f are apparently related. Indeed, we shall now show that F_1^f is a left adjoint of G_1.

It is clear that $F_1^f \cdot G_1 : \bar{S}^f \to \bar{S}^f$ is an identity functor. Less obvious is the composition in the reverse order. For each $\rho : C \times X \to Y$, let $\rho_c : X \to Y$ such that $\rho_c(x) = \rho(c, x)$.

Proposition 2.8. $G_1 F_1^f : \bar{S}_g^f \to \bar{S}_g^f$ is a functor. Furthermore, for each $\rho \in \bar{S}_g^f$ there is a morphism $\tau_1(\rho) \in \mathrm{Mor}\,(\rho, G_1 F_1^f(\rho))$, as shown in Fig. 2.3. $\tau_1(\rho)$ is given by $\tau_1(\rho) = \langle I_x, I_y, h_c \rangle$, where for $\rho : C \times X \to Y$, $h_c(c) = \rho_c$.

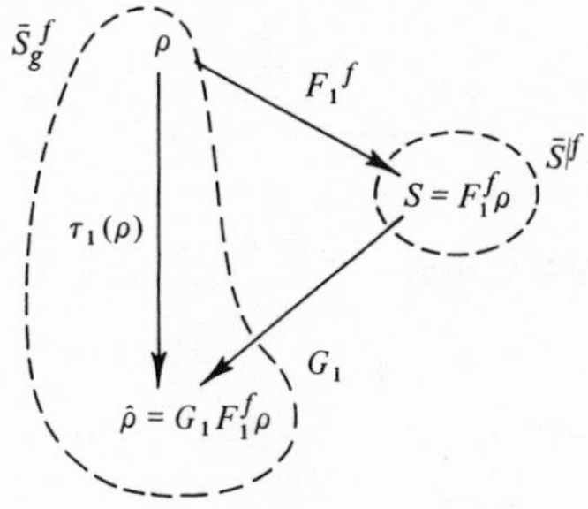

FIG. 2.3

PROOF. Let $\rho : C \times X \to Y$ and $\hat{\rho} = G_1 F_1^f \rho : \hat{C} \times X \to Y$. The definition of F_1^f implies that $F_1^f(\rho) = S$, where

$$S = \{(x, y) : (\exists c)(\rho(c, x) = y)\}$$

Hence, for every $c \in C$, we have $\rho_c \in \mathrm{Fun}\,(S) = \hat{C}$. Consequently, $h_c : C \to \hat{C}$ is properly defined. The definition of $\hat{\rho}$ implies that

$$\hat{\rho}(h_c \cdot c, I_x \cdot x) = \hat{\rho}(\rho_c, x) = \rho_c(x) = \rho(c, x) = I_y \rho(c, x)$$

Therefore,

$$\langle I_x, I_y, h_c \rangle \in \mathrm{Mor}\,(\rho, G_1 F_1^f \rho)$$

Q.E.D.

We can now establish the following proposition.

Proposition 2.9. $\tau_1 : I \to G_1 F_1^f$ is a natural transformation, where $I : \bar{S}_g^f \to \bar{S}_g^f$ is the identity functor, i.e., the diagram

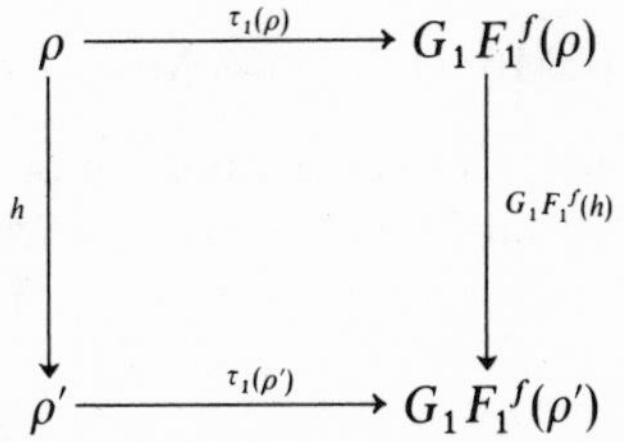

is commutative, where $\tau_1(\rho)$ is given in Proposition 2.8.

PROOF. Let $h = \langle h_x, h_y, h_c \rangle$, $\tau_1(\rho) = \langle I_x, I_y, h_c \rangle$, $\tau_1(\rho') = \langle I_{x'}, I_{y'}, h_{c'} \rangle$, and $G_1 F_1{}^f(h) = \langle h_x, h_y, h_c' \rangle$, where $\rho : C \times X \to Y$, $\rho' : C' \times X' \to Y'$, $G_1 F_1{}^f(\rho) = \hat{\rho} : \hat{C} \times X \to Y$, and $G_1 F_1{}^f(\rho') = \hat{\rho}' : \hat{C}' \times X' \to Y'$. Then it is sufficient to show that the diagram

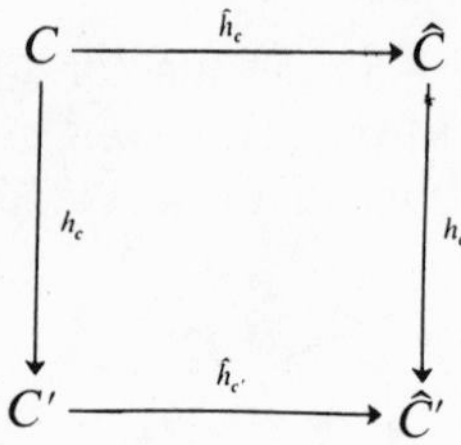

is commutative. The definitions of $\hat{h}_c$, $\hat{h}_{c'}$, and h_c' imply that $\hat{h}_c(c) = \rho_c$, $\hat{h}_{c'}(c') = \rho_{c'}'$, and

$$h_c'(\hat{c}) = \hat{c}' \leftrightarrow \hat{c}'(h_x \cdot x) = h_y \cdot \hat{c}(x) \qquad \text{for every} \quad x \in X$$

Therefore,

$$h_c' \cdot \hat{h}_c(c) = h_c'(\rho_c) \equiv \hat{c}'' \leftrightarrow \hat{c}''(h_x \cdot x) = h_y \rho_c(x) \qquad \text{for every} \quad x \in X$$

and

$$\hat{h}_c' \cdot h_c(c) = \rho'_{h_c(c)} \equiv \hat{c}' \leftrightarrow \hat{c}'(h_x \cdot x) = \rho'(h_c \cdot c, h_x \cdot x) \qquad \text{for every} \quad x \in X$$

Since $\rho'(h_c \cdot c, h_x \cdot x) = h_y \rho(c, x)$ for every $x \in X$, we have $\hat{c}'' = \hat{c}'$. Q.E.D.

Proposition 2.10. For each $\rho \in \text{Ob}\, \bar{S}_g{}^f$ and $S \in \text{Ob}\, \bar{S}^f$, let

$$\tau_2(\rho, S)(h) = G_1(h) \cdot \tau_1(\rho)$$

where $h \in \text{Mor}\,(F_1{}^f \rho, S)$. Then

$$\tau_2(\rho, S) : \text{Mor}\,(F_1{}^f \rho, S) \to \text{Mor}\,(\rho, G_1 S)$$

is a morphism in $\overline{\text{Set}}$, where $\overline{\text{Set}}$ is the category of sets. Furthermore, τ_2 is a natural transformation from $\text{Mor}\,(F_1{}^f -, -)$ to $\text{Mor}\,(-, G_1 -)$; that is,

$$\text{Mor}\,(F_1{}^f -, -), \text{Mor}\,(-, G_1 -) : (\bar{S}_g{}^f)^{\text{op}} \times \bar{S}^f \to \overline{\text{Set}}$$

$$\tau_2 : \text{Mor}\,(F_1{}^f -, -) \to \text{Mor}\,(-, G_1 -)$$

and the diagram

$$\begin{array}{ccc}
\operatorname{Mor}(F_1{}^f\rho', S) & \xrightarrow{\tau_2(\rho', S)} & \operatorname{Mor}(\rho', G_1 S) \\
\Big\downarrow{\scriptstyle \operatorname{Mor}(F_1{}^f f,\, g)} & & \Big\downarrow{\scriptstyle \operatorname{Mor}(f,\, G_1 g)} \\
\operatorname{Mor}(F_1{}^f\rho, S') & \xrightarrow{\tau_2(\rho, S')} & \operatorname{Mor}(\rho, G_1 S')
\end{array}$$

is commutative, where $f \in \operatorname{Mor}(\rho, \rho')$ and $g \in \operatorname{Mor}(S, S')$.

PROOF. Since $\tau_1(\rho) \in \operatorname{Mor}(\rho, G_1 F_1{}^f(\rho))$ and $G_1(h) \in \operatorname{Mor}(G_1 F_1{}^f \rho, G_1 S)$,

$$G_1(h) \cdot \tau_1(\rho) \in \operatorname{Mor}(\rho, G_1 S)$$

Hence,

$$\tau_2(\rho, S) \in \operatorname{Mor}(\operatorname{Mor}(F_1{}^f \rho, S), \operatorname{Mor}(\rho, G_1 S))$$

Let $h \in \operatorname{Mor}(F_1{}^f \rho', S)$ be an arbitrary element. Then

$$\begin{aligned}
\operatorname{Mor}(f, G_1 g)\tau_2(\rho', S)(h) &= \operatorname{Mor}(f, G_1 g) G_1(h)\tau_1(\rho') \\
&= G_1(g) G_1(h) \tau_1(\rho') f \\
&= G_1(g) G_1(h) G_1 F_1{}^f(f)\tau_1(\rho) \qquad \text{(from Proposition 2.9)} \\
&= G_1(g \cdot h \cdot F_1{}^f(f)) \cdot \tau_1(\rho) \\
&= \tau_2(\rho, S')(g \cdot h \cdot F_1{}^f(f)) \\
&= \tau_2(\rho, S') \operatorname{Mor}(F_1{}^f f, g)(h)
\end{aligned}$$

Q.E.D.

Proposition 2.11. For each $\rho \in \operatorname{Ob} \bar{S}_g{}^f$ and $S \in \operatorname{Ob} \bar{S}^f$, let

$$\tau_3(\rho, S)(h) = F_1{}^f(h)$$

where $h \in \operatorname{Mor}(\rho, G_1 S)$. Then

$$\tau_3(\rho, S) : \operatorname{Mor}(\rho, G_1 S) \to \operatorname{Mor}(F_1{}^f \rho, S)$$

is a morphism in $\overline{\text{Set}}$. Furthermore, $\tau_3 : \operatorname{Mor}(-, G_1 -) \to \operatorname{Mor}(F_1{}^f -, -)$ is a natural transformation; that is, the diagram

$$\begin{array}{ccc}
\operatorname{Mor}(\rho', G_1 S) & \xrightarrow{\tau_3(\rho', S)} & \operatorname{Mor}(F_1{}^f \rho', S) \\
\Big\downarrow{\scriptstyle \operatorname{Mor}(f, G_1 g)} & & \Big\downarrow{\scriptstyle \operatorname{Mor}(F_1{}^f f, g)} \\
\operatorname{Mor}(\rho, G_1 S') & \xrightarrow{\tau_3(\rho, S')} & \operatorname{Mor}(F_1{}^f \rho, S')
\end{array}$$

is commutative, where $f \in \operatorname{Mor}(\rho, \rho')$ and $g \in \operatorname{Mor}(S, S')$.

PROOF. If $h \in \operatorname{Mor}(\rho, G_1 S)$, then $F_1{}^f(h) \in \operatorname{Mor}(F_1{}^f \rho, F_1{}^f G_1 S)$. Since $F_1{}^f G_1 = I$ holds, $F_1{}^f(h) \in \operatorname{Mor}(F_1{}^f \rho, S)$. Consequently,

$$\tau_3(\rho, S) \in \operatorname{Mor}(\operatorname{Mor}(\rho, G_1 S), \operatorname{Mor}(F_1{}^f \rho, S))$$

Let $h \in \operatorname{Mor}(\rho', G_1 S)$ be an arbitrary element. Then

$$\begin{aligned}
\operatorname{Mor}(F_1{}^f f, g)\tau_3(\rho', S)(h) &= \operatorname{Mor}(F_1{}^f f, g) \cdot F_1{}^f(h) \\
&= g \cdot F_1{}^f(h) F_1{}^f(f) \\
&= (F_1{}^f G_1) g \cdot F_1{}^f(h) \cdot F_1{}^f(f) \\
&= F_1{}^f(G_1 g \cdot h \cdot f) \\
&= \tau_3(\rho, S')(G_1 g \cdot h \cdot f) \\
&= \tau_3(\rho, S') \operatorname{Mor}(f, G_1 g)(h)
\end{aligned}$$

Q.E.D.

Proposition 2.12. $\tau_2(\rho, S)$ is invertible such that

$$\tau_2(\rho, S) \cdot \tau_3(\rho, S) = I_{\operatorname{Mor}(\rho, G_1 S)}$$

and

$$\tau_3(\rho, S) \cdot \tau_2(\rho, S) = I_{\operatorname{Mor}(F_1{}^f \rho, S)}$$

Consequently, τ_2 is a natural isomorphism of $\operatorname{Mor}(F_1{}^f -, -) \cong \operatorname{Mor}(-, G_1 -)$, and hence, $F_1{}^f$ is a left adjoint of G_1.

PROOF. Let $h \in \operatorname{Mor}(F_1{}^f \rho, S)$ be arbitrary. Then

$$\begin{aligned}
\tau_3(\rho, S)\tau_2(\rho, S)(h) &= \tau_3(\rho, S) G_1(h) \tau_1(\rho) \\
&= F_1{}^f(G_1(h)\tau_1(\rho)) \\
&= F_1{}^f G_1(h) \cdot F_1{}^f \tau_1(\rho) \\
&= h
\end{aligned}$$

Let $h \in \operatorname{Mor}(\rho, G_1S)$ be arbitrary. Then

$$\begin{aligned}\tau_2(\rho, S)\tau_3(\rho, S)(h) &= \tau_2(\rho, S)F_1{}^f(h)\\ &= G_1(F_1{}^f(h)) \cdot \tau_1(\rho)\end{aligned}$$

Since the following diagram is commutative,

$$\begin{array}{ccc} \rho & \xrightarrow{\tau_1(\rho)} & G_1F_1{}^f(\rho) \\ {\scriptstyle h}\downarrow & & \downarrow{\scriptstyle G_1F_1{}^f(h)} \\ G_1S & \xrightarrow{\tau_1(G_1S)} & G_1F_1{}^fG_1S = G_1S \end{array}$$

$$\begin{aligned}\tau_2(\rho, S)\tau_3(\rho, S)(h) &= G_1(F_1{}^f(h)) \cdot \tau_1(\rho)\\ &= \tau_1(G_1S) \cdot h\\ &= h \qquad (\text{because } \tau_1(G_1S) = I) \qquad \text{Q.E.D.}\end{aligned}$$

It follows immediately from Proposition 2.12 that for each $\rho \in \operatorname{Ob} \bar{S}_g{}^f$

$$\operatorname{Mor}(\rho, G_1 -) \cong \operatorname{Mor}(F_1{}^f\rho, -) : \bar{S}^f \to \bar{S}^f$$

Recall that G_1 is a constructive functor, which represents a "realization procedure." The above relationship implies that the model relation between ρ and the global response G_1S constructed from $S \in \operatorname{Ob} \bar{S}^f$ is essentially the same as that between $F_1{}^f\rho$ and S.

3. CATEGORIES OF TIME SYSTEMS

In the preceding section, we were concerned with the general systems, and the results apply, of course, to more specialized structures such as time systems. Completely analogous statements could be given for $\bar{S}$ and $\bar{S}_g$ if they were the classes of time systems and initial-state responses, respectively, using the basic definitions given in Section 2. However, there are additional categories and functors that use explicitly the structure of a time system. These will be discussed in the present sections.

A new category of time systems will be introduced by the following proposition.

Proposition 3.1. Let $\operatorname{Ob} \bar{I}$ be the class of families of mappings $\bar{\rho} = \{\rho_t : C_t \times X_t \to Y_t \mid t \in T\}$, where C_t is an arbitrary set for each $t \in T$, and X_t and Y_t are restrictions of time objects X and Y such that

(i) the time object X satisfies the input consistency, i.e.,

$$(\forall x)(\forall x')(\forall t)(x, x' \in X \rightarrow x^t \cdot x_t' \in X)$$

(ii) $\bar{\rho}$ satisfies the realizability conditions, i.e.,

(P1) $(\forall t)(\forall t')(\forall c_t)(\forall x_{tt'})(\exists c_{t'})(\forall x_{t'})(\rho_{t'}(c_{t'}, x_{t'}) = \rho_t(c_t, x_{tt'} \cdot x_{t'}) \mid T_{t'})$

(P2) $(\forall t)(\forall c_t)(\exists c_0)(\exists x^t)(\forall x_t)(\rho_t(c_t, x_t) = \rho_0(c_0, x^t \cdot x_t) \mid T_t)$

For any pair $\bar{\rho}, \bar{\rho}' \in \text{Ob } \bar{I}$, let Mor $(\bar{\rho}, \bar{\rho}')$ be the set of all triplets of homomorphisms $h = \langle h_x, h_y, h_c \rangle$, $h_x : X \rightarrow X'$, $h_y : Y \rightarrow Y'$, $h_c : C_t \rightarrow C_t'$, such that h_x is onto, and the diagram

$$\begin{array}{ccc} C_t \times X_t & \xrightarrow{\rho_t} & Y_t \\ \downarrow{\scriptstyle h_c \times h_x} & & \downarrow{\scriptstyle h_y} \\ C_t' \times X_t' & \xrightarrow{\rho_t'} & Y_t' \end{array}$$

is commutative for every $t \in T$, i.e.,

$$h_y \cdot \rho_t(c_t, x_t) = \rho_t'(h_c \cdot c_t, h_x \cdot x_t)$$

(Ob $\bar{I}$, {Mor $(\bar{\rho}, \bar{\rho}')$}) is then a category, which we shall denote by $\bar{I}$ and refer to as the category of time-system responses.

It should be noticed that strictly speaking, a different function h_c should be defined for each $t \in T$. However, for the sake of simplicity in notation, we shall use the same symbol h_c.

It should be noticed also that the realizability conditions used in Proposition 3.1 are more restricted than in the general case; that is, the general realizability conditions require

(P2)′ $(\forall t)(\forall c_t)(\forall x_t)(\exists c_0)(\exists x^t)(\rho_t(c_t, x_t) = \rho_0(c_0, x^t \cdot x_t) \mid T_t)$

instead of (P2). It is apparent, however, that if $\bar{\rho}$ satisfies (P1) and (P2), $\bar{\rho}$ is realizable, because (P2) implies (P2)′. Conversely, if the state transition $\phi_{tt'} : C_t \times X_{tt'} \rightarrow C_{t'}$ associated with $\bar{\rho}$ satisfies the condition

$$C_t = \phi_{0t}(C_0 \times X^t)$$

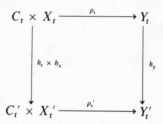

(P2) is obviously satisfied. As a matter of fact, (P1) and (P2) are the realizability conditions under the requirement that $C_t = \phi_{0t}(C_0 \times X^t)$. Since this requirement is quite convenient for the categorical arguments, it is adopted here for the realizability conditions rather than (P2)'. We shall show later that this restriction does not result in any serious loss of generality.

There exists a forgetful functor, which assigns an initial-state response to any time-system response. It is defined in the following proposition.

Proposition 3.2. Let $F_2 : \bar{I} \to \bar{S}_g$ be a mapping such that for every $\bar{\rho} \in \text{Ob}\, \bar{I}$

$$F_2(\bar{\rho}) = \rho (\in \text{Ob}\, \bar{S}_g)$$

such that

$$\rho = \rho_0$$

and for every $\langle h_x, h_y, h_c \rangle \in \text{Mor}\, (\bar{\rho}, \bar{\rho}')$

$$F_2(\langle h_x, h_y, h_c \rangle) = \langle h_x, h_y, h_c \rangle$$

F_2 is then a functor.

There exists also a constructive functor, which, to any initial-state response, assigns a time-system response.

For a given $\rho \in \text{Ob}\, \bar{S}_g$, let a relation $E_t \subset (C \times X^t) \times (C \times X^t)$ be defined by

$$((c, x^t), (c', x'^t)) \in E_t \leftrightarrow (\forall x_t)(\rho(c, x^t \cdot x_t) \mid T_t = \rho(c', x'^t \cdot x_t) \mid T_t)$$

and

$$(c, c') \in E_0 \leftrightarrow (\forall x)(\rho(c, x) = \rho(c', x))$$

Proposition 3.3. Let $G_2 : \bar{S}_g \to \bar{I}$ be a mapping such that for every $\rho \in \text{Ob}\, \bar{S}_g$

$$G_2(\rho) = \bar{\rho}$$

where $\bar{\rho} = \{\rho_t : C_t \times X_t \to Y_t\}$, $C_t = (C \times X^t)/E_t$, and

$$\rho_t(c_t, x_t) = \rho(c, x^t \cdot x_t) \mid T_t$$

where

$$c_t = [(c, x^t)]$$

and for every $\langle h_x, h_y, h_c \rangle \in \text{Mor}\, (\rho, \rho')$,

$$G_2(\langle h_x, h_y, h_c \rangle) = \langle h_x, h_y, \bar{h}_c \rangle$$

where for $c_t = [(c, x^t)]$,

$$\bar{h}_c(c^t) = [(h_c c, h_x x^t)] \in C_t'$$

G_2 is then a functor.

PROOF. It is easy to show that ρ_t is well defined, and the realizability conditions (P1) and (P2) are satisfied by $G_2(\rho)$ for any $\rho \in \text{Ob}\,\bar{S}_g$; i.e., $G_2(\rho) \in \text{Ob}\,\bar{I}$ is true.

For an arbitrary $\langle h_x, h_y, h_c \rangle \in \text{Mor}(\rho, \rho')$, let $\langle h_x, h_y, \bar{h}_c \rangle = G_2(\langle h_x, h_y, h_c \rangle)$, $\bar{\rho} = G_2(\rho)$, and $\bar{\rho}' = G_2(\rho')$. If $c_t = [(c, x^t)]$, then

$$\begin{aligned} h_y \cdot \rho_t(c_t, x_t) &= h_y(\rho(c, x^t \cdot x_t) \mid T_t) \\ &= (h_y \cdot \rho(c, x^t \cdot x_t)) \mid T_t \\ &= \rho'(h_c(c), h_x(x^t) \cdot h_x(x_t)) \mid T_t \\ &= \rho_t'([(h_c(c), h_x(x^t))], h_x(x_t)) \\ &= \rho_t'(h_c(c_t), h_x(x_t)) \end{aligned}$$

Hence, $\langle h_x, h_y, \bar{h}_c \rangle \in \text{Mor}(G_2(\rho), G_2(\rho'))$.

It is also easy to show that $G_2(h' \cdot h) = G_2(h') \cdot G_2(h)$, where $h \in \text{Mor}(\rho, \rho')$ and $h' \in \text{Mor}(\rho', \rho'')$ and that $G_2(I_\rho) = I_{G_2}(\rho)$. Q.E.D.

F_2 can be shown to be a left adjoint to G_2 by the same arguments used to show that $F_1{}^f$ is a left adjoint to G_1.

Proposition 3.4. $F_2 G_2 : \bar{S}_g \to \bar{S}_g$ is an identity functor.

Let $\bar{\rho} \in \text{Ob}\,\bar{I}$ be arbitrary. Since $\bar{\rho}$ satisfies condition (P2) of Proposition 3.1, there is a mapping $\phi : C_t \to (C \times X^t)/E_t$ such that

$$\phi(c_t) = [(c, x^t)] \leftrightarrow \rho_t(c_t, x_t) = \rho_0(c, x^t \cdot x_t) \mid T_t \qquad \text{for every} \quad x_t$$

Strictly speaking, ϕ depends on t. However, for notational convenience, we use ϕ for all t.

By using ϕ we can define a natural transformation $\mu_1 : I \to G_2 F_2$.

Proposition 3.5. For each $\bar{\rho} \in \text{Ob}\,\bar{I}$, let

$$\mu_1(\bar{\rho}) = \langle I_x, I_y, \phi \rangle$$

Then $\mu_1(\bar{\rho}) \in \text{Mor}(\bar{\rho}, G_2 F_2 \bar{\rho})$. Furthermore, μ_1 is a natural transformation; i.e., $\mu_1 : I \to G_2 F_2$ such that the diagram

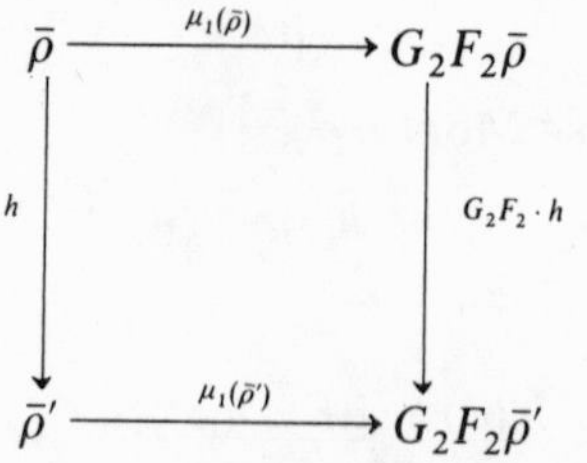

is commutative, where I is the identity functor from $\bar{I} \to I$.

PROOF. Let

$$\bar{\rho} = \{\rho_t : C_t \times X_t \to Y_t \text{ and } t \in T\}$$

and

$$G_2F_2\bar{\rho} = \{\hat{\rho}_t : \hat{C}_t \times X_t \to Y_t \text{ and } t \in T\}$$

where $\hat{C}_t = (C \times X^t)/E_t$. Then

$$\hat{\rho}_t([c, x^t], x_t) = \rho_0(c, x^t \cdot x_t) \mid T_t$$

is satisfied for all $[c, x^t] \in \hat{C}_t$. Let $\phi(c_t) = [c, x^t]$. Then

$$\hat{\rho}_t([c, x^t], x_t) = \rho_0(c, x^t \cdot x_t) \mid T_t = \rho_t(c_t, x_t)$$

that is, $\langle I_x, I_y, \phi \rangle \in \text{Mor}\,(\bar{\rho}, G_2F_2\bar{\rho})$.

Let $h = \langle h_x, h_y, h_c \rangle$, $G_2F_2h = \langle h_x, h_y, \bar{h}_c \rangle$, and $\langle I_x, I_y, \phi' \rangle \in \text{Mor}\,(\bar{\rho}', G_2F_2\bar{\rho}')$. For a given $c_t \in C_t$, let $\phi(c_t) = [(c, x^t)]$ and $\phi' h_c(c_t) = [c', x''^t]$, where $\rho_t' : C_t' \times X_t' \to Y_t'$. Then

$$\begin{aligned} &\rho_t(c_t, x_t) = \rho_0(c, x^t \cdot x_t) \mid T_t && \text{for every } x_t \\ &\quad \to \rho_t'(h_c \cdot c_t, h_x \cdot x_t) = \rho_0'(h_c \cdot c, h_x(x^t) \cdot h_x(x_t)) \mid T_t && \text{for every } x_t \\ &\quad \to \rho_t'(h_c \cdot c_t, x_t') = \rho_0'(h_c \cdot c, h_x(x^t) \cdot x_t') \mid T_t && \text{for every } x_t' \end{aligned}$$

On the other hand, $\phi' h_c(c_t) = [c', x''^t]$ implies that

$$\rho_0'(c', x''^t \cdot x_t') \mid T_t = \rho_t'(h_c \cdot c_t, x_t') \qquad \text{for every } x_t'$$

Consequently, $[c', x''^t] = [h_c \cdot c, h_x \cdot x^t]$; that is, $h_c \cdot \phi(c_t) = \phi' h_c(c_t)$ for every c_t. Therefore, $G_2F_2h\mu_1(\bar{\rho}) = \mu_1(\bar{\rho}') \cdot h$. Q.E.D.

Proposition 3.6. Let

$$\mu_2(\bar{\rho}, \rho) : \text{Mor}\,(F_2\bar{\rho}, \rho) \to \text{Mor}\,(\bar{\rho}, G_2\rho)$$

be such that for each $h \in \text{Mor}\,(F_2\bar{\rho}, \rho)$, $\mu_2(\bar{\rho}, \rho)(h) = G_2(h)\mu_1(\bar{\rho})$. Then μ_2 is a natural transformation; i.e.,

$$\mu_2 : \text{Mor}\,(F_2-, -) \to \text{Mor}\,(-, G_2-)$$

and the diagram

$$\begin{array}{ccc} \text{Mor}\,(F_2\bar{\rho}', \rho) & \xrightarrow{\mu_2(\bar{\rho}', \rho)} & \text{Mor}\,(\bar{\rho}', G_2\rho) \\ \Big\downarrow{\scriptstyle \text{Mor}\,(F_2f,\, g)} & & \Big\downarrow{\scriptstyle \text{Mor}\,(f,\, G_2g)} \\ \text{Mor}\,(F_2\bar{\rho}, \rho') & \xrightarrow{\mu_2(\bar{\rho}, \rho')} & \text{Mor}\,(\bar{\rho}, G_2\rho') \end{array}$$

is commutative, where $f \in \text{Mor}(\bar{\rho}, \bar{\rho}')$ and $g \in \text{Mor}(\rho, \rho')$, and

$$\text{Mor}(F_2 -, -), \text{Mor}(-, G_2 -): \bar{I}^{op} \times \bar{S}_g \to \bar{\text{S}}\text{et}$$

Proposition 3.7. Let

$$\mu_3(\bar{\rho}, \rho): \text{Mor}(\bar{\rho}, G_2\rho) \to \text{Mor}(F_2\bar{\rho}, \rho)$$

be such that for each $h \in \text{Mor}(\bar{\rho}, G_2\rho)$, $\mu_3(\bar{\rho}, \rho)(h) = F_2(h)$. Then μ_3 is a natural transformation; i.e.,

$$\mu_3: \text{Mor}(-, G_2 -) \to \text{Mor}(F_2 -, -)$$

and the diagram

$$\begin{array}{ccc} \text{Mor}(\bar{\rho}', G_2\rho) & \xrightarrow{\mu_3(\bar{\rho}', \rho)} & \text{Mor}(F_2\bar{\rho}', \rho) \\ \Big\downarrow {\scriptstyle \text{Mor}(f, G_2 g)} & & \Big\downarrow {\scriptstyle \text{Mor}(F_2 f, g)} \\ \text{Mor}(\bar{\rho}, G_2\rho') & \xrightarrow{\mu_3(\bar{\rho}, \rho')} & \text{Mor}(F_2\bar{\rho}, G_2\rho') \end{array}$$

is commutative, where $f \in \text{Mor}(\bar{\rho}, \bar{\rho}')$ and $g \in \text{Mor}(\rho, \rho')$, and

$$\text{Mor}(F_2 -, -), \text{Mor}(-, G_2 -): \bar{I}^{op} \times \bar{S}_g \to \bar{\text{S}}\text{et}$$

Proposition 3.8. $\mu_2(\bar{\rho}, \rho)$ is invertible such that $\mu_2^{-1}(\bar{\rho}, \mu) = \mu_3(\bar{\rho}, \mu)$. Consequently, F_2 is a left adjoint to G_2.

We have established so far the diagram:

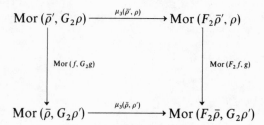

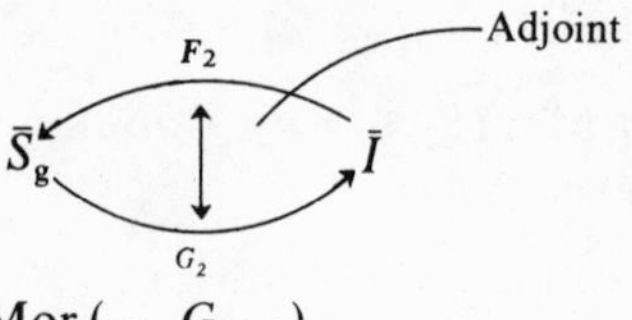

and $\text{Mor}(F_2 -, -) \cong \text{Mor}(-, G_2 -)$.

4. CATEGORIES OF DYNAMICAL SYSTEMS

Following the procedure from the preceding section, we shall introduce a category of dynamical systems and establish various functors that relate this category to the other categories considered so far. There is, of course, even more flexibility in defining a category of dynamical systems since it depends not only on the construction of the state objects C_t but the state-transition family $\bar{\phi}$ as well. The concept of a category of dynamical systems ought to be consistent with the basic approach adopted for the development of the present theory, namely, that the primary concept is that of a system as an input–output relation, while the concepts of state and state-transition family are derived, secondary concepts.

Before introducing a category of dynamical systems, the following observation is pertinent. Let $\{\phi_{tt'}: C_t \times X_{tt'} \to C_{t'}\}$ be a state-transition family of a dynamical system. In general, $\phi_{tt'}(C_t \times X_{tt'})$ is a proper subset of $C_{t'}$. However, we can make $\phi_{tt'}$ an onto funtion in a natural way. Let $\bar{C}_t = \phi_{ot}(C \times X^t)$ for every $t \in T$. We claim that $\phi_{tt'}(\bar{C}_t \times X_{tt'}) = \bar{C}_{t'}$. As a matter of fact,

$$\begin{aligned} c_{t'} \in \phi_{tt'}(\bar{C}_t \times X_{tt'}) &\leftrightarrow (\exists(c_t, x_{tt'}))(c_t \in \bar{C}_t \;\&\; c_{t'} = \phi_{tt'}(c_t, x_{tt'})) \\ &\leftrightarrow (\exists(c, x^t))(\exists x_{tt'})(c_t = \phi_{ot}(c, x^t) \;\&\; c_{t'} = \phi_{tt'}(c_t, x_{tt'})) \\ &\leftrightarrow (\exists(c, x^t \cdot x_{tt'}))(c_{t'} = \phi_{ot'}(c, x^t \cdot x_{tt'})) \\ &\leftrightarrow c_{t'} \in \bar{C}_{t'} \end{aligned}$$

Furthermore, the following is a direct consequence of the realization theorem:

$$\{(x_t, y_t) : (\exists c_t \in \bar{C}_t)(y_t = \rho_t(c_t, x_t))\} = \{(x_t, y_t) \mid (\exists c_t \in C_t)(y_t = \rho_t(c_t, x_t))\}$$

Consequently, we can always assume without losing any generality that every state transition of a dynamical system is an onto mapping.

Proposition 4.1. Let Ob $\bar{D}$ be the class of all pairs $(\bar{\rho}, \bar{\phi})$, where

$$\bar{\rho} = \{\rho_t : C_t \times X_t \to Y_t \;\&\; t \in T\}$$

$$\bar{\phi} = \{\phi_{tt'} : C_t \times X_{tt'} \to C_{t'} \;\&\; t, t' \in T\}$$

such that $\phi_{tt'}$ is onto and $\bar{\rho} = \{\rho_t\}$, and $\bar{\phi} = \{\phi_{tt'}\}$ satisfy the input state consistency, the semigroup property, and the system response consistency as defined in Chapter II. For any $(\bar{\rho}, \bar{\phi}), (\bar{\rho}', \bar{\phi}') \in$ Ob $\bar{D}$, let Mor $((\bar{\rho}, \bar{\phi}), (\bar{\rho}, \bar{\phi}))$ be the set of all triplets of homomorphisms $f = \langle h_x, h_y, h_c \rangle$ such that h_x is onto, and the diagram

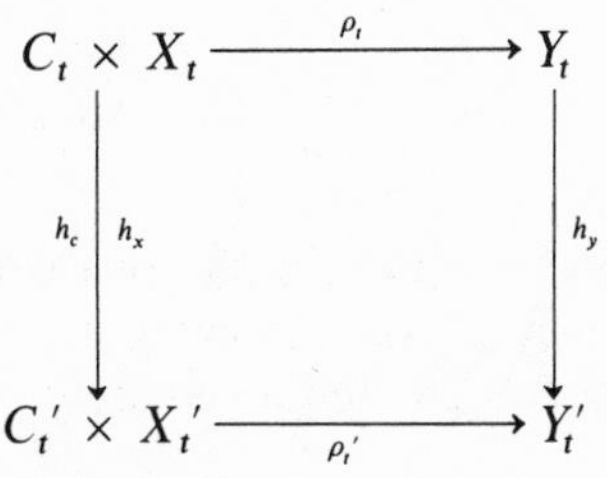

is commutative. Furthermore, let

$$\langle h_x, h_y, h_c\rangle \cdot \langle h_x', h_y', h_c'\rangle = \langle h_x \cdot h_x', h_y \cdot h_y', h_c \cdot h_c'\rangle$$

(Ob $\bar{D}$, {Mor $((\bar{\rho}, \bar{\phi}), (\bar{\rho}', \bar{\phi}'))$}) forms a category, which we shall denote by $\bar{D}$ and refer to as the category of dynamical systems.

In the definition of $\bar{D}$, a morphism $h = \langle h_x, h_y, h_c\rangle$ is not required to be a homomorphism with respect to $\bar{\phi}$. However, as we have shown in Section 1, a morphism h is also a homomorphism for $\bar{\phi}$ in the sense given by the following commutative diagram:

$$\begin{array}{ccc}
(C_t/E_t) \times X_{tt'} & \xrightarrow{\hat{\phi}_{tt'}} & (C_{t'}/E_{t'}) \\
\Big\downarrow \hat{h}_c \times h_x & & \Big\downarrow \hat{h}_c \\
(C_t'/E_t') \times X_{tt'}' & \xrightarrow{\hat{\phi}_{tt'}'} & (C_{t'}'/E_{t'}')
\end{array}$$

where $E_t \subset C_t \times C_t$ is an equivalence relation such that

$$(c_t, c_t') \in E_t \leftrightarrow (\forall x_t)(\rho_t(c_t, x_t) = \rho_t(c_t', x_t))$$

$$\hat{h}_c : (C_t/E_t) \to C_t'/E_t'$$

and

$$\hat{\phi}_{tt'} : (C_t/E_t) \times X_{tt'} \to (C_{t'}/E_{t'})$$

are extensions of $h_c : C_t \to C_t'$ and $\phi_{tt'} : C_t \times X_{tt'} \to C_{t'}$ such that

$$\hat{h}_c([c_t]) = [h_c(c_t)]$$

$$\hat{\phi}_{tt'}([c_t], x_{tt'}) = [\phi_{tt'}(c_t, x_{tt'})]$$

For notational convenience, we shall write $\hat{C}_t = C_t/E_t$. Then it is clear that for a given $\bar{\rho} \in \text{Ob}\,\bar{I}$ a mapping $\hat{\rho}_t : \hat{C}_t \times X_t \to Y_t$ can be defined for each $t \in T$ as follows:

$$\hat{\rho}_t([c_t], x_t) = \rho_t(c_t, x_t)$$

Moreover, it is easy to show that $\bar{\hat{\rho}} = \{\hat{\rho}_t : t \in T\} \in \text{Ob}\,\bar{I}$.

We can now establish some functors that relate the category of dynamical systems with the previously introduced categories of systems. First of all, a forgetful functor relating $\bar{D}$ with the immediately preceding $\bar{I}$ is given in the following proportion.

Proposition 4.2. Let $F_3 : \bar{D} \to \bar{I}$ be a mapping such that for all $(\bar{\rho}, \bar{\phi}) \in \text{Ob}\, \bar{D}$,

$$F_3(\bar{\rho}, \bar{\phi}) = \bar{\hat{\rho}}$$

and all morphisms

$$F_3(\langle h_x, h_y, h_c \rangle) = \langle h_x, h_y, \hat{h}_c \rangle$$

where $\bar{\hat{\rho}}$ and $\hat{h}_c$ are defined above. F_3 is then a functor.

In order to introduce a constructive functor between $\bar{I}$ and $\bar{D}$, we should observe that for each

$$\bar{\hat{\rho}} = \{\hat{\rho}_t : \hat{C}_t \times X_t \to Y_t \,\&\, t \in T\}$$

which is generated by

$$\bar{\rho} = \{\rho_t : C_t \times X_t \to Y_t \,\&\, t \in T\}$$

condition (P1) of Proposition 3.1 specifies a unique function $\hat{\phi}_{tt'} : \hat{C}_t \times X_{tt'} \to \hat{C}_{t'}$ for each $t, t' \in T$. Then a constructive functor can be given by $\bar{\hat{\rho}}$ and $\bar{\hat{\phi}} = \{\hat{\phi}_{tt'} : \hat{C}_t \times X_{tt'} \to \hat{C}_{t'}\}$.

Proposition 4.3. Let G_3 be a mapping $G_3 : \bar{I} \to \bar{D}$ such that for all $\bar{\rho} \in \text{Ob}\, \bar{I}$,

$$G_3(\bar{\rho}) = (\bar{\hat{\rho}}, \bar{\hat{\phi}})$$

For all $\langle h_x, h_y, h_c \rangle \in \text{Mor}\,(\bar{\rho}, \bar{\rho}')$,

$$G_3(\langle h_x, h_y, h_c \rangle) = \langle h_x, h_y, \hat{h}_c \rangle$$

where $\hat{h}_c([c_t]) = [h_c c_t]$. G_3 is then a functor.

Let $\bar{I}_r$ be a full subcategory of $\bar{I}$ such that

$$\bar{\hat{\rho}} = \{\hat{\rho}_t : \hat{C}_t \times X_t \to Y_t \,\&\, t \in T\} \in \text{Ob}\, \bar{I}_r \leftrightarrow [\bar{\hat{\rho}} \in \bar{I} \text{ and } \bar{\hat{\rho}} \text{ is reduced}]$$

As the construction of the functor F_3 shows, the range of F_3 is $\bar{I}_r$. If we consider $F_3{}^r$ as a functor from $\bar{D}$ to $\bar{I}_r$ and if the restriction of G_3 to $\bar{I}_r$ is represented by $G_3{}^r$, we can show by using the following propositions that $F_3{}^r$ is a left adjoint to $G_3{}^r$. Let $\eta : C_t \to \hat{C}_t$ be the natural mapping for every $t \in T$; i.e., $\eta(c_t) = [c_t]$. It is clear from the construction of $F_3{}^r$ and $G_3{}^r$ that

$$F_3{}^r G_3{}^r = I : \bar{I}_r \to \bar{I}_r$$

Proposition 4.4. For each $(\bar{\rho}, \bar{\phi}) \in \text{Ob}\, \bar{D}$, let $v_1(\bar{\rho}, \bar{\phi}) = \langle I_x, I_y, \eta \rangle$. Then

$$v_1(\bar{\rho}, \bar{\phi}) \in \text{Mor}\,((\bar{\rho}, \bar{\phi}), G_3{}^r F_3{}^r(\bar{\rho}, \bar{\phi}))$$

Furthermore, v_1 is a natural transformation; i.e., $v_1 : I \to G_3{}^r F_3{}^r$, and the diagram

$$
\begin{array}{ccc}
(\bar{\rho}, \bar{\phi}) & \xrightarrow{\nu_1(\bar{\rho}, \bar{\phi})} & G_3{}^r F_3{}^r(\bar{\rho}, \bar{\phi}) \\
\downarrow{\scriptstyle h} & & \downarrow{\scriptstyle G_3{}^r F_3{}^r h} \\
(\bar{\rho}', \bar{\phi}') & \xrightarrow{\nu_1(\bar{\rho}', \bar{\phi}')} & G_3{}^r F_3{}^r(\bar{\rho}', \bar{\phi}')
\end{array}
$$

is commutative.

Proposition 4.5. For each $(\bar{\rho}, \bar{\phi}) \in \operatorname{Ob} \bar{D}$ and $\bar{\bar{\rho}}_i \in \operatorname{Ob} \bar{I}_r$, let

$$\nu_2((\bar{\rho}, \bar{\phi}), \bar{\bar{\rho}}_i) : \operatorname{Mor}(F_3{}^r(\bar{\rho}, \bar{\phi}), \bar{\bar{\rho}}_i) \to \operatorname{Mor}((\bar{\rho}, \bar{\phi}), G_3{}^r \bar{\bar{\rho}}_i)$$

be such that for any $h \in \operatorname{Mor}(F_3{}^r(\bar{\rho}, \bar{\phi}), \bar{\bar{\rho}}_i)$

$$\nu_2((\bar{\rho}, \bar{\phi}), \bar{\bar{\rho}}_i)(h) = G_3{}^r(h)\nu_1(\bar{\rho}, \bar{\phi})$$

Then ν_2 is a natural transformation; i.e., $\nu_2 : \operatorname{Mor}(F_3{}^r -, -) \to \operatorname{Mor}(-, G_3{}^r -)$, and the diagram

$$
\begin{array}{ccc}
\operatorname{Mor}(F_3{}^r(\bar{\rho}', \bar{\phi}'), \bar{\bar{\rho}}_i) & \xrightarrow{\nu_2((\bar{\rho}', \bar{\phi}'), \bar{\bar{\rho}}_i)} & \operatorname{Mor}((\bar{\rho}', \bar{\phi}'), G_3{}^r \bar{\bar{\rho}}_i) \\
\downarrow{\scriptstyle \operatorname{Mor}(F_3{}^r f, g)} & & \downarrow{\scriptstyle \operatorname{Mor}(f, G_3{}^r g)} \\
\operatorname{Mor}(F_3{}^r(\bar{\rho}, \bar{\phi}), \bar{\bar{\rho}}_i') & \xrightarrow{\nu_2((\bar{\rho}, \bar{\phi}), \bar{\bar{\rho}}_i')} & \operatorname{Mor}((\bar{\rho}, \bar{\phi}), G_3{}^r \bar{\bar{\rho}}_i')
\end{array}
$$

is commutative, where $f \in \operatorname{Mor}((\bar{\rho}, \bar{\phi}), (\bar{\rho}', \bar{\phi}'))$ and $g \in \operatorname{Mor}(\bar{\bar{\rho}}_i, \bar{\bar{\rho}}_i')$.

Proposition 4.6. For each $(\bar{\rho}, \bar{\phi}) \in \operatorname{Ob} \bar{D}$ and $\bar{\bar{\rho}}_i \in \operatorname{Ob} \bar{I}_r$, let

$$\nu_2((\bar{\rho}, \bar{\phi}), \bar{\bar{\rho}}_i) : \operatorname{Mor}((\bar{\rho}, \bar{\phi}), G_3{}^r \bar{\bar{\rho}}_i) \to \operatorname{Mor}(F_3{}^r(\bar{\rho}, \bar{\phi}), \bar{\bar{\rho}}_i)$$

be such that for any $h \in \operatorname{Mor}((\bar{\rho}, \bar{\phi}), G_3{}^r \bar{\bar{\rho}}_i)$

$$\nu_3((\bar{\rho}, \bar{\phi}), \bar{\bar{\rho}}_i)(h) = F_3{}^r(h)$$

Then ν_3 is a natural transformation; i.e., $\nu_3 : \operatorname{Mor}(-, G_3{}^r -) \to \operatorname{Mor}(F_3{}^r -, -)$, and the diagram

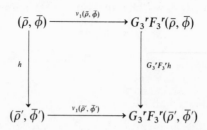

$$\begin{array}{ccc} \operatorname{Mor}((\bar{\rho}', \bar{\Phi}'), G_3{}^r\bar{\bar{\rho}}_i) & \xrightarrow{v_3((\bar{\rho}', \bar{\Phi}'), \bar{\bar{\rho}}_i)} & \operatorname{Mor}(F_3{}^r(\bar{\rho}', \bar{\Phi}'), \bar{\bar{\rho}}_i) \\ \Big\downarrow \scriptstyle\operatorname{Mor}(f, G_3{}^r g) & & \Big\downarrow \scriptstyle\operatorname{Mor}(F_3{}^r f, g) \\ \operatorname{Mor}((\bar{\rho}, \bar{\Phi}), G_3{}^r\bar{\bar{\rho}}_i') & \xrightarrow{v_3((\bar{\rho}, \bar{\Phi}), \bar{\bar{\rho}}_i')} & \operatorname{Mor}(F_3{}^r(\bar{\rho}, \bar{\Phi}), \bar{\bar{\rho}}_i') \end{array}$$

is commutative, where $f \in \operatorname{Mor}((\bar{\rho}, \bar{\Phi}), (\bar{\rho}', \bar{\Phi}'))$ and $g \in \operatorname{Mor}(\bar{\bar{\rho}}_i, \bar{\bar{\rho}}_i')$.

Finally we have the following proposition.

Proposition 4.7. $v_2((\bar{\rho}, \bar{\Phi}), \bar{\bar{\rho}}_i)$ is invertible such that

$$v_2^{-1}((\bar{\rho}, \bar{\Phi}), \bar{\bar{\rho}}_i) = v_3((\bar{\rho}, \bar{\Phi}), \bar{\bar{\rho}}_i)$$

Consequently, $F_3{}^r$ is a left adjoint to $G_3{}^r$; i.e.,

$$\operatorname{Mor}(F_3{}^r -, -) \cong \operatorname{Mor}(-, G_3{}^r -) : \bar{D}^{\text{op}} \times \bar{I}_r \to \overline{\text{Set}}$$

In summary, we have shown in this chapter the relations shown in Fig. 4.1, where I_n is the inclusion functor and $H_3 : \bar{I} \to \bar{I}_r$ is given by

$$H_3(\bar{\rho}) = \bar{\bar{\rho}}$$

and

$$H_3(\langle h_x, h_y, h_c \rangle) = \langle h_x, h_y, \hat{h}_c \rangle$$

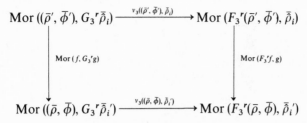

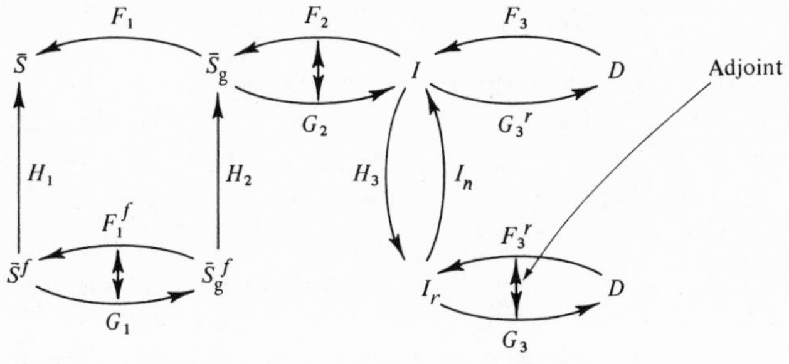

FIG. 4.1

Appendix I

REFERENCES AND HISTORICAL ACCOUNT

Since the present book is a monograph on a new theory, there are two types of references: the first type [1–19] is on the results from other fields that are used at strategic points in the presentation to indicate potential applications and the relationship of the new theory to other fields, while the second type [20–41] is on the earlier and subsequent related work. The first type is cited throughout the main text, and both types are included in this appendix.

The program for the development of a mathematical theory of general systems as described in Chapter I was first conceived in 1960 [20]. Listing the references over such a long period, one cannot but take a historical perspective which, like any history writing, cannot help but be of a personal character too. We shall face this problem squarely and present the references in the form of a historical account to show how the development of our theory has been affected by the other work. Chronologically, the period before 1960 should be distinguished since the earlier work had a direct influence to various degrees on the formulation of our program.

Three principal schools of thoughts, which shaped the formulation of the program and, indeed, in response to which the entire effort was initiated, are associated with the names of Ludwig von Bertalanffy [21], [22], Norbert Wiener [23], [24], and Herbert Simon [25], [26]. We refer to "the schools" to indicate that a plethora of scientists have contributed to the development of theories usually associated with the name of the initiator or the main protagonist. For example, Anatol Rappoport [27] and Kenneth Boulding [28] have contributed much to general systems theory in the Bertalanffy sense, while Ross Ashby [29] has done much to interpret cybernetics in the Wiener sense.

We shall start with von Bertalanffy because his was the first influence as far as our effort is concerned; what preceded his pronouncements on a general systems theory is up to the historians to discover. Von Bertalanffy proposed general systems theory as a philosophy of science. Such a theory should be concerned with the basic processes which are universal in the sense that they encompass the phenomena from any number of scientific disciplines. The theory also must be truly interdisciplinary, rejecting confinement to a disciplinary view regardless of how well a discipline is developed. In particular, he singled out mathematics among the disciplinary fields and urged that the development of a mathematical theory of general systems theory would be contradictory to the basic notion of a general systems theory [22].

It is the suggestion that a general systems theory is desirable that prompted our interest. But as a result of dissatisfaction with the views and approach proposed earlier our program was initiated. There is nothing confining in being explicit, precise, and rigorous, i.e., mathematical. On the contrary, the danger in developing a theory, and in particular of the interdisciplinary kind, is actually in being vague, imprecise, and open to any or all interpretations. Interdisciplinary considerations suffer precisely from a lack of context in which the transfer of knowledge can take place, and vagueness cannot but retard, rather than promote, progress in interdisciplinary efforts. This is probably why many accomplished scientists, even with philosophical inclinations (e.g., Monod [30]), have rejected general systems theory as being nonoperational, i.e., without power of explanation. While von Bertalanffy proposed a *theory-of-general-systems* meaning of systems, which will reflect the universal laws or principles valid for biological, social, physical, and any other phenomena, we are interested in a *general theory of systems* that deals with general properties of formal relationships between objects of concern and as such needs no reference to any discipline, since a formal theory is by nature interdisciplinary.

The second influence was that of Norbert Wiener for two reasons. First, he demonstrated that the interdisciplinary problems can be treated mathematically, and second, he pointed out the fundamental distinction and importance of control processes everywhere in nature and, fortunately, coined the term "cybernetics" to be used as a banner for the field of study concerned with control communication systems. Unfortunately, the banner was carried by a mixed crowd of researchers, and the major impact of cybernetics is yet to be felt. We have made certain steps toward a general theory of certain goal-seeking, i.e., cybernetic systems [31]. The present book provides enough of a foundation for a more complete general theory of cybernetics.

The influence of Herbert Simon on our work was more indirect. His deep insight and brilliant exemplification of how complex systems "really" work (as opposed to an idealized view often associated with optimality ideas), being of social, political, or "artificial" variety, have provided a wealth of paradigms which served as the food for the development of a formal theory. Indeed, the formalization of all these realistic images and scenarios presented a challenge to which we responded perhaps more explicitly in [31] than in this book.

What about subsequent developments? We shall mention some of the trends which either seem most relevant or have received attention.

First, since the mid-sixties a trend has appeared aimed at integrating various specialized, already well-developed theories in engineering [32–34] with the objective of developing a more abstract theory of specialized kinds of systems (e.g., algebraic theory of difference-equation systems [34]). These theories, by definition, lead toward the kind of theory and level of abstraction we are using, but as yet have to be much more fully developed in order to reach the domain of general systems theory. Most complete among them is that of Wymore. However, since his basic formalism requires so much mathematical structure, it cannot deal successfully with either large-scale or goal-seeking systems and, unless properly modified, cannot be elevated to the general systems level of abstraction.

The second trend is the application of our framework. Macko [35] has considered the question of natural states, while Windeknecht [36] has taken our notion of a time system in an attempt to develop a general theory of dynamical processes, which turned out to be quite limited in scope in spite of the appearance of a detailed analysis.

The third trend represents attempts to develop alternative approaches. Notable among them is the work by Klir [37] and Goguen [38]. It is too early to say how well formulated these approaches are and how they are related with the earlier or other on-going efforts.

Finally, a wealth of references can be found in [39] and [40], which the reader interested in an overview of the alternative approaches could consult.

References

1. Mesarovic, M. D., and Eckman, D. P., On some basic concepts of the general systems theory, *Proc. 3rd Internat. Conf. Cy.*, Namure, Belgium, 1961.
2. Nerode, A., Linear automaton transformation, *Proc. Amer. Math. Soc.* **9**, 1958.
3. Hajek, O., "Dynamical Systems in the Plane." Academic Press, New York and London, 1968.
4. Gill, A., "Introduction to the Theory of Finite-State Machines." McGraw-Hill, New York, 1962.

5. Brockett, R. W., "Finite-Dimensional Linear Systems." Wiley, New York, 1970.
6. Mikusinski, J., "Operational Calculus," Chapter II. Pergamon, Oxford, 1959.
7. MacLane, S., and Birkhoff, G., "Algebra," Chapter V. Macmillan, New York, 1971.
8. Lee, E. B., and Markus, L., "Foundations of Optimal Control Theory." Wiley, New York, 1967.
9. Taylor, A. E., "Introduction to Functional Analysis." Wiley, New York, 1963.
10. Kalman, R. E., Ho, Y. C., and Narendra, K. S., "Controllability of linear dynamical systems," *in* "Contributions to Differential Equations," Vol. 1. Wiley, New York, 1963.
11. Brockett, R. W., and Mesarovic, M. D., The reproducibility of multivariable systems, *Math. Anal. Appl.* **10**, 1965.
12. Hartmanis, J., and Stearns, R. E., "Algebraic Structure Theory of Sequential Machines." Prentice-Hall, Englewood Cliffs, New Jersey, 1966.
13. Bushaw, D., A stability criterion for general systems, *J. Math. Systems Theory* **1**, 1967.
14. Nemytskii, V. V., and Stepanov, V. V., "Qualitative Theory of Differential Equations." Princeton Univ. Press, Princeton, New Jersey, 1960.
15. Thom, R., "Stabilité Structurelle et Morphogénèse." Benjamin, New York, 1972.
16. Zubov, V. I., "Methods of A. M. Lyapunov and Their Application." Noordhoff, Amsterdam, 1964.
17. Krohn, K. B., and Rhodes, J. L., "Algebraic theory of machines," *in* Proc. Symp. Math. Theory of Automata. Wiley, New York, 1963.
18. Smullyan, R. M., "Theory of Formal Systems." Princeton Univ. Press, Princeton, New Jersey, 1961.
19. Davis, M., "Computability and Unsolvability." McGraw-Hill, New York, 1958.
20. Yoshii, S., "General Stability of Sets," M.S. thesis, Case Western Reserve Univ., Cleveland, Ohio, 1971.
21. von Bertalanffy, L., An outline of general system theory, *Brit. J. Philos. Sci.*, **1**, 134–164, 1950. Reprinted in "General System Theory: Foundations, Development, Applications." George Braziller, New York, 1968.
22. von Bertalanffy, L., "General Systems Theory," Chapters 1 and 8. George Braziller, New York, 1968.
23. Wiener, N., "Cybernetics or Control and Communication in the Animal and the Machine." M.I.T. Press, Cambridge, Mass., and Wiley, New York, 1961.
24. Wiener, N., "The Human Use of Human Beings; Cybernetics and Society." Houghton Mifflin, Boston, 1950.
25. Simon, H. A., "Administrative Behavior." Free Press, New York, 1957.
26. Simon, H. A., "The Sciences of the Artificial." M.I.T. Press, Cambridge, Mass., 1969.
27. Mesarovic, M. D., "Views on general systems theory," *in* "Proc. 2nd Systems Symp., Cast Inst. Tech.," Wiley, New York, 1964.
28. Boulding, K., "General systems theory-skeleton of science," *in* "General Systems Yearbook," Vol. 1. Univ. of Michigan Press, Ann Arbor, Michigan, 1956.
29. Ashby, W. R., "An Introduction to Cybernetics," 3rd ed. Wiley, New York, 1958.
30. Monod, Jacques, "Le Hasard et la Necessite." Editions Du Seuil, Paris, 1970.
31. Mesarovic, M. D., Macko, D., and Takahara, Y., "Theory of Hierarchical Multilevel Systems." Academic Press, New York and London, 1970.
32. Volterra, V., "Theory of Functionals and of Integral and Integro-Differential Equations." Dover, New York, 1959.
33. Wymore, A. W., "A Mathematical Theory of Systems Engineering—The Elements." Wiley, New York, 1967.
34. Kalman, R. E., Falk, P. L., and Arbib, M. A., "Topics in Mathematical System Theory." McGraw-Hill, New York, 1969.

35. Macko, D., Natural states and past-determinism of general time systems, *Inform. Sci.* **3**, 1971.
36. Windeknecht, T. G., "General Dynamical Processes: A Mathematical Introduction." Academic Press, New York and London, 1971.
37. Klir, G. J., "An Approach to General Systems Theory," Preview and Chapters 1, 7, 9, 10, and 11. Van Nostrand, New York, 1969.
38. Goguen, J. A., Realization is universal, *J. Math. Systems Theory* **6**, No. 4, January 1973.
39. Klir, G. J., "Trends in General Systems Theory." Wiley, New York, 1972.
40. Mesarovic, M. D., "Views on general systems theory," *in* "Proc. 2nd Systems Symp. Cast Inst. Tech." Wiley, New York, 1964.
41. Pareigis, B., "Categories and Functions." Academic Press, New York and London, 1970.

Appendix II

ALTERNATIVE BASIS FOR MATHEMATICAL GENERAL SYSTEMS THEORY

The starting point for our theory is the notion of a system introduced on the set-theoretic level.

There are, of course, a number of other possible approaches, which are either formulated on a lower level of abstraction (i.e., use richer mathematical structure) or define the basic concept of a system from a different angle. We have considered a number of these alternatives and find them lacking in some of the crucial requirements for the foundation of a general systems theory and inferior to the approach we have adopted. Here is briefly the rationale why we have not taken some of the most obvious alternative candidates.

1. AXIOMATIC LOGIC STRUCTURES

The abstract mathematical structures used in the axiomatic logic systems are apparently too specialized to be used to define the basic notion of a system. However, one can use logic in the way it is used in metamathematics, i.e., to consider the statements *about* the system and deductions that can be made about systems properties and behavior. This is an attractive approach which we have considered. Specifically, let F be a (formal) language and f a set of (well-formed) sentences in F, expressing the observed facts or assumed properties of the system. Let f be "exhaustive" as far as our knowledge of the system is concerned; i.e., it contains all known or presumed facts about the system. Then we have the following concept, termed *linguistic definition* of a system [20].

A system is a proper set of well-formed sentences.

Again, such a set *refers* to a system rather than *defining* it as such, but since by definition it encodes all that we know about the system, it is essentially one and the same thing.

The relationship between the system as a relation on abstract sets and as defined by the set of statements is obvious. In the language of mathematical logic, the former is a model, a "realization" of the latter. Model theory, then, can be used to explore these relationships still further.

The approach to general systems theory via statements about the systems behavior is attractive because in practice that is often all one knows in a definite way about the system. This is the very first step in formalizing the observations, i.e., the firsthand experience. However, we have not pursued the linguistic approach in this book since it deals with the systems properties we decided to consider in a too indirect fashion.†

2. TOPOLOGICAL, FUNCTIONAL ANALYSIS, AND QUANTITATIVE APPROACHES

The relationship between the system as a relation on abstract sets and as

Apparently, a systems theory can be developed starting from more specific notions defined on the sets with richer mathematical structure, e.g., in terms of mappings on topological spaces, on function spaces, etc. Still further, one could start from the continuous and discrete-time dynamical systems in an attempt to unite these two theories. Indeed, this is the approach via abstraction mentioned in Chapter I and Appendix I. However, none of these approaches is consistent with the objectives we have stated for our program, in particular with respect to generality and ability to deal with poorly structured, uncertain situations. Furthermore, attempts in that direction have shown that one gets entangled in a wealth of purely technical problems with little, if any, conceptual interests for the systems theory per se.

3. ALGEBRAIC SYSTEMS THEORY

Attempts have also been made to develop a pure algebraic systems theory. Our approach contains within itself what is best of these attempts. Indeed, we have used algebraic structures whenever conceptually required, and a good portion of specific mathematical developments in this book is algebraic. However, for the introduction of the basic concepts—the very foundation

† Klir [39] seems to have picked up the linguistic approach starting from what he calls "traits of a system."

itself—the set-theoretic framework is definitely preferable to being restricted in use to only algebraic structures. Actually, starting from the set-theoretic level, it is more appropriate for certain class of problems to go to topological rather than algebraic structures (see, e.g., Chapter IX on stability).

While using algebraic structures where appropriate, we have avoided paying too much attention to some specialized algebraic-type questions of limited generality, as , e.g., the problems involving only the sets of finite cardinality. (However, see Chapter IX for the consideration of such a case in our framework.) Actually, the hardest and most intriguing questions in general systems theory involve precisely the situations where sets of finite, continuum, and higher cardinality are involved. Attempts to restrict algebraic systems theory solely to the "discrete-time functions," i.e., sets that are at most denumerable, interesting as they are in themselves, miss the main purpose and avoid the true challenge of a general, algebraically based systems theory.

4. MORE RESTRICTED NOTION OF A SYSTEM

Last, but not least, objections can be raised to the concept of a system as a set-theoretic relation on the ground that more restrictions ought to be introduced for an "entity" to be called a system, even if the set-theoretic level is used. Prime candidate for such an additional necessary specification for a system is the requirement that a state space and associated state-transition and output functions have to be given before a relation can be qualified for consideration as a system. While such a requirement (i.e., that a relation is a system *only* if state-transition and output functions are given) might seem reasonable in principle, some major difficulties appear in further development and application when such restrictions are introduced on the primary concept of a system for the following reasons:

(i) There is a class of real-life systems, in particular in biological and social sciences, where the system under consideration is goal seeking and can be described formally only as such. Such a system can still be represented as a system in our sense, but any attempts to introduce states and state transition cannot but introduce arbitrary and unwarranted assumptions of such magnitude that the validity of the model becomes questionable. Although we do not consider goal-seeking systems as such in this book, our treatment allows such a possibility, i.e., does not prevent future developments in this direction (see Appendix III).

(ii) In applications, a system is very often defined by means of a family of subsystems and interactions among them. Even if the state transitions for individual subsystems are given, the state transition of the overall system, as a rule, is very difficult and cumbersome if not impossible to define. Yet a combination of systems is certainly a system, and the assertion that the overall system to be considered as such must have a state transition is of a very limited utility. No such difficulties appear using our notion of a system.

(iii) The starting point for any modeling is the observations and the assumption about the existence of relationships between them. The primary concept of a system ought to be definable just with that much data. Whether such a relationship can be described as a transition in a state space is a point that needs to be proven. Even if this is possible, a state space is not unique, which indicates the secondary nature of the concept of state. These points are well documented in the realization theory developed in this book.

Appendix III

OPEN SYSTEMS AND GOAL-SEEKING SYSTEMS

Two important concepts, which are not treated in this book, yet which are fundamental for the general systems theory, are the notions of an open system and a goal-seeking system. One of the reasons for such an omission is the fact that the input–output, terminal approach as presented in this book had to be developed first and can serve as the starting point for theories of open and goal-seeking systems. We shall indicate here the direction for such a development.

1. OPEN SYSTEMS

A system is, in general, a relation rather than a function, and in such a general case it is not possible to determine what will be the output for any one input that is either observed or assumed. This situation can be remedied as shown in this book by the introduction of a global object or an initial state object, a procedure which, in principle, is always possible. However, the introduction of a global state object in an arbitrary fashion might lead to the following difficulties:

(i) Although the existence of the global state is assured, the information needed to determine the global state of the system in any specific instance might not be available in practice. It is not possible then to determine the output when an input is given, and it is better to face that fact at the very formulation of the problem.

(ii) The introduction of the global state might lead to an inappropriate (or even incorrect) interpretation of what we know about the system; e.g., it might require that an anticipatory system be changed into nonanticipatory or that a system about which we have only probabilistic information be presented as a deterministic system, etc. In either of these cases, it is better then to describe the system so as to reflect in the best possible way our understanding of the behavior of the real system and to obtain the predictability of the output in an indirect way.

Intuitively, a system that cannot be represented (satisfactorily) as a function (i.e., one cannot state, under all conditions, what will the output be) represents an *open system*. In order to formalize the notion of an open system, it is simplest to start by assuming that the system has two components in the input object, $X = M \times U$, i.e.,

$$S \subset M \times U \times Y$$

Assume, then, that one can determine (or detect) what the m-component of the input will be (or is) while one can say, at most, to which subset, $U_m \subset U$, the u-component will belong. This is a typical case of an uncertain situation. M represents the measurable, directly observable, or controlled inputs, while U represents the inputs on which the information is provided only indirectly if at all. For any given $m \in M$, the most one can say about the output is that it will belong to $Y_m = S(m, U_m)$. Such a system represents an open system in a general sense.

The concept of an open system as introduced above is fully consistent with the traditional notion. Classically, a system is considered as open if

(i) either there is a throughput of energy or information that cannot be controlled or even precisely (directly) observed,

(ii) or the system cannot be described as state determined—usually in reference to a given framework (e.g., when the ordinary differential equations are used, the equations depend explicitly upon time). It is quite apparent that both of these cases are covered by the concept of an open system introduced here.

Let us briefly sketch how the "predictability" for an open system can be regained in an indirect way. Two such methods will be briefly discussed. In both of these, one essentially tries to solve the problem of a nonpredictable description by considering the *subsets* rather than elements of the output object.

(a) Almost Predictable Outputs

For any given $x \in X$, there is defined a subset $Y_x \subset Y$ such that

$$y \in Y_x \leftrightarrow (\exists x)((x, y) \in S)$$

where $X = M \times U$.

The set Y_x consists of all outputs that can occur as a response to the given x.

Denote by $\Pi(Y)$ the power set of Y. It is possible under certain circumstances to define a new system

$$S' \subseteq X \times \Pi(Y)$$

such that

$$(x, \hat{Y}) \in S' \leftrightarrow \hat{Y} = Y_x$$

where $\hat{Y} \in \Pi(Y)$.

Apparently, S' is a functional system

$$S' : X \to \Pi(Y)$$

Then one can talk about the predictability only in the sense of the subsets of Y. If for any $x \in X$, the set Y_x is small enough in a given sense, the description of S by S' might be satisfactory. This apparently depends upon the nature of the system S and the criteria introduced in Y to evaluate the "size" of Y_x.

(b) Almost Predictable Systems

In many practical cases, the predictability in the systems behavior is absent not only because we do not know how the system will respond but also because we might not be certain about the effect of the environment. Formally, this means that we cannot specify uniquely which input will be applied. However, in most cases, it is possible to delineate a subset of X which describes the conditions of the systems operation. For example, in a multi-input system, some of the inputs can be under our control, while no choice exists for the remaining inputs (even if they can be observed). With any selection of the inputs under our control, a subset of the total inputs appears as possible to affect the system.

The predictability can then be achieved by passing to the power sets in both input and output objects.

Let $\Pi(X)$ be the power set of the input X, and the conditions at the input during an observation can be described by a subset of $X \subset X$. It is possible then to define a functional system

$$S_\Pi : \Pi_s(X) \to \Pi(Y)$$

such that for any $\hat{X} \in \Pi_s(X)$

$$S_\Pi(\hat{X}) = \hat{Y} \leftrightarrow \hat{Y} = \{Y_x : x \in \hat{X}\}$$

A more detailed description of S can be given by the introduction of some additional structure in both $\Pi(X)$ and $\Pi(Y)$. These considerations lead directly to fuzzy and probabilistic descriptions of systems.

2. GOAL-SEEKING SYSTEMS

The system is described, in this case, indirectly in reference to a decision-making problem. Essentially, one defines a system $S \subseteq X \times Y$ so that the pair (x, y) belongs to S if and only if y is a solution of a decision problem specified by x. As an illustration of what is meant by a decision problem, consider the following two situations.

(a) General Optimization Problem

Let $g: X \to V$ be a function from an arbitrary set X into a set V; the set V is linearly or partially ordered by the relation $\leq$. A general optimization problem is then the following:

Given a subset $X^f \subseteq X$, find $x \in X^f$ such that for all x in X^f

$$g(\hat{x}) \leq g(x) \tag{A.1}$$

The set X is the *decision set*, while the set X^f is the set of feasible decisions. The function g is the *objective function*, and the set V is the *value set*. A general optimization problem is specified by a pair (g, X^f). An element $\hat{x}$ of X^f satisfying Eq. (A.1) for all x in X^f is a *solution* of the optimization problem specified by the pair (g, X^f).

Often the function g is specified by two functions,

$$P: X \to Y \qquad \text{and} \qquad G: X \times Y \to V$$

$$g(x) = G(x, P(x))$$

In this case, the function P is referred to as an *outcome function* or model (of a controlled process), and the function G is referred to as a *performance* or *evaluation function*; an optimization problem may then be specified by a triplet (P, G, X^f) or a pair (P, G) if $X^f = X$.

Reference to P as a "model of the controlled process" implies that the optimization problem specified by the triplet (P, G, X^f) is actually defined in reference to a system to be controlled and whose image is represented by P. In general, however, nothing has to be assumed about the relationship between the model P and the actual process; even the existence of the actual process as an entity might be just an assumption used to define an optimization problem so as to be able to specify a system as decision making.

(b) General Satisfaction Problem

Let X and Ω be arbitrary sets; let g be a function from $X \times \Omega$ into a set V, which is linearly ordered by $\leq$, and let τ be a function from Ω into V; that is, $g: X \times \Omega \to V$ and $\tau: \Omega \to V$. A satisfaction problem is then the following:

Given a subset $X^f \subseteq X$, find $\hat{x}$ in X^f such that for all ω in Ω,

$$g(\hat{x}, \omega) \leq \tau(\omega) \tag{A.2}$$

The set Ω is referred to as an *uncertainty set*; τ is a *tolerance function*; and the inequality, Eq. (A.2), over Ω is the *satisfaction criterion*. The other elements of the satisfaction problem have the same interpretation as in the optimization problem. A quadruple (g, τ, X^f, Ω) specifies a satisfaction problem, and any $\hat{x}$ in X^f satisfying Eq. (A.2) for all ω in Ω is a solution of the satisfaction problem specified by (g, τ, X^f, Ω).

The uncertainty set Ω is sometimes referred to as a *disturbance set*, for it represents the set of all possible effects that can influence the performance. When the objective function g is given in terms of an outcome function $P: X \times \Omega \to Y$ and a performance function $G: X \times \Omega \times Y \to V$,

$$g(x, \omega) = G(x, \omega, P(x, \omega))$$

then the set Ω represents the set of all possible effects that can influence the outcome of a given decision x. Notice that, due to the level of abstraction, Ω covers both the so-called parametric and structural uncertainties. The function τ then specifies an "upper limit" of the tolerable or acceptable performance. A decision x is considered satisfactory if it yields a performance that does not exceed the specified tolerance $\tau(\omega)$ for any outcome of the uncertainty ω from the given set Ω.

In a more general formulation of the satisfaction problem, the ordering $\leq$ may be replaced by any desired relation $R \subseteq V \times V$. A general satisfaction problem is then to find $\hat{x}$ in X^f such that for all ω in Ω

$$g(\hat{x}, \omega) R \tau(\omega)$$

where R is a given "satisfaction" relation in the value set V; that is, a decision $\hat{x}$ in X^f is satisfactory if for any ω in Ω, the performance $g(\hat{x}, \omega)$ is R-related to the tolerance $\tau(\omega)$.

We can now introduce a formal general systems concept of a decision-making system.

A system $S \subseteq X \times Y$ is a decision-making system if there is given a family of decision problems D_x, $x \in X$, with the solution set Z and a mapping $T: Z \to Y$ such that for any x in X and y in Y, the pair (x, y) is in the system S if and only if there exists $z \in Z$ such that z is a solution of D_x and $T(z) = y$.

In most considerations (but not always!), the output itself is a solution of the given decision problem; i.e., $Z = Y$, and T is identity.

In conclusion, let us offer the following remarks regarding the notion of a decision-making system.

(i) A constructive specification in terms of a set of equations might be given for a decision system, particularly if there exists an analytical solution to the corresponding decision problem in the sense that for any input x in X, there exists an analytical algorithm that determines the output $y = S(x)$. However, such an algorithm need not always exist. In general, it is only required that there is a well-defined decision problem in order to specify a system as decision making; therefore, no requirements on the existence of an algorithm in one form or another to obtain the solution of the decision problem involved are postulated.

(ii) Any input–output system can be described as a decision system and conversely. A system may be viewed as a decision system purely on the grounds of expediency in studying its behavior. Actually, both input–output and decision-making specifications can be used alternatively, depending upon one's interest. A notable example of this can be found in classical physics, where a given phenomenon can be described either on the basis of a law or a variational principle. Other examples can be found in the so-called behavioristic approach in psychology.

(iii) An optimization problem is apparently a special case of the satisfaction problem; for example, let Ω be a unit set, $\{\omega\}$, and define $\tau(\omega)$ as the minimum of g over $X^f \times \{\omega\}$. Conversely, one might argue that the satisfaction problem can be reformulated as an optimization problem by a proper selection of a new performance function. This, of course, is to some extent a question of interpretation and aesthetics. For conceptual reasons which we shall not elaborate here, we prefer, however, to make a distinction.

Finally, we can consider the notion of a goal-seeking system. In general, the concept of a goal and a goal-seeking activity might have to be left unformalized, since one has to refrain from describing purposeful behavior when both a precise meaning of the goal and the processes in terms of which it can be achieved are not explicitly stated. To initiate formalization, it should be assumed at least that the "state" when a goal is achieved is recognizable. An example from psychology might help here. Reaching the "state" of "being happy" might be a goal for a man, yet its meaning is probably known only to the person involved, and the methods of achieving it are not known even to him beforehand. In pursuing such a goal, one usually tries strategies that have some prospects of leading to fulfillment of the goal. He can try to get more education, make a fortune, get married, or try any other combination of strategies, but none of these strategies will necessarily lead him up to the goal, nor would he know with certainty which one is preferable. He can very well learn that his goal-seeking behavior has failed only after several attempts have been made.

If a formalization of the goal-seeking behavior is possible, it will inevitably lead to a general decision-type situation. We assume, therefore, that formalized goal seeking is the same concept as general decision making. Formalized goal is defined by a decision problem, and a goal being achieved means that the corresponding decision problem is solved. It should be emphasized again that the goal-seeking problem can be formalized without giving a solution method for the associated decision problem.

Appendix IV

BASIC NOTIONS IN CATEGORY THEORY

We are using in this book standard notions of categorical algebra [41]. For reference purposes, definitions of the key concepts used are given below.

1. Let $\overline{X}$ be a class of objects, X, Y, together with two functions as follows:

(i) A function assigning to each pair (X, Y) of objects of $\overline{X}$ a set $\text{Mor}(X, Y)$. An element $f \in \text{Mor}(X, Y)$ in this set is called a morphism $f: X \to Y$ of $\overline{X}$, with domain X and codomain Y.

(ii) A function assigning to each triple (X, Y, Z) of objects of $\overline{X}$ a function

$$\text{Mor}(Y, Z) \times \text{Mor}(X, Y) \to \text{Mor}(X, Z)$$

For morphisms $g: Y \to Z$ and $f: X \to Y$, this function is written as $(g, f) \to g \cdot f$, and the morphism $g \cdot f: X \to Z$ is called the composite of g with f. The class $\overline{X}$ with these two functions is called a category when the following two axioms hold.

Associativity. If $h: Z \to W$, $g: Y \to Z$, and $f: X \to Y$ are morphisms of $\overline{X}$ with the indicated domains and codomains, then

$$h \cdot (g \cdot f) = (h \cdot g) \cdot f$$

Identity. For each object Y of $\overline{X}$, there exists a morphism $I_y: Y \to Y$ such that

$$I_y \cdot f = f \qquad \text{for} \quad f: X \to Y$$

and

$$g \cdot I_y = g \quad \text{for} \quad g: Y \to Z$$

2. The class of objects of $\overline{X}$ will be denoted as Ob $\overline{X}$.

3. A morphism $f: X \to Y$ in a category $\overline{X}$ is called invertible in $\overline{X}$ if there exists a morphism $g: Y \to X$ in $\overline{X}$ with both $g \cdot f = I_x$ and $f \cdot g = I_y$. g will be denoted by f^{-1} since if g exists, it is unique.

4. If $\overline{X}$ and $\overline{X}'$ are two categories, a functor $F: \overline{X} \to \overline{X}'$ is a pair of functions, an object function and a mapping function. The object function assigns to each object X of the first category $\overline{X}$ an object $F(X)$ of $\overline{X}'$; the mapping function assigns to each morphism $f: X \to Y$ of the first category a morphism $F(f): F(X) \to F(Y)$ of the second category $\overline{X}'$. These functions must satisfy two requirements:

$$F(I_x) = I_{F(X)} \qquad \text{for each identity } I_x \text{ of } \overline{X}$$

$$F(g \cdot f) = F(g) \cdot F(f) \qquad \text{for each composite } g \cdot f \text{ defined in } \overline{X}$$

5. To each category $\overline{X}$ we can construct another category $\overline{Y}$ as follows: Objects of $\overline{Y}$ are all the objects of $\overline{X}$; morphisms of $\overline{Y}$ are $f^{op}: Y \to X$, one for each morphism $f: X \to Y$ in $\overline{X}$; composites $f^{op} \cdot g^{op} = (g \cdot f)^{op}$ in $\overline{Y}$ are defined whenever $g \cdot f$ is defined. $\overline{Y}$ will be denoted by $\overline{X}^{op}$.

6. If F, $G: \overline{X} \to \overline{X}'$ are functors, a natural transformation $\tau: F \to G$ from F to G is a function which assigns to each object X of $\overline{X}$ a morphism $\tau(X): F(X) \to G(X)$ of $\overline{X}'$ in such a way that every morphism $f: X \to Y$ of $\overline{X}$ yields a commutative diagram

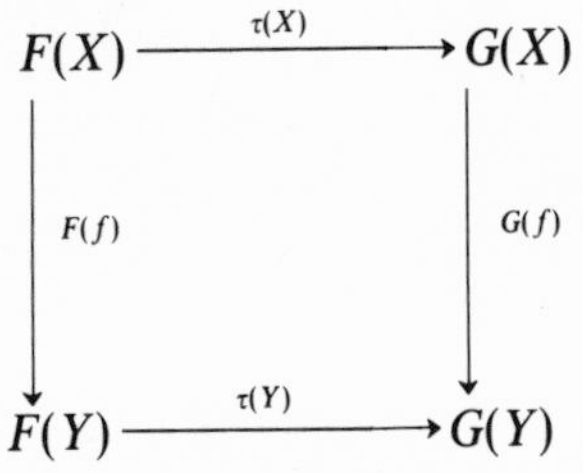

7. To each pair of categories $\overline{X}$ and $\overline{X}'$, we can construct another category $\overline{X} \times \overline{X}'$ called product category as follows: An object of $\overline{X} \times \overline{X}'$ is an ordered pair (X, X') of objects of $\overline{X}$ and $\overline{X}'$, respectively. A morphism $(X, X') \to (Y, Y')$ with the indicated domain and codomain is an ordered pair (f, f') of morphisms $f: X \to Y$, $f': X' \to Y'$. The composite of morphisms is defined termwise.

8. If $F:\overline{X} \to \overline{X}'$ is a functor, we can construct another functor

$$\text{Mor}(F-, -):\overline{X}^{op} \times \overline{X}' \to \overline{\text{Set}}$$

$\text{Mor}(F-, -)(X, X') \equiv \text{Mor}(FX, X')$ is the set of all morphisms $FX \to X'$; $\text{Mor}(F-, -)(f^{op}, g) \equiv \text{Mor}(Ff, g)$ of a morphism $(f^{op}, g):(Y, X') \to (X, Y')$ is a mapping $\text{Mor}(Ff, g):\text{Mor}(FX, X') \to \text{Mor}(FY, Y')$ such that for every $h \in \text{Mor}(FY, X')$

$$\text{Mor}(Ff, g)(h) = g \cdot h \cdot Ff$$

9. A natural transformation $\tau:F \to G$ from a functor $F:\overline{X} \to \overline{X}'$ to a functor $G:\overline{X} \to \overline{X}'$ is termed a natural isomorphism if and only if for each X, $\tau(X)$ is invertible in $\overline{X}'$. If τ is a natural isomorphism, we shall write $F \cong G$.

10. Each pair of functors $F:\overline{X} \to \overline{X}'$ and $G:\overline{X}' \to \overline{X}$ is called a pair of adjoint functors iff $\text{Mor}(F-, -) \cong \text{Mor}(-, G-):\overline{X}^{op} \times \overline{X}'$; F is called left adjoint to G, and G is called right adjoint to F.

INDEX

B
C 6
D 7
E 8
F 9
G 0
H 1
I 2
J 3